Selected Titles in This Series

176 **E. V. Shikin, Editor,** Some Questions of Differential Geometry in the Large

175 **R. L. Dobrushin, R. A. Minlos, M. A. Shubin, and A. M. Vershik, Editors,** Contemporary Mathematical Physics (F. A. Berezin Memorial Volume)

174 **A. A. Bolibruch, A. S. Merkur′ev, and N. Yu. Netsvetaev, Editors,** Mathematics in St. Petersburg

173 **V. Kharlamov, A. Korchagin, G. Polotovskiĭ, and O. Viro, Editors,** Topology of Real Algebraic Varieties and Related Topics

172 **K. Nomizu, Editor,** Selected Papers on Number Theory and Algebraic Geometry

171 **L. A. Bunimovich, B. M. Gurevich, and Ya. B. Pesin, Editors,** Sinai's Moscow Seminar on Dynamical Systems

170 **S. P. Novikov, Editor,** Topics in Topology and Mathematical Physics

169 **S. G. Gindikin and E. B. Vinberg, Editors,** Lie Groups and Lie Algebras: E. B. Dynkin's Seminar

168 **V. V. Kozlov, Editor,** Dynamical Systems in Classical Mechanics

167 **V. V. Lychagin, Editor,** The Interplay between Differential Geometry and Differential Equations

166 **O. A. Ladyzhenskaya, Editor,** Proceedings of the St. Petersburg Mathematical Society, Volume III

165 **Yu. Ilyashenko and S. Yakovenko, Editors,** Concerning the Hilbert 16th Problem

164 **N. N. Uraltseva, Editor,** Nonlinear Evolution Equations

163 **L. A. Bokut′, M. Hazewinkel, and Yu. G. Reshetnyak, Editors,** Third Siberian School "Algebra and Analysis"

162 **S. G. Gindikin, Editor,** Applied Problems of Radon Transform

161 **K. Nomizu, Editor,** Selected Papers on Analysis, Probability, and Statistics

160 **K. Nomizu, Editor,** Selected Papers on Number Theory, Algebraic Geometry, and Differential Geometry

159 **O. A. Ladyzhenskaya, Editor,** Proceedings of the St. Petersburg Mathematical Society, Volume II

158 **A. K. Kelmans, Editor,** Selected Topics in Discrete Mathematics: Proceedings of the Moscow Discrete Mathematics Seminar 1972–1990

157 **M. Sh. Birman, Editor,** Wave Propagation. Scattering Theory

156 **V. N. Gerasimov, N. G. Nesterenko, and A. I. Valitskas,** Three Papers on Algebras and Their Representations

155 **O. A. Ladyzhenskaya and A. M. Vershik, Editors,** Proceedings of the St. Petersburg Mathematical Society, Volume I

154 **V. A. Artamonov et al.,** Selected Papers in K-Theory

153 **S. G. Gindikin, Editor,** Singularity Theory and Some Problems of Functional Analysis

152 **H. Draškovičová et al.,** Ordered Sets and Lattices II

151 **I. A. Aleksandrov, L. A. Bokut′, and Yu. G. Reshetnyak, Editors,** Second Siberian Winter School "Algebra and Analysis"

150 **S. G. Gindikin, Editor,** Spectral Theory of Operators

149 **V. S. Afraĭmovich et al.,** Thirteen Papers in Algebra, Functional Analysis, Topology, and Probability, Translated from the Russian

148 **A. D. Aleksandrov, O. V. Belegradek, L. A. Bokut′, and Yu. L. Ershov, Editors,** First Siberian Winter School "Algebra and Analysis"

147 **I. G. Bashmakova et al.,** Nine Papers from the International Congress of Mathematicians, 1986

146 **L. A. Aĭzenberg et al.,** Fifteen Papers in Complex Analysis

(*Continued in the back of this publication*)

Some Questions of Differential Geometry in the Large

American Mathematical Society
TRANSLATIONS
Series 2 • Volume 176

Some Questions of Differential Geometry in the Large

E. V. Shikin
Editor

American Mathematical Society, Providence, Rhode Island
in cooperation with
MIR Publishers, Moscow, Russia

Translated by R. S. Wadhwa from a collection of original Russian manuscripts.

The present translation is published under an agreement between MIR Publishers and the American Mathematical Society.

1991 *Mathematics Subject Classification.* Primary 53Cxx; Secondary 53Axx, 51Fxx.

ABSTRACT. This collection contains articles that present results obtained recently by geometers of Russia and Ukraine. The disparity in statements of problems and in methods of their solutions is unified by the fact that all articles belong to the field of mathematics known as geometry in the large. Papers in the collection deal with various questions related to the structure, symmetries, and embeddings of submanifolds in Euclidean and pseudo-Euclidean spaces. The collection offers a review of problems facing specialists on geometry in the large, and of the latest research in the field. It is useful to researchers and graduate students working in differential geometry.

Library of Congress Cataloging-in-Publication Data
Some questions of differential geometry in the large / E. V. Shikin, editor: [translated by R. S. Wadhwa from a collection of original Russian manuscripts].
p. cm. — (American Mathematical Society translations: ser. 2, v. 176)
Includes bibliographical references.
ISBN 0-8218-7506-X (alk. paper)
1. Geometry, Differential. I. Shikin, E. V. II. Series.
QA3.A572 ser. 2, vol. 176
[QA641]
510 s—dc20 96-21901
[516.3′6] CIP

∞ The paper used in this book is acid-free and falls within the guidelines
established to ensure permanence and durability.
♲ Printed on recycled paper.

10 9 8 7 6 5 4 3 2 1 01 00 99 98 97 96

Contents

Preface

This collection contains articles that present results obtained recently by geometers of Russia and Ukraine. Even a cursory glance at the titles shows the disparity of their content. Nevertheless, this disparity in statements of problems and in methods of their solutions is unified by the fact that all articles belong to the field of mathematics long known as *geometry in the large*. For the reader to gain a better understanding of the nature of the results in this collection, their depth, and the methods developed, we believe it is advisable to comment briefly on the content of each article.

The paper "Three-manifolds with Cr-structure" gives a complete description of Cr-manifolds. When we say that a smooth manifold has a Cr-structure, we mean that it can be decomposed into parts, each being a Seifert bundle determined by a certain simplicial space (i.e., a recipe for how the whole should be assembled), so that we have an action on the manifold of a sheaf of groups with tori as fibres. A Cr-structure also means that there is an F-structure on the manifold, with polarization in the sense of Cheeger or Gromov. Hence we see that on a closed manifold there is a Cr-structure of a family of collapsing metrics with bounded curvature and vanishing volume. On the other hand, the property of possessing a Cr-structure is homotopy invariant for closed three-manifolds with nonpositive sectional curvature. A complete description of three-manifolds with a Cr-structure is used to show that the homotopy equivalence of complete Riemannian manifolds with nonpositive sectional curvature, one of which has a Cr-structure, implies their diffeomorphism (not assuming that the manifolds are compact).

The paper "Algebraic surfaces with an infinite set of skew symmetry planes" discusses the geometric theory of invariants of groups generated by reflections. Let G be an infinite group generated by skew reflections with respect to hyperplanes in the space E^m, N the set of all directions of symmetry, and let the μ_j-planes Π^{μ_j} be linear spans of $G(u)$-orbits of vectors $u \in N$. Then $\Pi^{\mu_j} = \Pi^{d_f} \oplus \Pi^{\gamma_j}$ (symmetry planes conjugate to the vectors Π^{d_j}, of a noncylindrical hypersurface F which is invariant under G, passing through Π^{γ_j}). A complete solution to the problem of mutual position of four γ_j-planes Π^{γ_j} $(\gamma_0 \geq \gamma_1 \geq \gamma_2 \geq \gamma_3)$ is given when $\Pi^{\gamma_j} \bigcap \Pi^{\gamma_{j'}} = 0$, $j \neq j'$, $\dim(\Pi^{\gamma_0} + \Pi^{\gamma_1} + \Pi^{\gamma_2}) = \gamma_0 + \gamma_1 + \gamma_2$, and a partial one is given when $\dim(\Pi^{\gamma_0} + \Pi^{\gamma_1} + \Pi^{\gamma_2}) < \gamma_0 + \gamma_1 + \gamma_2$. Equations of special surfaces whose symmetry groups do not admit extension are obtained.

Existence in the Euclidean space E^n of curves and discontinua with unusual geometric and analytic properties is proved in the paper "Curves and discontinua with paradoxical geometric properties." In particular, it is established that there exist (1) a homeomorphism $f : E^n \mapsto E^n$, $n \geq 2$, such that for any straight line $l \subset E^n$ the projection of the curve $f(l)$ onto any hyperplane $P \subset E^n$ is the whole hyperplane, and (2) a curve $L \subset E^n$, homeomorphic to the straight line and differentiable almost everywhere, such that for any unbounded regular curve γ with bounded curvature, the intersection $\gamma \bigcap L$ has infinite Lebesgue measure standardly defined on γ, and the intersection $Q \bigcap L$ for any plane $Q \subset E^n$ is of infinite Lebesgue measure defined in the standard fashion on Q. Existence of the above curves is related to the existence of discontinua nicely embedded in E^n in the topological sense and having unexpected geometric properties.

The paper "Structure of the neighborhood of an isolated zero of the Lipschitz–Killing curvature on a surface which is not 0-tight" generalizes the concept of canonical neighborhood introduced by N. V. Efimov for a two-dimensional surface in three-dimensional Euclidean space E^3 to the case of an m-dimensional surface in E^{m+p}. It is proved that an isolated zero of the Lipschitz–Killing curvature has a canonical neighborhood on the surface. Moreover, it is established that the surface is decomposable by a hyperplane into more than two components.

The paper "Space-like convex surfaces in pseudo-Euclidean space" discusses elements of the intrinsic and extrinsic geometry of convex surfaces in pseudo-Euclidean space. Theorems on the uniqueness of closed surfaces in hyperbolic and pseudo-Euclidean spaces, and analogs of the Cauchy and Minkowski theorems for polyhedra in pseudo-Euclidean space, are proved. The uniqueness theorems and the maximum principle are generalized to convex polyhedra with boundary in pseudo-Euclidean space. The essential difference of the Euclidean case from the hyperbolic one is discussed.

The paper "A small parameter in the theory of isometric imbedding of two-dimensional Riemannian manifolds in Euclidean spaces" discusses the use of small parameters in solving problems of regular realization of two-dimensional metrics in Euclidean spaces of small dimensions. The small parameter method in the theory of regular isometric imbeddings in three- and four-dimensional Euclidean spaces originated in two-manifolds with negative curvature and has now an almost twenty five year old history. With the aid of a small (numerical or functional) parameter, the existence of regular realization in the space E^3 of arbitrary compact domains, horocycles and extending strips of such manifolds is proved. The last result is an application of the small parameter method to the proof of the existence of regular isometric imbeddings in E^3 of complete open simply connected Riemannian manifolds with negative curvature. The same method was used to prove that a regular imbedding of any compact part of a two-manifold with alternating curvature in the space E^4 is possible, to solve the problem of local regular realization in E^3 of Riemannian two-manifolds with nonnegative curvature and of certain manifolds of alternating curvature, and to construct an analytic isometric imbedding in the large in E^3 of arbitrary analytic two-manifolds without conjugate points.

With all this variety in its papers, the collection offers a review of problems facing specialists on geometry in the large, and of the latest research in the field.

E. Shikin

Amer. Math. Soc. Transl.
(2) Vol. **176**, 1996

Three-manifolds with Cr-structure

S. V. Buyalo

§1. Statement of the problem

When we say that a Cr-structure exists on a smooth manifold, we mean that the manifold can be decomposed into parts, each of which is a Seifert foliation defined by a certain simplicial space (assembly of parts) so that the manifold is subjected to the action of a bundle of groups whose fibers are tori. The exact definition of the Cr-structure will be given in §2. It was shown in [**1**] that if a Cr-structure exists on a manifold, an F-structure with a polarization in the sense of Cheeger and Gromov [**2**] also exists on it. It follows from this and from results in [**2**] (Theorems 3.1 and 5.2) that a manifold with a Cr-structure contains a family of Riemannian metrics collapsing with a limited curvature and a vanishing volume.

On the other hand, it was shown in [**1, 3**] that a closed manifold of dimension ≤ 4 with a Riemannian metric of a nonpositive sectional curvature whose inscribed circle (defined in 1.3) has a very small radius as compared with the maximum absolute values of curvatures, possesses a Cr-structure. The metric of such a manifold is constructed in a very special way, namely it locally contains a Euclidean cofactor which is compatible with its Cr-structure (see 1.4 for definition). It is also generally known that a Riemannian manifold whose inscribed circle has a very small radius in comparison with the maximum absolute value of curvature possesses an F-structure [**4**]. Moreover, it is well known that the property of "possessing a Cr-structure" is homotopically invariant, at least for closed three-manifolds with nonpositive sectional curvature [**5**].

This paper consists of two parts. In the first part, we shall give a complete description of three-manifolds possessing a Cr-structure. Each such manifold is conceived as follows.

1.1. THEOREM. *Let M be a connected three-manifold having a Cr-structure. Then we have*

• *a manifold V presented as a topological sum $\bigsqcup_s V_s$ of a set of compact three-manifolds V_s with a boundary*;

1991 *Mathematics Subject Classification.* Primary 53C15; Secondary 53C12.

• *a collar K of the boundary ∂V, i.e., $K = \bigsqcup_{s,\alpha} K_{s,\alpha}$, where $K_{s,\alpha}$ is a collar of the component of the boundary of the manifold V_s; and*

• *an involution h: $K \to K$ without fixed points,*

such that the manifold M is homeomorphic to V/h.

In this case, the following facts hold.

1.1.1. Each manifold V_s is a Seifert foliation.

1.1.2. The collar $K_{s,\alpha}$ of each component of the boundary ∂V_s consists of integral fibers and is homeomorphic to $[0,1] \times F_{s,\alpha}$, where $F_{s,\alpha}$ is either the torus T^2 or the Klein bottle.

The component of the boundary $\partial K_{s,\alpha}$ lying inside V_s is called the boundary component for K and the other component is called the boundary component.

1.1.3. The involution h transposes the boundary components of the boundary ∂K with the boundary components, and the fiber q_x of the Seifert foliation for each point $x \in K$ intersects transversally with the image $h(q_{h^{-1}(x)})$ of the fiber $q_{h^{-1}(x)}$.

Hence it can be expected that the concept of a manifold with a Cr-structure is a natural generalization to any dimension of a three-dimensional graph-manifold in which Seifert foliations of appropriate dimensions serve as the "building blocks."

Another result presented in this paper concerns a description of the phenomenon of rigidity for three-dimensional Riemannian manifolds of nonpositive sectional curvature having a Cr-structure. To formulate this problem exactly, we use the following definitions 1.2–1.4.

1.2. The class α of freely homotopic loops in the topological space X is said to be peripheral (see [**6**]) if for any compact set $R \subset X$ there exists a loop from α that does not intersect R. An element of the fundamental group $\pi_1(X, x)$ is said to be peripheral if its free homotopy class is peripheral.

1.2.1. Note that although the fundamental group $\pi_1(X,x)$ is homotopy invariant, the property of the element $\alpha \in \pi_1(X,x)$ to be (or not to be) peripheral is not homotopically invariant.

Example: the spaces $X = (0,1) \times S^1$ and $Y = S^1$ are homotopy equivalent, but all elements in $\pi_1(X,x)$ are peripheral while those in $\pi_1(Y,y)$ are not.

1.3. Let M be a Riemannian manifold, and $i_M : M \to R$ its injectivity radius. The quantity $\sigma(M) = \sup_{x \in M} i_M(x)$ will be called the radius of the sphere inscribed in the Riemannian manifold M.

1.4. The Riemannian manifold M locally contains a Euclidean cofactor [**1**] if there exists an everywhere open dense manifold $M' \subset M$ such that for each point $x \in M'$ there exists a neighborhood that is isometric to the metric product $V \times (-\varepsilon, \varepsilon)$, $\varepsilon > 0$.

If the manifold M has a Cr-structure and each interval $v \times (-\varepsilon, \varepsilon)$, $v \in V$, is in a certain fiber of this Cr-structure, we shall say that the Riemannian metric on M is compatible with its Cr-structure.

1.5. THEOREM. *Let M_0 and M_1 be three-dimensional complete Riemannian manifolds with nonpositive sectional curvatures, such that*

(a) *the fundamental group $\pi_1(M_0)$ does not contain nontrivial peripheral elements,*

(b) *a* Cr-*structure exists on M_0, and*

(c) *the fundamental groups* $\pi_1(M_0)$ *and* $\pi_1(M_1)$ *are isomorphic.*

Then the manifold M_1 *contains an absolutely convex open subset* M_1^* *having the following properties*:

1. *A* Cr-*structure exists on* M_1^* *and is compatible with its metric induced by the inclusion* $M_1^* \subset M$.
2. M_1^* *is diffeomorphic both to the manifold* M_0 *and to the manifold* M_1. *In particular, the manifolds* M_0 *and* M_1 *are diffeomorphic, and a* Cr-*structure exists on* M_1.
3. *If the radius of the inscribed sphere* $\sigma(M_1)$ *is finite, then* $M_1^* = M_1$.

1.6. Remarks. 1.6.1. NP-graph manifolds in Schroeder's sense [**7**] have a Cr-structure (see [**5**]). It can be shown that their fundamental group does not contain nontrivial peripheral elements. However, as is seen from the definition of an NP-graph manifold and Theorem 1.1, the class of manifolds M_0 satisfying conditions (a) and (b) of Theorem 1.5 is much wider than the class of NP-manifolds. Thus, Theorem 1.5 is a generalization of Theorem 2 in [**7**]. Unfortunately, the effects associated with the possibility that $M_1^* \neq M_1$ are not taken into account in [**7**] (see 1.6.3 and [**5**] in this connection).

1.6.2. If condition (a) is omitted, the statement of Theorem 1.5 no longer holds. Example: Let F_0 be a pierced torus T^2, and let F_1 be a three times pierced sphere S^2. The manifolds $M_0 = F_0 \times S^1$ and $M_1 = F_1 \times S^1$ are three-dimensional and homotopy equivalent, carry complete metrics with a limited nonpositive sectional curvature and a finite volume, and possess, as can be verified easily, a Cr-structure (caution: the number of blocks from which these manifolds are glued as in Theorem 1 is infinite). However, the group $\pi_1(M_0)$ (as well as $\pi_1(M_1)$) contains nontrivial peripheral elements, and the manifolds M_0 and M_1 are obviously not homeomorphic.

1.6.3. The condition of finiteness of the radius of the inscribed sphere $\sigma(M_1)$ in Theorem 1.5 can be slightly relaxed: for the equality $M_1^* = M_1$, it is sufficient that the injectivity radius of the manifold M_1 be bounded along any geodesic. However, in the absence of any such condition, the Riemannian metric for M_1 may not be compatible with any Cr-structure on M_1. An example of this is discussed in [**5**].

1.6.4. Apparently, the conclusion of Theorem 1.5 remains valid even without the assumption that the manifold M_0 has a complete metric with nonpositive curvature. Of course, in this case we must assume instead of condition (c) that the manifolds M_0 and M_1 are homotopy equivalent.

§2. Cr-structure on a smooth manifold

In this section, we shall define a Cr-structure.

2.1. We say that the bijection γ: $Y \to Y$ induces automorphism of the action ξ: $G \times Y \to Y$ of the group G on the set Y if there exists an automorphism γ_*: $G \to G$ such that $\gamma \circ \xi = \xi \circ (\gamma_* \times \gamma)$.

2.2. The simplicial space S is said to be graded if the following three conditions are satisfied.

2.2.1. To each vertex $s \in S$ there corresponds a natural number $\operatorname{rk}(s)$, called its rank.

2.2.2. Simplices from S do not contain different vertices of the same rank. In particular, the relation $\mathrm{rk}(s) \leq \mathrm{rk}(s')$ for the vertices s and s' of one simplex transforms S into an ordered graded space: we assume that $s \leq s'$.

2.2.3. If $s_1 \leq \cdots \leq s_k$, then $s_1, \dots, s_k$ are the vertices of one simplex.

We shall denote by $\mathrm{ske}_0(S)$ the zero-dimensional skeleton, i.e., the set of vertices of S.

2.3. Let M be a smooth n-manifold. Assume that for each vertex s of the graded simplicial space S we have a smooth imbedding $\varphi_s : V_s \to M$ for the compact connected n-manifold V with the boundary (which may be empty). Together with the set $\{\varphi_s : s \in \mathrm{ske}_0(S)\}$, the space S is called the Cr-structure $\mathcal{S} = (S, \{\varphi_s\})$ on the manifold M if the following seven conditions are satisfied.

2.3.1. *Local finiteness of a covering.* The sets $U_s := \varphi_s(\mathrm{Int}(V_s))$, $s \in \mathrm{ske}_0(S)$, cover the manifold M, and this covering is locally finite.

2.3.2. *The structure of pure polarizations (or Seifert foliations).* Let a normal covering $\psi_s : \widetilde{V}_s \to V_s$ with a finite number of sheets be defined. In this case, $\widetilde{V}_s = V_s' \times T^{\mathrm{rk}(s)}$ is acted upon by the torus $T^{\mathrm{rk}(s)}$ identically on the first cofactor and by translation on the second, and the group of slips of the cover ψ_s induces a group of automorphisms of this action.

Thus, each manifold V_s has a foliation $\mathcal{T}_s$ with compact fibers of the type $\psi_s(v \times T^{\mathrm{rk}(s)})$, $v \in V_s'$. In the notation used in [**2**], the manifold V_s has the F-structure with a pure polarization of rank $\mathrm{rk}(s)$. In the terminologgy of [**8**], the manifold V_s has the structure of an injective Seifert foliation. It was mentioned in [**8**] that the fibers of the foliation $\mathcal{T}_s$ are Euclidean space forms whose topological form may change from point to point. The fundamental groups of these fibers are crystallographic groups, and this explains the origin of the term Cr-structure.

It also follows from the property 1.3.2 that on the manifold V_s there acts a locally constant bundle of groups P_s with fibers $(P_s)_y$, $y \in V_s$, which are isomorphic to the torus $T^{\mathrm{rk}(s)}$. The orbits of this action are fibers of the foliation $\mathcal{T}_s$. For $y \in V_s$, we denote by t_y the tangent space to the fiber q_y of $\mathcal{T}_y$ at the point y, and by $\tau_{s,y}$ the subspace of the tangent space T_xM of the type $d\varphi_s(ty)$, $x := \varphi_s(y)$. Thus, $\dim(\tau_{s,y}) = \mathrm{rk}(s)$.

2.3.3. *Injectivity relative to foliation.* If $\tau_{s,y} = \tau_{s,y'}$, then $y = y'$.

For a point $x \in M$, we denote by $t(x)$ (or by $\mathrm{ti}(x)$) the collection of all subspaces $\tau_{s,y} \subset T_xM$, where $s \in \mathrm{ske}_0(S)$, $y \in V_s$ (resp. $y \in \mathrm{Int}(V_s)$).

2.3.4. *Maximality.* Any set of subspaces from $t(x)$ lies in a certain subspace from $t(x)$; any set of subspaces from $\mathrm{ti}(x)$ lies in a certain subspace from $\mathrm{ti}(x)$.

2.3.5. *Local compatibility.* The inclusion $\tau_{s,y} \subset \tau_{s',y'}$ implies that $s \leq s'$, and that for the internal points $y \in V_s$ and $y' \in V_{s'}$ the homeomorphism

$$\varphi_s^{-1} \circ \varphi_{s'}|W' : W' \to W$$

of their appropriate neighborhoods induces the inclusion $\xi_y^{y'} : (P_s)_y \to (P_{s'})_{y'}$ of the fibers of the bundles P_s and $P_{s'}$, which commutes with the local actions on the neighborhoods of W and W'. If $\tau_{s,y} \subset \tau_{s',y'} \subset \tau_{s'',y''}$, we obtain $\xi_{y'}^{y''} \circ \xi_y^{y'} = \xi_y^{y''}$.

2.3.6. *Global compatibility.* It follows from $s \leq s'$ that

(i) $U_{s'} \subset U_s$,

(ii) if $\varphi_{s'}(q') \cap U_s$ is not empty (q' is a fiber of $\mathcal{T}_{s'}$), then $\varphi_{s'}(q') \subset U_s$, and

(iii) if $\varphi_{s'}(q') \cap \varphi_s(B)$ is not empty (q' is a fiber of $\mathcal{T}_{s'}$, and B is a component of the boundary ∂V_s), then there exist a vertex $s'' \geq s'$ and a fiber q'' of $\mathcal{T}_{s''}$ lying on $\partial V_{s''}$, for which $\varphi_{s'}(q') \subset \varphi_{s''}(q'') \subset \varphi_s(B)$.

2.3.7. π_1-*injectivity.* Each mapping φ_s induces a monomorphism of fundamental groups.

2.4. Remarks. 2.4.1. The boundaries of manifolds used for determining the Cr-structure and the restriction of the corresponding mappings to the boundary need not be smooth: all that is required is that these restrictions must be local inclusions that are smooth together with the boundary along the fibers of the existing foliations.

2.4.2. In the definition of the Cr-structure given in [**1, 5**] less stringent requirements are imposed than in the property 2.3.6 of global compatibility, i.e., it is sufficient that conditions (i) and (ii) be satisfied. Although condition (iii) in 2.3.6 imposes more stringent requirements, the results obtained in [**1, 5**] also remain valid under condition (iii). This will be explained in the Appendix (see §6). On the other hand, condition (iii) is very useful for proving Theorem 1.1.

2.4.3. It can easily be proved (see [**1**]) that the simplicial space S is connected if and only if the manifold M is connected. In what follows, we shall always assume that the manifold M is connected.

2.5. Mapping of s and y. Reduced Cr-structure. Let $\mathcal{S} = (S, \{\varphi_s\})$ be a Cr-structure on the manifold M. In this case, the equality $\tau(x) = \tau_{s(x),y(x)}$, where $\tau(x)$ is a subspace from the collection $\mathrm{ti}(x)$ such that all subspaces from $\mathrm{ti}(x)$ lie in this subspace, defines the mappings $s : M \to \mathrm{ske}_0(S)$ and $y : M \to \bigsqcup_s \mathrm{Int}(V_s)$ (see [**1**] for details). For an arbitrary Cr-structure $\mathcal{S}$, we have in general $s(M) \neq \mathrm{ske}_0(S)$.

We shall call the Cr-structure $\mathcal{S} = (S, \{\varphi_s\})$ a reduced structure if $s(M) = \mathrm{ske}_0(S)$. Any Cr-structure $\mathcal{S}$ can always be modified in such a way as to obtain a reduced Cr-structure $\mathcal{S}$. For this purpose, it is sufficient to assume that $\hat{S}$ is a complete subspace in S with a set of vertices $s(M)$. It can easily be seen that $\hat{\mathcal{S}} := (S, \{\varphi_s : s \in \mathrm{ske}_0(\hat{S})\})$ is a reduced Cr-structure. Hence in what follows we shall always assume that the Cr-structure under consideration is a reduced one.

2.5.1. LEMMA. *If for a ceratin vertex $s \in \mathrm{ske}_0(S)$ the boundary of the manifold V_s is empty, the mapping $\varphi_s : V_s \to M$ is a diffeomorphism.*

PROOF. The image $\varphi_s(V_s)$ is open and closed in M. Since M is connected, $\varphi_s(V_s) = M$. Among other things, M is compact and φ_s is a finite sheeted covering. Since the Cr-structure under consideration is reduced, there exists a point $x \in M$ for which $s(x) = s$. In this case, the inverse image $\varphi_s^{-1}(x) \subset V_s$ consists of a single point $y(x)$ in view of the property of injectivity relative to fibration. Hence φ_s is a diffeomorphism. This proves the lemma.

It follows from the property 2.3.2 that under the conditions of Lemma 2.5.1, the manifold M is an injective Seifert foliation whose fibers are compact Euclidean space forms of dimension $\mathrm{rk}(s)$. In this case, $(\{s\}, \varphi_s)$ is a Cr-structure on M with a unique vertex. In particular, this situation is observed when the simplicial space S contains a vertex of rank $n := \dim M$. In this case, M is a compact n-dimensional Euclidean space form.

Below we always assume that unless stated otherwise, the boundary ∂V_s of any vertex $s \in \mathrm{ske}_0(S)$ is not empty. In particular, $1 \leq \mathrm{rk}(s) \leq n-1$.

2.6. Structure of the vertices of rank $n-1$**.** Let $s \in \mathrm{ske}_0(S)$ be a vertex of rank $n-1$. In this case, the manifold M is diffeomorphic to the product $[0,1] \times T^{n-1}$. The group of covering transformations $\psi_s : \widetilde{V}_s \to V_s$ preserves this structure of the product, inducing automorphisms of the natural action of the torus T^{n-1} on $\widetilde{V}_s$. Hence the following possibilities exist for the manifold V_s.

2.6.1. V_s is diffeomorphic to the product $[0,1] \times F_s$, where F_s is a compact $(n-1)$-dimensional Euclidean space form.

2.6.2. The boundary ∂V_s is bounded and V_s itself is a foliation with compact fibers $F_s(t)$ over the I-orbital image $[0 \leq t \leq 1/2]$, where $\{0\}$ is a boundary, $\{1/2\}$ is the point of reflection, and all fibers are compact Euclidean $(n-1)$-dimensional space forms that are homeomorphic for $0 \leq t < 1/2$ and cover the fiber $F_s(1/2)$ with two sheets.

In the case 2.6.1, the vertex s will be called a jumper, while for 2.6.2 it will be called a stopper.

2.7. Locally maximal and locally minimal vertices. The vertex $s \in \mathrm{ske}_0(S)$ is said to be locally maximal (locally minimal) if there does not exist a vertex $s' \in \mathrm{ske}_0(S)$ with $s' > s$ (resp. $s' < s$)). We denote by $N(S)$ the set of all locally maximal vertices in the space S, and by $n(S)$ the set of all locally minimal vertices.

2.7.1. LEMMA. *Suppose that $s \in N(S)$. In this case, the mapping $\varphi_s : V_s \to M$ is an imbedding. If the image $\varphi_s(V_s)$ intersects any other image $\varphi_{s'}(V_{s'})$, then $s' < s$.*

PROOF. Suppose that the condition $\varphi_s(y) = x = \varphi_s(y')$ is satisfied for distinct points y, $y' \in V_s$. According to the property of injectivity relative to foliation, the subspaces $\tau_{s,y}$ and $\tau_{s,y'}$ in T_xM do not coincide. By the maximality property, they are contained in a certain subspace $\tau_{s_0,z}$, which is different from each of these subspaces. It follows from the property of local compatibility that $s_0 > s$. However, this contradicts the above condition. Hence φ_s is an imbedding. The second statement of the lemma is proved similarly. This completes the proof of the lemma.

We define an n-dimensional manifold $\mathcal{W}$ as a disjoint sum $\bigsqcup\{V_s : s \in N(S)\}$. The formula $\Phi|V_s = \varphi_s$ for $s \in N(S)$ defines the immersion $\Phi : \mathcal{W} \to M$. It follows from Lemma 2.7.1 that Φ is an imbedding. Hence below we assume that $\mathcal{W}$ is a submanifold in M and Φ is an inclusion.

We define an n-dimensional manifold $\mathcal{V}$ as the disjoiont sum $\bigsqcup\{V_s : s \in n(S)\}$. The formula $\varphi|V_s = \varphi_s$ for $s \in n(S)$ defines the immersion $\varphi \colon \mathcal{V} \to M$.

2.7.2. LEMMA. *The image $\varphi(\mathrm{Int}(\mathcal{V}))$ coincides with M.*

PROOF. Suppose that $x \in M$. In view of the property 2.3.1, there exists a vertex $s' \in \mathrm{ske}_0(S)$ for which $x \in U_{s'}$. Since the simplicial space S is graded and its vertices have limited ranks, there obviously exists a simplicial path in S that joins the vertex s' with any vertex $s \in n(S)$ and does not have "local maxima". In other words, if $s_0 := s'$, and $s_1, \ldots, s_k := s$ are the vertices successively encountered in it, then $s_0 \geq s_1 \geq \cdots \geq s_k$.

In this case $s' \geq s$, and, according to the condition of global compatibility, $U_{s'} \subset U_s$. Thus, $x \in \varphi(\mathrm{Int}(\mathcal{V}))$. The lemma is proved.

§3. Structure of a three-manifold having a Cr-structure

In this section, we prove Theorem 1.1. Let M be a connected three-dimensional smooth manifold with a (reduced) Cr-structure $\mathcal{S} = (S, \{\varphi_s\})$. As before, we assume that the simplicial space S contains more than one vertex and that the boundary ∂V_s of each manifold V_s is not empty. Thus, if S is connected (see 2.4.3), the set $N(S)$ of locally maximal vertices coincides with the set $\{s \in \mathrm{ske}_0(S) : \mathrm{rk}(s) = 2\}$, while the set $n(S)$ of locally minimal vertices coincides with the set $\{s \in \mathrm{ske}_0(S) : \mathrm{rk}(s) = 1\}$. Each connected component V_s of the manifold $\mathcal{V} = \bigsqcup\{V_s : \mathrm{rk}(s) = 1\}$ has a finite number of sheets and is normally covered by the product $V_s' \times S'$, where V_s' is a compact surface with boundary. According to Lemma 2.7.2, the image $\varphi(\mathrm{Int}(\mathcal{V}))$, where $\varphi : \mathcal{U} \to M$ is an imbedding, coincides with M.

The set $J = \{y \in \mathcal{V} : \varphi^{-1}(\varphi(y)) = y\}$ is called the set of injectivity of the mapping φ. Its complement $\mathcal{V} \setminus J$ is denoted by K. If all this is taken into consideration, Theorem 1.1 obviously follows from the following theorem.

3.1. THEOREM. *Each connected component K_α of the set K is homeomorphic to the product $\{0,1\} \times F_\alpha$, where F_α is either the torus T^2, or the Klein bottle whose one component of the boundary ∂K_α coincides with a component of the boundary $\partial\mathcal{V}$, while the other lies inside $\mathcal{V}$, i.e., is the boundary component for the set K. In this case, $\partial\mathcal{V} \subset K$. There exists an involution $h : K \to K$ without stationary points, for which the statement* 1.1.3 *of Theorem* 1.3 *is valid as well as the equality $\varphi \cdot h = \varphi|K$. In this case, the induced natural mapping $\overline{\varphi} : \mathcal{V}/h \to M$ is a homeomorphism.*

3.1.1. COROLLARY. *Let $s \in \mathrm{ske}_0(S)$ be a vertex of rank* 1. *In this case, the connected compact surface V_s' with boundary in the decomposition $V_s = V_s' \times S^1$ is different from a disk.*

It should be recalled that according to Lemma 2.7.1, the manifold $\mathcal{W} := \sqcup\{V_s : \mathrm{rk}(s) = 2\}$ can be treated as a submanifold in M. It can be naturally expected that there exists a relation between the noninjectivity set K and the manifold $\mathcal{W}$ viz. $\varphi^{-1}(\mathcal{W}) = K$. This is actually the case if the nonessential vertices are removed first.

3.2. Nonessential vertices. The vertex $s \in \mathrm{ske}_0(S)$ is said to be nonessential if after it and its open star are removed from S, the remaining subspace in S and the set of mappings $\{\varphi_s\}$ still continue to define a Cr-structure on M.

3.2.1. LEMMA. *Let $s_0 \in \mathrm{ske}_0(S)$ be a vertex of rank* 2, *such that at all points $x \in U_{s_0}$, all the possible subspaces in T_xM of the type $\tau_{s,y}$, where $s \prec s_0$ and $y \in \varphi_s^{-1}(x)$, coincide. In this case, s_0 is an nonessential vertex.*

PROOF. Suppose that the inequality $\varphi_s(V_s) \cap V_{s_0} \neq \emptyset$ holds for the vertex $s \in \mathrm{ske}_0(S)$, $s \neq s_0$. In this case, we have $s \prec s_0$ according to Lemma 2.7.1 and, in particular, $\mathrm{rk}(s) = 1$. In view of the property of global compatibility, $U_{s_0} \subset U_s$. It follows from the condition of the lemma and the property of local compatibility that the number of such vertices s does not exceed one. Hence the subspace in

S left after the removal of the vertex s_0 from S together with its open star is a connected graded simplicial space.

Let us prove that the condition $V_{s_0} \subset U_s$ is satisfied for $s \prec s_0$. Indeed, it was mentioned above that $U_{s_0} \subset U_s$. Suppose that $x \in v_{s_0} \backslash U_{s_0}$. If $x \in U_s$, then by the property 2.3.1 there exists a vertex $s' \in \mathrm{ske}_0(S)$ with $x \in U_{s'}$. Hence the vertex s' differs from each of the vertices s_0, s. However, this contradicts the proved uniqueness of the vertex s satisfying the condition $\varphi_s(V_s) \cap V_{s_0} \neq \emptyset$. Thus, $V_{s_0} \subset U_s$.

Removal of the vertex s_0 could violate only the properties 2.3.1 and 2.3.4 of the Cr-structure. However, it follows from the uniqueness of the vertex s with the property $\varphi_s(V_s) \cap V_{s_0} \neq \emptyset$ and the inclusion $V_{s_0} \subset U_s$ that this does not happen. The lemma is proved.

In what follows, we shall assume that the Cr-structure under consideration does not contain nonessential vertices.

3.3. Noninjectivity set K and submanifold $\mathcal{W}$. The following statement serves as an important step toward the proof of Theorem 3.1.

3.3.1. PROPOSITION. $\varphi^{-1}(\mathcal{W}) = K$.

In this subsection, we shall prove only the inclusion $K \subset \varphi^{-1}(\mathcal{W})$. The proof of the reverse inclusion requires a more detailed analysis of the images obtained from the mapping φ of the boundary components $\partial\mathcal{V}$ and will be given in 3.4.3.

PROOF OF THE INCLUSION $K \subset \varphi^{-1}(\mathcal{W})$. Let $y \in K$, $x = \varphi(y)$. Then there exists a point $y' \in \mathcal{V}$, $y' \neq y$, such that $\varphi(y') = x$. Thus, $x \in \varphi_s(V_s) \cap \varphi_{s'}(V_{s'})$, where s, $s' \in \mathrm{ske}_0(S)$ are vertices of rank 1 which may coincide. From the properties of local compatibility and injectivity relative to foliation, we have $\tau_{s,y} \neq \tau_{s',y'}$. According to the property of maximality, there exists a subspace of the type $\tau_{s_0,y_0} \subset T_xM$ containing the subspaces $\tau_{s,y}$ and $\tau_{s',y'}$. Hence $\mathrm{rk}(s_0) = 2$ and $x = \varphi_{s_0}(y) \in \mathcal{W}$. This means that $K \subset \varphi^{-1}(\mathcal{W})$.

3.3.2. LEMMA. *The set $\varphi(K)$ is closed in M.*

PROOF. Let $x_0 \in M$ be the limit point for a converging sequence of points $x_i \in \varphi(K)$. In this case, there exist points y_i, $y_i' \in \mathcal{V}$, $y_i \neq y_i'$, with $\varphi(y_i) = x_i = \varphi(y_i')$ for each i. Since the covering of the manifold M by the sets U_s is locally finite and each manifold V_s is compact, it can be assumed that the sequence y_i (resp. y_i') converges at the point y_0 (resp. y_0'). Since φ is an immersion, $y_0 \neq y_0'$. However, in accordance with the continuity condition, $\varphi(y_0) = x_0 = \varphi(y_0')$. This means that $x_0 \in \varphi(K)$. The lemma is proved.

3.4. Structure of the image $\varphi(\partial\mathcal{V})$ of the boundary $\partial\mathcal{V}$ of the manifold $\mathcal{V}$.

3.4.1. PROPOSITION. *The image $\varphi(\partial\mathcal{V})$ of the boundary $\partial\mathcal{V}$ of the manifold $\mathcal{V}$ is a smooth submanifold in M, each of whose components in M is two-sided and is either the torus T^2 or the Klein bottle. The contraction of the mapping φ to the boundary, i.e.,*

$$\varphi|\partial\mathcal{V} : \partial\mathcal{V} \to \varphi(\partial\mathcal{V}),$$

is a covering. In this case, the image $\varphi(B)$ of each component B of the boundary $\partial\mathcal{V}$ coincides with a certain component of the boundary $\partial\mathcal{W}$.

The proof of this proposition is based on the following lemma.

3.4.2. LEMMA. *Let $s \in \mathrm{ske}_0(S)$ be a vertex of rank 1 and let the point $x_0 \in M$ lie in $\varphi_s(\partial V_s)$. Then there exists a unique vertex $s_0 \succ s$ for which $x_0 \in \partial V_{s_0}$.*

PROOF. We use the mapping $s : M \to \mathrm{ske}_0(S)$ described in 2.5. According to its definition, $x_0 \in U_{s(x_0)}$. Let the points $y_0 \in \partial V_s$ and $y_0' \in \mathrm{Int}(V_{s(x_0)})$ be such that $\varphi_s(y_0) = x_0 = \varphi_{s(x_0)}(y_0')$. According to the properties of local compatibility and injectivity relative to foliation, the subspaces τ_{s,y_0} and $\tau_{s(x_0),y_0'}$ of the tangent space $T_{x_0}M$ do not coincide. By the maximality property, there exists a vertex $s_0 \in \mathrm{ske}_0(S)$ such that $\tau_{s,y_0} \cup \tau_{s(x_0),y'} \subset \tau_{s_0,z}$ for a certain point $z \in v_{s_0}$. In particular, $\mathrm{rk}(s_0) = \dim \tau_{s_0,z} = 2$. According to the local compatibility, $s \prec s_0$. The fiber q_z of the foliation $\mathcal{T}_{s_0}$ passes through the point x_0. It follows from the condition of global compatibility that $q_z \subset \partial V_{s_0}$. This means that $x_0 \in \partial V_{s_0}$. The uniqueness of such a vertex s_0 follows from Lemma 2.7.1. The lemma is proved.

PROOF OF PROPOSITION 3.4.1. Let B be a component of the boundary of the manifold $\mathcal{U}$ and $x_0 \in \varphi(B)$. According to Lemma 3.4.2, $x_0 \in \partial V_{s_0}$, where $\mathrm{rk}(s_0) = 2$. In particular, the component B_{s_0} of the boundary ∂V_{s_0} containing the point x_0 is a fiber of the foliation $\mathcal{T}_{s_0}$. According to Lemma 2.7.1 and the property of global compatibility, $B_{s_0} \subsetneq \varphi(B)$. Since the image $\varphi(B)$ is connected, the above arguments (and the uniqueness of the vertex s_0) show that the component B_{s_0} passes through each point of the image $\varphi(B)$, i.e., $\varphi(B) = B_{s_0}$. This proves the last statement in Proposition 3.4.1. The remaining statements now follow from the description of the structure of rank $n-1$ vertices (see 2.6). This proves Proposition 3.4.1.

The proof of Proposition 3.3.1 can now be completed easily by 3.4.3.

PROOF OF THE INCLUSION $\varphi^{-1}(\mathcal{W}) \subset K$. Let $s_0 \in \mathrm{ske}_0(S)$ be a rank 2 vertex. Since s_0 is an essential vertex, there exist in accordance with Lemma 3.2.1 a point $x \in U_{s_0}$ and different points y, $y' \in \mathcal{V}$, for which $\varphi(y) = x = \varphi(y')$. Hence the set $\varphi(K) \cap U_{s_0}$ is not empty. According to Lemma 2.7.1 and Proposition 3.4.1, the image $\varphi(\partial\mathcal{V})$ of the boundary $\partial\mathcal{V}$ does not pass through the internal points of the set V_{s_0}. Hence the set $\varphi(K) \cap U_{s_0}$ is open in U_{s_0}. By Lemma 3.3.2, this set is closed in U_{s_0}. It follows from the connection of U_{s_0} that $\varphi(K) \cap U_{s_0} = U_{s_0}$. Applying Lemma 3.3.2 once again, we obtain $\varphi(K) \cap V_{s_0} = V_{s_0}$. Hence $\varphi^{-1}(V_{s_0}) \subset K$. This proves the inclusion $\varphi^{-1}(\mathcal{W}) \subset K$ as well as Proposition 3.3.1.

3.5. Boundary, boundary, and internal points of the noninjectivity set K. It follows from Lemma 3.3.2 that the boundary of the noninjectivity set K is the set $\mathrm{Fr}\, K = K \cap \overline{J}$.

3.5.1. PROPOSITION. *The boundary $\mathrm{Fr}\, K$ of the set K coincides with the set $\mathrm{Int}(\mathcal{V}) \cap \varphi^{-1}(\partial\mathcal{W})$. In particular, $\mathrm{Fr}\, K$ is a submanifold in $\mathrm{Int}(\mathcal{V})$, with each connected component homeomorphic to the torus T^2 or the Klein bottle.*

3.5.2. PROPOSITION. *Each connected component of the set K is a three-submanifold in $\mathcal{V}$ with a nonempty boundary that must inevitably contain the boundary component for K.*

Before proving these propositions, let us prove two lemmas.

3.5.3. LEMMA. *Let the point $y \in K$ be a boundary point for the set K. Then $\varphi^{-1}(\varphi(y))\backslash y \subset \partial\mathcal{V}$.*

PROOF. Suppose that the point $y' \in \mathcal{V}$, $y' \neq y$, is such that $\varphi(y') = \varphi(y)$. If $y' \in \mathrm{Int}(\mathcal{V})$, then since φ is an immersion, any small neighborhood of the point y in $\mathcal{V}$ lies in the set K. However, this contradicts the statement that y is a boundary point in K. The lemma is proved.

3.5.4. LEMMA. *The boundary of the set K does not intersect the boundary $\partial\mathcal{V}$.*

PROOF. Suppose that there exists a point $y \in \partial\mathcal{V}$, which is a boundary point for K. Then, in view of Lemma 3.5.3, the point $x := \varphi(y)$ does not lie in $\varphi(\mathrm{Int}(\mathcal{V}))$. However, this contradicts Lemma 2.7.2. Hence the lemma is proved.

PROOF OF PROPOSITION 3.5.1. According to Lemma 3.5.4, $\mathrm{Fr}\, K \subset \mathrm{Int}(\mathcal{V})$. By Proposition 3.3.1, $K = \varphi^{-1}(\mathcal{W})$. Hence the boundary $\mathrm{Fr}\, K$ is contained in $\varphi^{-1}(\mathcal{W})$ and clearly does not intersect $\varphi^{-1}\,\mathrm{Int}(\mathcal{W})$. This means that $\mathrm{Fr}\, K \subset \varphi^{-1}(\partial\mathcal{W})$.

Suppose that the point $y \in \mathrm{Int}(\mathcal{V}) \cap \varphi^{-1}(\partial\mathcal{W})$. Then $y \in K$. We assume that the point y lies in K together with a certain neighborhood in $\mathcal{V}$. Then the point $\varphi(x)$ lies in $\varphi(x) = \mathcal{W}$ together with a certain neighborhood in M. However, this contradicts the statement that $\varphi(y) \in \partial\mathcal{W}$. Thus, $y \in \mathrm{Fr}\, K$, and this proves Proposition 3.5.1.

PROOF OF PROPOSITION 3.5.2. Suppose that K_α is a connected component of the set K. In view of Lemma 3.5.4, K cannot lie in $\partial\mathcal{V}$. Hence it follows from Propositions 3.3.1 and 3.4.1 that K_α is a three-dimensional submanifold in $\mathcal{V}$. It remains for us to verify that the boundary of this submanifold contains a component which is a boundary component of the set K.

If this is not so, then K_α coincides with some connected component V_s of the manifold $\mathcal{V}$, where $s \in \mathrm{ske}_0(S)$ is a vertex of rank 1. Hence the set $v_s \cap J$ is empty. This contradicts the statement that the Cr-structure under consideration is a reduced structure in $\mathcal{S}$. The proposition is proved.

3.5.5. COROLLARY. *The mapping $\varphi|K : K \to \mathcal{W}$ is a finite sheeted covering.*

The fact that $\varphi|K$ is a covering follows from Propositions 3.3.1, 3.3.2 and the properties of the mapping φ. That this covering is finite sheeted follows from the local finiteness of the covering.

3.6. Fibration of the noninjectivity set K into two-dimensional tori and the Klein bottle. Let Σ be a foliation on the manifold $\mathcal{V}$ formed by foliations $\mathcal{T}_s$, where s runs through all the vertices of rank 1 in the space S. Thus, fibers of the foliation Σ are circles. Similarly, we define the foliation $\mathcal{T}$ on the manifold $\mathcal{W}$

as composed of foliations $\mathcal{T}_s$ where s runs through all the vertices of rank 2 in the space S.

It follows from Corollary 3.5.5 that the mapping $\varphi|K : K \to \mathcal{W}$ and the foliation $\mathcal{T}$ on $\mathcal{W}$ induce a foliation $\widetilde{\mathcal{T}}$ on K, each of whose fibers is either the torus or the Klein bottle. It follows from the properties of the Cr-structure that each fiber in the foliation $\widetilde{\mathcal{T}}$ is composed of the fibers of the foliation Σ.

The foliation $\widetilde{\mathcal{T}}$ on K has codimension 1, and all its fibers are compact. Hence the space of all fibers of the foliation $\widetilde{\mathcal{T}}$ lying in the connected component K_α of the set K has a natural structure of the connected one-dimensional orbital image Q_α (see [**15,** Theorem 5.9]). The one-sheeted fibers in K_α have the mapping points in Q_α corresponding to them. According to Proposition 3.5.2, K_α always contains a boundary fiber. Hence K_α contains not more than one one-sheeted fiber, while Q_α, which is treated as a topological space, is homeomorphic to the segment.

3.6.1. LEMMA. *The foliation $\widetilde{\mathcal{T}}$ on K does not have unilateral fibers.*

PROOF. Let us assume that the converse is true. In this case, there exists a connected component K_α of the set K containing one-sheeted fibers. Hence the boundary C_α of the manifold K_α is connected and, in accordance with Proposition 3.5.2, is a boundary component of the set K. In particular, C_α lies on the boundary $\overline{J}\backslash J$ of the injectivity set. The image $V_s = \varphi(K_\alpha)$ has a nonempty connected boundary B_s. Hence the rank 2 vertex s is a stopper (see 2.6). Since $C_\alpha \subset \overline{J} \cap \operatorname{Int}(\mathcal{V})$, the restriction $\varphi|C_\alpha$ is injective. On the other hand, C_α lies in the injectivity set K and hence there exists a connected component K_β of K such that $\varphi(C_\alpha) = B_s = \varphi(B_s)$, where $B_\beta \subset K_\beta$ is a component of the boundary $\partial\mathcal{V}$ (see Lemma 3.5.3). In particular, $K_\beta \neq K_\alpha$. According to Proposition 3.5.2, the boundary of K_β also has a boundary component $C_\beta \subset \operatorname{Int}(\mathcal{V})$. Since $\varphi|K : K \to \mathcal{W}$ is a covering and $\varphi(K_\beta) \cap U_s \neq \emptyset$, it follows that $\varphi|K_\beta : K_\beta \to V_s$ is a covering. Since the vertex s is a stopper, $\varphi(C_\beta) = B_s = \varphi(C_\alpha)$. This contradicts Lemma 3.5.3. The lemma is proved.

3.6.2. It follows from Lemma 3.6.1 that each connected component K_α of the set K is homeomorphic to the product $[0,1] \times F_\alpha$, where F_α is either the torus T^2 or the Klein bottle. One of the components $\{0\} \times F_\alpha$, $\{1\} \times F_\alpha$ of the boundary ∂F_α is a boundary component of the set K.

3.6.3. LEMMA. *Let K_α be a connected component of the set K. The restriction of the mapping φ to each connected component of the boundary ∂K_α is an imbedding.*

PROOF. Let $C_\alpha = \operatorname{Fr} K_\alpha$ be the boundary component of the boundary ∂K_α. Since $C_\alpha \subset \overline{J} \cap \operatorname{Int}(\mathcal{V})$, we obtain in the same way as in the proof of Lemma 3.6.1 that the restriction of φ to C_α is an imbedding.

Let $s \in \operatorname{ske}_0(S)$ be a rank 2 vertex for which $\varphi(K_\alpha) = V_s$. For any point $x \in \partial V_s$ in the manifold V_s there exists a smooth path $\gamma : [0,1] \to V_s$ with the beginning $x = \gamma(0)$ and the end $\gamma(1) \in \partial V_s$, which intersects each leaf of the foliation $\mathcal{T}_s$ transversally. Any lift of this path (the path $\tilde{\gamma} : [0,1] \to K_\alpha$) satisfying the condition $\varphi \circ \tilde{\gamma} = \gamma$ also intersects transversally the leaves of the foliation $\widetilde{\mathcal{T}}$ that it encounters. Hence the beginning and end of the path $\tilde{\gamma}$ lie in different components of the boundary of the manifold K_α. Since $\varphi'|C_\alpha$ is an imbedding, it becomes obvious that the restriction φ to the other component of the boundary ∂K_α is also an imbedding. The lemma is proved.

3.7. Proof of Theorem 3.1.

3.7.1. LEMMA. *Let $s \in \mathrm{ske}_0(S)$ be a rank 2 vertex. If s is a jumper, the inverse image $\varphi^{-1}(V_s)$ consists of two connected components of the manifold K. If s is a stopper, the inverse image $\varphi^{-1}(V_s)$ is a connected component of the manifold K.*

PROOF. The inverse image $\varphi^{-1}(V_s)$ consists of the connected components of the manifold K, each of which is homeomorphic to the product $[0,1] \times F_{s,\alpha}$, where $F_{s,\alpha}$ is either the torus T^2 or the Klein bottle. According to Lemma 3.6.3, the restrictions of φ to $\{0\} \times F_{s,\alpha}$ and $\{1\} \times F_{s,\alpha}$ are injective.

If the vertex s is a jumper, the boundary components $\{0\} \times F_{s,\alpha}$ and $\{1\} \times F_{s,\alpha}$ of each connected component $K_{s,\alpha}$ of the set $\varphi^{-1}(V_s)$ are mapped into different components of the boundary ∂V_s. Hence the restriction $\varphi|K_{s,\alpha}$ is injective, and the set $\varphi^{-1}(V_s)$ contains at least two components. If the set contained more than two components, by Proposition 3.5.2 there would exist two components of the set $\mathrm{Fr}\, K$ mapped by φ onto the same component of the boundary ∂V_s. This contradicts Lemma 3.5.3.

If the vertex s is a stopper, the boundary ∂V_s is connected. Arguments similar to those presented above show that the set $\varphi^{-1}(V_s)$ cannot contain more than one connected component. This proves the lemma.

3.7.2. LEMMA. *Let K_α be a connected component of the manifold K. Then the boundary ∂K_α consists of two components, one of which coincides with a certain component of the boundary $\partial \mathcal{V}$, while the other lies inside $\mathcal{V}$.*

PROOF. According to 3.6.2 and Proposition 3.5.2, it is sufficient to prove that both components of the boundary of the manifold K_α cannot lie inside $\mathcal{V}$, i.e., cannot be boundary components for the set K_α in $\mathcal{V}$.

The image $\varphi(K_\alpha)$ coincides with the manifold V_s, where the vertex $s \in \mathrm{ske}_0(S)$ has rank 2. If s is a jumper, then according to Lemma 3.7.1 the set $\varphi^{-1}(\varphi(K_\alpha)) = \varphi^{-1}(V_s)$ contains a component $K_{\alpha'} \neq K_\alpha$ which also has a boundary component of the boundary. If s is a stopper, then according to Lemma 3.7.1 the set $\varphi^{-1}(\varphi(K_\alpha))$ coincides with K_α. In the latter case, the images for φ corresponding to different components of the boundary ∂K_α coincide.

In both cases, Lemma 3.5.3 forbids both components of the boundary ∂K_α to lie inside $\mathcal{V}$. The lemma is proved.

3.7.3. LEMMA. *The covering $\varphi|K : K \to \mathcal{W}$ is two-sheeted.*

PROOF. Let K_α be a connected component of the manifold K. In this case, $\varphi(K_\alpha) = V_s$, where $s \in \mathrm{ske}_0(S)$ is a vertex of rank 2.

If s is a jumper, then according to Lemma 3.7.1 the inverse image $\varphi^{-1}(\varphi(K_\alpha))$ consists of two connected components. It was shown in the proof of Lemma 3.7.1 that the restriction of φ to each component is injective. Hence $\varphi|\varphi^{-1}(V_s) : \varphi^{-1}(V_s) \to V_s$ is a two-sheeted covering.

If the vertex s is a stopper, then according to Lemma 3.7.1 the inverse image $\varphi^{-1}(\varphi(K_\alpha))$ coincides with K_α. In this case, φ maps both components of the boundary ∂K_α onto the (connected) boundary ∂V_s. It follows from Lemma 3.6.3 that in this case also, the covering $\varphi|\varphi^{-1}(V_s) = \varphi|K_\alpha : K_\alpha \to V_s$ is two-sheeted. The lemma is proved.

In order to complete the proof of Theorem 3.1, it remains to prove the existence of the involution $h\colon K \to K$ which does not contain stationary points and transposes the boundary components of $\operatorname{Fr} K$ and the components of the boundary ∂K lying in $\partial\mathcal{V}$. For this involution $\varphi \circ h = \varphi|K$, and the induced mapping $\overline{\varphi} : \mathcal{V}/h \to M$ is a homeomorphism.

According to Lemma 3.7.3, the covering $\varphi|K : K \to \mathcal{W}$, being two-sheeted, is normal. The group of its deck transformations is isomorphic to $\mathbf{Z}_2$ and contains the involution $h : K \to K$, which possesses the required properties. This proves Theorem 3.1.

3.8. Absence of disks.

PROOF OF COROLLARY 3.1.1. Suppose that there exists a vertex $s \in \operatorname{ske}_0(S)$ of rank 1 for which $\widetilde{V}_s = D^2 \times s^1$. In this case, the manifold V_s has a connected boundary which is homeomorphic either to the torus T^2 or to the Klein bottle, and the imbedding $\partial V_s \hookrightarrow V_s$ is not injective on the level of fundamental groups. According to Proposition 3.4.1, there exists a rank 2 vertex s_0 for which $\varphi(\partial V_s) = B_{s_0}$, where B_{s_0} is a component of the boundary ∂V_{s_0}. It follows from Proposition 3.3.1 that $\partial V_s \subset K$, while according to Lemma 3.6.3 the mapping $\varphi|\partial V_s : \partial V_s \to B_{s_0}$ is a homeomorphism. Since the inclusion homomorphisms $\pi_1(\partial V_s) \to \pi_1(V_s)$ is not injective, the homomorphism $\pi_1(B_{s_0}) \to \pi_1(M)$ induced by the inclusion $B_{s_0} \hookrightarrow M$ is also not injective. Recalling the structure of the manifold V_{s_0} (see 2.6), we find that the homomorphism $\pi_1(V_{s_0}) \to \pi_1(M)$ induced by the imbedding $\varphi_{s_0} : V_{s_0} \to M$ is not injective. This contradicts the π_1-noninjectivity, and the corollary is proved.

The rest of this paper is devoted to a proof of Theorem 1.5.

§4. Auxiliary remarks

In this section, we present the information from [**1, 3, 5**] that is necessary for proving Theorem 1.5.

4.1. Commensurable groups. Simplicial space sp Cab **of commensurable abelian subgroups.** 4.1.1. Let Γ be a group. Subgroups a and a' of Γ are said to be commensurable if their intersection $a \cap a'$ has finite index in both a and a'.

We denote by Cab the set of classes of commensurable Abelian positive rank subgroups of the group Γ. The inclusion relation induces a partial order on Cab: the element α precedes the element α' (notation: $\alpha \le \alpha'$) if there exist subgroups $a \in \alpha$ and $a' \in \alpha'$ such that $a \subset a'$.

4.1.2. The set Cab can be identified with the set of vertices of the ordered simplicial space sp Cab: the condition $\alpha_1 \le \cdots \le \alpha_k$ is satisfied if and only if

$\alpha_1, \ldots, \alpha_k$ form an ordered set of vertices from sp Cab. If each element $\alpha \in$ Cab is assigned a rank $\mathrm{rk}(\alpha) := \mathrm{rk}(a)$, where $a \in \alpha$, then sp Cab is transformed into a graded simplicial space (see 2.2).

4.1.3. The group Γ acts by conjugations on the set of its subgroups: $(\gamma, a) \mapsto \gamma a \gamma^{-1}$. This action is transformed to the set Cab and preserves its partial order. In its turn, the action of the group Γ on Cab is continued to the action of Γ by monotonic simplicial homeomorphisms on the simplicial space sp Cab.

4.1.4. THEOREM (see Theorem 2.4 in [**5**]). *Each* Cr-*structure* $\mathcal{S}$ *on a connected manifold* M *canonically defines a nonempty, complete, connected and* Γ-*invariant subspace* $C(\mathcal{S})$ *of the simplicial space* sp Cab *of the fundamental group* $\Gamma := \pi_1(M)$.

4.2. Euclidean planes and Euclidean spheres at infinity for an Hadamard manifold. The simplicial space sp Es. It should be recalled that an Hadamard manifold is a complete simply connected Riemannian manifold of nonpositive sectional curvature. A detailed presentation of all the pertinent information about Hadamard manifolds can be found in [**9**].

If M is a complete Riemannian manifold of nonpositive sectional curvature, its universal covering X is an Hadamard manifold. The fundamental group Γ acts on X by isometries discretely and freely in such a way that the projection $\pi : X \to X/\Gamma$ is a universal covering of the manifold $M = X/\Gamma$.

4.2.1. A completely geodesic subspace $E \subset X$ isometric to the Euclidean subspace $\mathbf{R}^k$, $k > 0$, is called a Euclidean k-plane. Its boundary $\partial_\infty E$ at infinity is isometric in Tietze's metric to the absolute $\partial_\infty X$ of a unit sphere in $\mathbf{R}^k$ (see [**9**]).

4.2.2. We denote by Es the collection of all subsets in $\partial_\infty X$ of the type $\partial_\infty E$ (Euclidean spheres), where E is a Euclidean plane in X of positive dimension. The inclusion relation introduces a partial order in Es, which enables us, as in 4.1.2, to consider Es as a set of vertices of the ordered simplicial space sp Es. Each vertex $l \in$ Es has rank $\mathrm{rk}(l) := \dim E$, where E is a Euclidean plane in X with $\partial_\infty E = l$. In particular, sp Es is a graded simplicial space.

4.2.3. The isometries $X \to X$ of the Hadamard manifold X induce the isometries $\partial_\infty X \to \partial_\infty X$ (in Tietze's metric). Hence the group Γ acts on sp Es through monotonic simplicial homeomorphisms (see [**1**] for details).

4.2.4. The isometry $\gamma \in \Gamma$ is said to be hyperbolic if its displacement function $\delta_\gamma(x) = \rho(x, \gamma(x))$ (ρ is the distance in X) attains a minimum that is positive. A hyperbolic isometry must have an invariant straight line in X, while the set formed by all its invariant straight lines is convex and splits metrically as $D \times \mathbf{R}$.

4.2.5. If X/Γ is a compact manifold, then $\Gamma \setminus \mathrm{id}$ consists of hyperbolic isometries (see [**9**]). In a more general form, the statement is also valid if $\Gamma \setminus \mathrm{id}$ does not contain peripheral elements (see [**6**]).

4.2.6. THEOREM (see Theorem 3.1, Corollary 3.5, and Remark 3.7 in [**5**]). *Let* C *be a complete subspace of the simplicial space* sp Cab *of the fundamental group* Γ *of the complete Riemannian manifold* $M = X/\Gamma$ *with nonpositive sectional curvature, and let each vertex in* C *have a representative that is an Abelian subgroup in* Γ *consisting of hyperbolic isometries.*

Then C *can be identified canonically with the complete subspace of the simplicial space* sp Es *of Euclidean spheres of the absolute* $\partial_\infty X$.

4.3. Convex shells Conv(l) **of the vertices of the space** sp Es **and conditions for the existence of a Cr-structure.**

4.3.1. Let $M = X/\Gamma$ be a complete Riemannian manifold of nonpositive sectional curvature. For a "Euclidean sphere at infinity" $l \in$ Es, we denote by Conv(l) the subset in X consisting of all Euclidean k-planes ($k = \text{rk}(l)$) the boundary of each of which coincides with l in $\partial_\infty X$. The set Conv(l) is a closed convex subset in X which is isometric to the metric product $D \times \mathbf{R}^k$, where all Euclidean k-planes $d \times \mathbf{R}^k \subset D \times \mathbf{R}^k$ satisfy the condition $\partial_\infty(d \times \mathbf{R}^k) = l$ (see [**1**] for details).

4.3.2. For the subset A of the space on which the group Γ acts, we shall denote its stabilizer $\{\gamma \in \Gamma : \gamma(A) = A\}$ by Stab(A).

4.3.3. Theorem (see Proposition 4.2 in [**5**]). *Let $M = X/\Gamma$ be a complete three-manifold with nonpositive sectional curvature satisfying the following conditions.*

(a) *The simplicial space* sp Cab *of commensurable Abelian subgroups in Γ contains a nonempty complete connected Γ-invariant subspace $\mathcal{C}$.*
(b) *The action of the group Γ in $\mathcal{C}$ does not have finite orbits.*
(c) *Each vertex in $\mathcal{C}$ has a representative in the form of an Abelian subgroup in Γ consisting of hyperbolic isometries.*

In accordance with Theorem 4.2.6, we identify $\mathcal{C}$ with a complete subspace of the simplicial space sp Es. *Then,*

1. *For any vertex $l \in \text{ske}_0(\mathcal{C})$, there exists a free Abelian group $a_l \in l$ which acts on the set $\text{Conv}(l) = D \times R^k$, $k = \text{rk}(l)$, by isometries of the type* (id, shift).
2. *For any vertex $l \in \text{ske}_0(\mathcal{C})$, the set* Conv($l$) *contains a Euclidean hyperplane.*
3. *The collection $\{\text{Conv}(l) : l \in \text{ske}_0(\mathcal{C})\}$ is closed relative to intersections: if the set $\bigcap_\alpha \text{Conv}(l_\alpha)$ is not empty for the tuple $\{l_\alpha\} \subset \text{ske}_0(\mathcal{C})$, there exists a vertex $l \in \text{ske}_0(\mathcal{C})$ such that $\bigcap_\alpha l_\alpha \subset l$ and $\bigcap_\alpha \text{Conv}(l_\alpha) = \text{Conv}(l)$.*
4. *The set $\text{Conv}(\mathcal{C}) := \bigcup\{\text{Conv}(l) : l \in \text{ske}_0(\mathcal{C})\}$ is convex, and its covering by the sets* Conv(l), *$l \in \text{ske}_0(\mathcal{C})$, is locally finite.*

4.3.4. Theorem (see §3 in [**1**]). *Let $M = X/\Gamma$ be a complete n-dimensional ($n \geq 2$) Riemannian manifold of nonpositive sectional curvature. Suppose that the corresponding simplicial space* sp Es *contains a complete Γ-invariant subspace $\mathcal{C}$ having properties* (1)–(4) *in Theorem* 4.3.3. *Moreover, assume that the group* Stab(l) *acts on the set* Conv(l) *uniformly for each vertex $l \in \text{ske}_0(\mathcal{C})$, and that the Γ-invariant convex set* Conv(l) *is open in X.*

Then the set $M^* := \text{Conv}(\mathcal{C})/\Gamma$ is an absolutely convex open subset in M having Cr-*structure $\mathcal{S}^*$ such that its metric induced by the inclusion $M^* \subset M$ is matched with $\mathcal{S}^*$.*

4.3.5. Remark. Let us briefly describe the method used for obtaining the Cr-structure $\mathcal{S}^*$. Its simplicial space S^* is defined as $\mathcal{C}/\Gamma$. For the sake of simplicity, we assume that each vertex l is nondegenerate. In other words, the convex set Conv(l) is n-dimensional (for a degenerate vertex l, the set Conv(l) is a Euclidean hyperplane according to statement (2) of Theorem 4.3.3). Let $l \in \mathcal{C}$ and $s = \Gamma(l) \in S^*$. We assume that $V_s := \text{Conv}(l)/\text{Stab}(l)$. According to Theorem 4.3.4 and the assumption about nondegeneracy, V_s is a compact n-manifold with boundary (which may be empty; in this case, $\text{Conv}(l) = X$). The immersion $\varphi_s : V_s \to M$

is induced by the inclusions $\mathrm{Conv}(l) \subset X$ and $\mathrm{Stab}(l) \subset \Gamma$. It has been proved in §3 of [**1**] that $\mathcal{S}^* = (S^*, \{\varphi_s : s \in \mathrm{ske}_0(S^*)\})$ is a Cr-structure on the manifold M^* under these conditions. (See also §6.)

4.4. Complex Γ-invariant sets. Before concluding this section, let us consider two simple statements about convex sets.

4.4.1. LEMMA. *Let $M = X/\Gamma$ be an open manifold with nonpositive sectional curvature, and let A be a nonempty convex Γ-invariant subset of the Hadamard manifold X. Then for any connected component B of the complement $X \backslash A$, the stabilizer* $\mathrm{Stab}(B)$ *consists of peripheral elements.*

PROOF. Let $\gamma \in \mathrm{Stab}(B)$, and let K be a compact in M. We can find a metric tubular neighborhood V of the set A such that its image U contains K in the case of the universal covering $\pi : X \to X/\Gamma$. Since sectional curvatures are nonpositive, the set V is convex. Hence the set $B \backslash V$ is connected. Since points from V are at a limited distance from A, $B \backslash V$ is not empty. Let $x \in B \backslash V$. Since $\gamma(V) = V$, this means that $\gamma(x) \in B \backslash V$. Hence we can find a path in $B \backslash V$ connecting the points x and $\gamma(x)$. Its image in M will be a loop lying in $M \backslash U \subset M \backslash K$ and representing the free homotopy class of the element γ. The lemma is proved.

4.4.2. COROLLARY. *Let $M = X/\Gamma$ be an open manifold of nonpositive sectional curvature and let A be a convex noncompact Γ-invariant subset of the Hadamard manifold X such that A/Γ is compact in M. Then the group Γ contains nontrivial peripheral elements.*

PROOF. Let B be a connected component of the complement $X \backslash A$. Since A/Γ is compact and A is not compact, the stabilizer of the boundary $\overline{B} \backslash B \subset A$ is nontrivial. Hence the stabilizer $\mathrm{Stab}(B) \subset \Gamma$ is also nontrivial. The corollary now follows from the preceding lemma.

§5. Proof of Theorem 1.5 on homotopy equivalent three-manifolds with nonpositive curvature

Let us briefly outline the strategy to be used for proving Theorem 1.5 in 5.1 and 5.3. This theorem is proved for the cases when the manifold M_0 is flat and when it is closed, respectively. These cases can be covered easily, using well-known facts. In 5.2.3, we shall also consider a simple case when the action of the group $\Gamma_0 := \pi_1(M_0)$ on the simplicial space $\mathcal{C}_0$ has a finite orbit. Thus, the main difficulties are associated with the case of noncompact manifolds M_0 and M_1. Two things have to be considered in this case, viz. the representation of the required Abelian subgroups in Γ_0, $\Gamma_1 := \pi_1(M_1)$ by hyperbolic isometries (see Theorems 4.2.6 and 4.3.3), and the uniformity of the action of the group $\mathrm{Stab}(l)$ on the set $\mathrm{Conv}(l)$ (see Theorem 4.3.4). For the manifold M_0, these conditions are satisfied in view of condition (a) of Theorem 1.5 (see 5.2). Hence Theorems 4.3.3 and 4.3.4 enable us to find an absolutely convex submanifold $M_0^* \subset M_0$ with a Cr-structure $\mathcal{S}_0^*$. The diffeomorphism of manifolds M_0^* and M_0 can be derived easily from the absence of nontrivial peripheral elements in Γ_0 (see 5.4). Since the latter property is not homotopy invariant, the existence of the isomorphism $\Gamma_0 \to \Gamma_1$ does not mean that the fundamental group Γ_1 of the manifold M_1 does not have nontrivial peripheral elements.

However, the isomorphism $\Gamma_0 \to \Gamma_1$ induces the isomorphism of the corresponding simplicial spaces of commensurable Abelian subgroups, and thus the space sp Cab for the group Γ_1 contains a nonempty, connected, complete and Γ_1-invariant subspace $\mathcal{C}_1$ isomorphic to $\mathcal{C}_0$. Condition (a) of Theorem (1.5) is replaced for the group Γ_1 by arguments from [**7**] (see 5.6). This enables us to apply Theorems 4.3.3 and 4.3.4 to the space $\mathcal{C}_1$.

5.1. If the manifold M_0 is flat, then, since the group Γ_0 does not have nontrivial peripheral elements, it follows from Theorem 3.3.3 in [**10**] that M_0 is closed. Theorem 1.5 now follows easily from results obtained in Chapter 3 of [**10**].

Below we assume that the manifold M_0 is not flat.

5.2. Uniformity of action of vertex stabilizers. The manifold M_0 has a Cr-structure. Hence, according to Theorem 4.1.4, there exists a nonempty, connected, complete, and Γ_0-invariant subspace $\mathcal{C}$ of the simplicial space sp Cab of the fundamental group Γ_0.

5.2.1. Since the group Γ_0 does not have nontrivial peripheral elements, it acts on the universal covering X_0 of the manifold M_0 through hyperbolic isometries (see 4.2.5). Hence, in accordance with Theorem 4.2.6, we can identify $\mathcal{C}_0$ with a complete subspace of the simplicial space sp Es constructed from the absolute $\partial_\infty X_0$.

5.2.2. LEMMA. *For any vertex $l \in \operatorname{ske}_0(\mathcal{C})$, the group* Stab$(l)$ *acts on the set* Conv(l) *uniformly.*

PROOF. To begin with, we assume that

(*) a certain Abelian group $a \in l$ acts on the set $\operatorname{Conv}(l) = D \times \mathbf{R}^k$, $k = \operatorname{rk}(l)$, through isometries of the type (id, shift).

Then a loop of fixed length from the free homotopy class of any element $\gamma \in a$ passes through each point of the set $\pi(\operatorname{Conv}(l)) \subset M_0$ (here, $\pi := x_0 \to M_0$ is a universal covering). Since Γ_0 does not contain nontrivial peripheral elements, it follows that the set $\pi(\operatorname{Conv}(l))$ is bounded.

Now we use the construction presented in §3 of [**3**]. Let $\operatorname{Gr}_k(X_0) \to X_0$ be a Grassmannian foliation of k-planes in tangent spaces to X_0. The action of the group Γ_0 on X_0 is continued to the action on $\operatorname{Gr}_k(X_0)$, which is totally disconnected and free. It was shown in §3 of [**3**] that under the assumption (*), the following imbedding holds:

$$\varphi : \operatorname{Conv}(l)/\operatorname{Stab}(l) \to \operatorname{Gr}_k(X_0)/\Gamma_0 = \operatorname{Gr}_k(M_0)$$

and $\operatorname{pr} \circ \varphi = \pi|\operatorname{Conv}(l)$, where $\operatorname{pr} : \operatorname{Gr}_k(M_0) \to M_0$ is the natural projection. In particular, the image φ is closed. Since the foliation $\operatorname{Gr}_k(M_0) \to M_0$ has compact fibers, the boundedness of the set $\pi(\operatorname{Conv}(l))$ means that the image φ is bounded. Hence this image is compact as well as the manifold $\operatorname{Conv}(l)/\operatorname{Stab}(l)$.

It remains for us to prove that the assumption (*) is satisfied. Since $n = 3$, we can use Lemma 3.4 from [**3**]. According to this lemma, if condition (*) is not satisfied, then $k = 1$ and either $\operatorname{Conv}(l) = X_0$, or $\operatorname{Conv}(l) = D \times \mathbf{R}$, where D is a two-dimensional metric disk. Any group $a \in l$ is an infinite cyclic group, acts on

Conv(l) through isometries of the type (γ', shift), and has an invariant straight line of the type $x \in \mathbf{R} \subset D \times \mathbf{R}$.

If D is a disk, then $\mathrm{Conv}(l)/\mathrm{Stab}(l)$ is obviously compact. We assume that $\mathrm{Conv}(l) = X_0$. In this case, since X_0 is not flat, the group $\mathrm{Stab}(l)$ coincides with Γ_0. It follows from condition (a) of Theorem 1.5 and Corollary 4.4.2 that the group $\mathrm{Stab}(l)$ does not have invariant straight lines of the type $x \times \mathbf{R}$ in $X_0 = D \times \mathbf{R}$. Hence for any group $a \in l$, there exists an isometry $\gamma \in \mathrm{Stab}(l)$ such that the invariant straight lines for the groups $\gamma a \gamma^{-1} \in l$ and a do not coincide. But it can easily be seen that in this case (see 4.8.2 in [**5**]) that a certain subgroup a' with index ≤ 2 in $a \cap \gamma a \gamma^{-1}$ belongs to the class l and has the property (*). The lemma is proved.

5.2.3. We prove Theorem 1.5 for the case when the action of the group Γ_0 on the simplicial space $\mathcal{C}_0$ has a finite orbit.

In this case, the group $\mathrm{Stab}(l)$ has a finite index in Γ_0 for a certain vertex $l \in \mathrm{ske}_0(\mathcal{C}_0)$. The set $\mathrm{Conv}(l) \subset X_0$ is convex, noncompact, and $\mathrm{Stab}(l)$-invariant. According to Lemma 5.2.2, $\mathrm{Conv}(l)/\mathrm{Stab}(l)$ is compact. If the manifold M_0 is open, its finite sheeted covering $M_0' := X_0/\mathrm{Stab}(l)$ is also open. Then, in accordance with Corollary 4.4.2, the fundamental group of the manifold M_0', and hence the manifold M_0 also, has nontrivial peripheral elements, which is impossible. Hence the manifolds M_0 and M_0' are closed. In this case, the $\mathrm{Stab}(l)$-invariant convex set $\mathrm{Conv}(l)$ coincides with X_0. Since X_0 is not flat, $\mathrm{rk}(l) = 1$ and $\mathrm{Stab}(l) = \Gamma_0$. In other words, the manifold M_0 is a compact Seifert fibration. It is well known [**11**] that in this case the manifold M_1 is diffeomorphic to M_0, and the metric of M_1 locally contains a Euclidean cofactor matched with the corresponding trivial Cr-structure (see [**5**]). Theorem 1.5 is proved for this case.

5.2.4. Below we assume that the action of the group Γ_0 on $\mathcal{C}_0$ does not have finite orbits. Since the fundamental group of a flat manifold contains a normal free Abelian subgroup with finite index, it follows in particular that the manifold M_1 is not flat.

Thus, in view of 5.2.1, Lemma 5.2.2, and Theorem 4.3.3, all conditions of Theorem 4.3.4 are satisfied for the manifold M_0, except possibly for the condition of openness of the set $\mathrm{Conv}(\mathcal{C}_0)$ in X_0.

5.3. Proof of Theorem 1.5 for the case when the manifold M_0 is closed. In this case, the Γ_0-invariant set $\mathrm{Conv}(\mathcal{C}_0)$ coincides with X_0 and we can use Theorem 4.3.4. Since the action of the group Γ_0 on the simplicial space $\mathcal{C}_0$ does not have finite orbits, the simplicial space $S^* = \mathcal{C}_0/\Gamma_0$ of the Cr-structure $\mathcal{S}_0^*$ on M_0 contains vertices of rank 1 as well as of rank 2. It can be derived easily from the property of π_1-injectivity of Cr-structure and Theorem 1.1 that the manifold M_0 contains an incompressible surface. The fundamental group Γ_0 has no finite order elements, and the universal covering X_0 of the manifold M_0 is diffeomorphic to the Euclidean space $\mathbf{R}^3$. Hence the manifold M_0 is P^2-irreducible and quite large in the orientable case. The manifold M_1 is homotopy equivalent to the manifold M_0 and is closed together with it. Hence, according to [**12**], the manifolds M_0 and M_1 are diffeomorphic. The statement about the matching of the metric of M_1 with its Cr-structure follows from [**5**]. Theorem 1.5 is proved for this case.

Below we assume that the manifolds M_0 and M_1 are open.

5.4. Diffeomorphism of the manifolds M_0^* and M_0. In this case, we establish the diffeomorphism of the manifolds $M_0^* := \mathrm{Conv}(\mathcal{C})/\Gamma_0$ and M_0 (see Theorem 4.3.4) and consider the relation between these manifolds depending on the behavior of the injectivity radius. The results of this subsection will also be applied to the manifold M_1.

The set $\mathrm{Conv}(\mathcal{C}_0) = \bigcup\{\mathrm{Conv}(l) : l \in \mathrm{ske}_0(\mathcal{C}_0)\}$ considered in Theorem 4.3.3 is convex and Γ_0-invariant. Since the group Γ_0 does not have nontrivial peripheral elements, Lemma 4.4.1 leads to the following corollaries.

5.4.1. COROLLARY. *For any connected component B of $X_0 \setminus \mathrm{Conv}(\mathcal{C}_0)$, the stabilizer* $\mathrm{Stab}(B)$ *is trivial.*

5.4.2. COROLLARY. *If the injectivity radius of the manifold M_0 is bounded along any geodesic, then* $\mathrm{Conv}(\mathcal{C}_0) = X_0$*, and hence $M_0^* = M_0$.*

PROOF. We assume that $\mathrm{Conv}(\mathcal{C}_0) \neq X_0$. In this case, there exists a geodesic $h\colon [0,\infty) \to X_0$ which has a vertex $h(0)$ at the boundary of the set $\mathrm{Conv}(\mathcal{C}_0)$ and is an outward normal to it. Since the set $\mathrm{Conv}(\mathcal{C}_0)$ is convex and Γ_0-invariant, while the sectional curvatures are nonpositive, the displacement function δ_γ of any isometry $\gamma \in \Gamma_0$ is convex and does not decrease along the geodesic h. If in this case $\gamma \neq \mathrm{id}$, then according to Corollary 5.4.1 the geodesics h and $\gamma(h)$ lie in different connected components of the complement $X_0 \setminus \mathrm{Conv}(\mathcal{C}_0)$. It follows from the formula for the first variation of a geodesic that the derivative of the function δ_γ along the geodesic h is strictly positive at the initial moment.

Let δ_0 be a supremum of the injectivity radius of the manifold M_0 along the geodesic $\pi \circ h$. Since $\delta_0 < \infty$, the tuple $A = \{\gamma \in \Gamma_0\colon \delta_\gamma(h(0)) \leq 2\delta_0\}$ of isometries of the group Γ_0 is finite. Hence it follows from the above that there exists $t_0 > 0$ such that $\delta_\gamma(h(t_0)) > 2\delta_0$ for all $\gamma \in A$. But then this inequality is valid for all $\gamma \in \Gamma_0 \setminus \mathrm{id}$. This contradicts the fact that the injectivity radius of the manifold M_0 does not exceed Δ_0 at the point $\pi \circ h(t_0)$. Hence $\mathrm{Conv}(\mathcal{C}_0) = X_0$. The corollary is proved.

5.4.3. LEMMA. *The set* $\mathrm{Conv}(\mathcal{C}_0)$ *is open, and each connected component of its boundary is a Euclidean 2-plane in X_0.*

PROOF. Let $l \in \mathrm{ske}_0(\mathcal{C}_0)$. The set $\mathrm{Conv}(l)$ is closed and the stabilizer $\mathrm{Stab}(l)$ acts uniformly on it. Hence the stabilizer of any connected component of $X_0 \setminus \mathrm{Conv}(l)$ is not trivial. It follows from Corollary 5.4.1 that none of the components of $X_0 \setminus \mathrm{Conv}(l)$ is a component of $X_0 \setminus \mathrm{Conv}(\mathcal{C}_0)$. But then it follows from statements (2) and (3) of Theorem 4.3.3 that the set $\mathrm{Conv}(l)$ lies inside $\mathrm{Conv}(\mathcal{C}_0)$. Hence the set $\mathrm{Conv}(\mathcal{C}_0)$ is open.

It should be noted now that each connected component of the boundary of the set $\mathrm{Conv}(l)$ for any vertex $l \in \mathrm{ske}_0(\mathcal{C}_0)$ is a Euclidean 2-plane. This is obvious for a rank-2 vertex since the set $\mathrm{Conv}(l)$ is isometric to the metric product of the segment (probably degenerate) and the Euclidean 2-plane. For a vertex of rank one, it follows from statement (3) of Theorem 4.3.3 and the fact that $\mathrm{Conv}(l)$ lies inside $\mathrm{Conv}(\mathcal{C}_0)$.

Moreover, it also follows from the above-mentioned statement (3) that the connected components of the boundaries of sets $\mathrm{Conv}(l)$ either coincide or do not intersect. Hence each connected component of the boundary of the set $\mathrm{Conv}(\mathcal{C}_0)$, being limiting for the set of disjunctive Euclidean 2-planes, is itself a Euclidean 2-plane. The lemma is proved.

5.4.4. PROPOSITION. *The manifolds* M_0^* *and* M_0 *are diffeomorphic.*

PROOF. It follows from Lemma 5.4.3 that the closed convex and Γ_0-invariant set $A := \mathrm{Conv}(\mathcal{C}_0)$ is a three-manifold in X_0 with a smooth boundary. Using the geodesics in $X_0 \setminus A$ orthogonal to A at the points on the boundary, we can easily construct a Γ_0-equivariant diffeomorphism of the manifold X_0 on any metric tubular open neighborhood U of A, and hence the diffeomorphism $M_0 \to U/\Gamma_0$. Since the boundary of the manifold A/Γ_0 is smooth, we can easily construct the diffeomorphism $U/\Gamma_0 \to \mathrm{Conv}(\mathcal{C}_0)/\Gamma_0 = M_0^*$ by using the collar theorem (see, for example, Theorem 5.9 in [**13**]). The proposition is proved.

5.5. Removal of rings. In this subsection, the simplicial space $\mathcal{C}_0$ is slightly modified in such a way that its isomorphic simplicial space $\mathcal{C}_1$ in the group Γ_1 can be realized geometrically. For this purpose, we must remove the vertices $l \in \mathcal{C}_0$ of rank 1 for which the manifold $\mathrm{Conv}(l)/\operatorname{Stab}(l)$ is covered by the product of a ring and a circle.

5.5.1. PROPOSITION. *The simplicial space* $\mathcal{C}_0 \subset \mathrm{sp}\,\mathrm{Cab}$ *can be changed so that it possesses the following properties.*

(a) $\mathcal{C}_0$ *is a nonempty, connected, complete, and* Γ_0*-invariant subspace in* $\mathrm{sp}\,\mathrm{Cab}$.
(b) *The action of the group* Γ_0 *on* $\mathcal{C}_0$ *does not have finite orbits.*
(c) *The group* $\operatorname{Stab}(l)$ *for each rank* 1 *vertex* $l \in \mathrm{ske}_0(\mathcal{C}_0)$ *contains a finite index normal subgroup* b_l *isomorphic to* $F_k \times \mathbf{Z}$, *where* F_k *is a free group with* $k \geq 2$ *generators, and the subgroup* $a_l = 1 \times \mathbf{Z} \subset b_l$ *belongs to the class* l.
(d) *Each vertex* $l \in \mathrm{ske}_0(\mathcal{C}_0)$ *of rank* 2 *has exactly two adjacent vertices.*

5.5.2. The change mentioned in Proposition 5.5.1 consists in the following. First, we eliminate all vertices $l \in \mathrm{ske}_0(\mathcal{C}_0)$ of rank 1 for which $\mathrm{Conv}(l) = \mathrm{Conv}(l')$, where l' is a rank 2 vertex adjacent to l. Second, we eliminate all rank 2 vertices $l \in \mathrm{ske}_0(\mathcal{C}_0)$ for which the set $\mathrm{Conv}(l)$ lies inside the set $\mathrm{Conv}(l')$ for a certain vertex $l' \in \mathrm{ske}_0(\mathcal{C}_0)$. It can easily be shown with the help of Lemmas 5.5.3 and 5.5.4 that the properties (a)–(d) are satisfied.

5.5.3. LEMMA. *The rank* 1 *vertex* $l \in \mathrm{ske}_0(\mathcal{C}_0)$ *has only one adjacent vertex* l' *in* $\mathcal{C}_0$ *if and only if the set* D *in the metric decomposition* $\mathrm{Conv}(l) = D \times \mathbf{R}$ *is isometric to the strip* $[a, b] \times \mathbf{R}$ *which may degenerate into a straight line. In this case,* $\mathrm{Conv}(l) = \mathrm{Conv}(l')$.

PROOF. Let $D = [a, b] \times \mathbf{R}$ and let $l' \in \mathrm{ske}_0(\mathcal{C}_0)$ be a vertex adjoining l. In this case, $\mathrm{rk}(l') = 2$, $l' \supset l$ and $\mathrm{Conv}(l') \subset \mathrm{Conv}(l)$. Since the set $\mathrm{Conv}(l')$ is isometric to the metric product $[c, d] \times \mathbf{R}^2$ and consists of all Euclidean 2-planes with a common boundary in $\partial_\infty X$, it follows from the equality $\mathrm{Conv}(l) = [a, b] \times \mathbf{R}^2$ that $\mathrm{Conv}(l) = \mathrm{Conv}(l')$. Hence any two vertices adjoining l in $\mathcal{C}_0$ coincide.

Now we observe that since $\mathrm{Conv}(l)$ lies inside $\mathrm{Conv}(\mathcal{C}_0)$ (Lemma 5.4.3), each boundary component of $\mathrm{Conv}(l)$ is a boundary component of $\mathrm{Conv}(l')$ for a vertex l' adjoining l (see statements (2) and (3) in Theorem 4.3.3). Hence, if the vertex l has only one adjoining vertex, the boundary of $\mathrm{Conv}(l)$ cannot have more than two components. Since the group $\mathrm{Stab}(l)$ acts uniformly on $\mathrm{Conv}(l)$ (Lemma 5.2.2), this can happen only in the case when $\mathrm{Conv}(l) = D' \times \mathbf{R}^2$, where D' is a segment (possibly degenerate). The lemma is proved.

5.5.4. LEMMA. *The rank* 2 *vertex* $l \in \mathrm{ske}_0(\mathcal{C}_0)$ *has only one adjacent vertex* l' *for which* $\mathrm{Conv}(l') \neq \mathrm{Conv}(l)$ *if and only if the set* $\mathrm{Conv}(l)$ *lies inside the set* $\mathrm{Conv}(l')$.

PROOF. Suppose that the vertex l has only one adjoining vertex l'. Since $l \supset l'$, we have $\mathrm{Conv}(l) \subset \mathrm{Conv}(l')$. We assume that the set $\mathrm{Conv}(l)$ touches the boundary $\partial\,\mathrm{Conv}(l')$. In this case, using the fact that $\mathrm{Conv}(l')$ lies inside $\mathrm{Conv}(\mathcal{C}_0)$ (Lemma 5.4.3) as well as statements (2) and (3) of Theorem 4.3.3, we can find a rank 1 vertex l'' differing from l' such that $l'' \subset l$. This contradicts our assumption. Hence $\mathrm{Conv}(l)$ lies inside $\mathrm{Conv}(l')$.

Conversely, suppose that $\mathrm{Conv}(l)$ lies inside $\mathrm{Conv}(l')$ for the rank 2 vertex $l \in \mathrm{ske}_0(\mathcal{C}_0)$. Then, according to statement (3) of Theorem 4.3.3, $l' \subset l$. If $l'' \neq l'$ for the rank 1 vertex $l'' \subset l$, it follows from statements (2) and (3) of Theorem 4.3.3 that $\mathrm{Conv}(l'') = \mathrm{Conv}(l)$. The theorem is proved.

5.5.5. LEMMA. *Let* $l \in \mathrm{ske}_0(\mathcal{C}_0)$ *be a rank* 1 *vertex having at least two adjoining vertices in* $\mathcal{C}_0$. *Then the group* $\mathrm{Stab}(l)$ *contains a normal finite index subgroup* b_l *isomorphic to* $F_k \times \mathbf{Z}$, *where* F_k *is a free group with* $k \geq 2$ *generators, the subgroup* $a_l = 1 \times \mathbf{Z}$ *belonging to the class* l. *The manifold* $\mathrm{Conv}(l)/b_l$ *is diffeomorphic to the product* $W_l \times S$, *where* W_l *is a closed orientable surface from which a finite number of open disks have been removed.*

PROOF. It follows from statement (2) of Theorem 4.3.3 and Lemma 5.5.3 that the set $\mathrm{Conv}(l)$ is three-dimensional. Then, in accordance with 3.4 in [**1**] (see also Remark 4.3.5 and the definition of the Cr-structure), the group $\mathrm{Stab}(l)$ contains a subgroup b_l having all the properties mentioned in the lemma with the possible exception of the estimate $k \geq 2$. But if $k = 1$, then since b_l acts uniformly on $\mathrm{Conv}(l)$, the set $\mathrm{Conv}(l)$ is isometric to $D \times \mathbf{R}^2$, where D is a segment, possibly degenerate. However, this contradicts the condition in view of Lemma 5.5.3. The lemma is proved.

PROOF OF PROPOSITION 5.5.1. According to Lemmas 5.5.3 and 5.5.4, the operations described in 5.5.2 are reduced to the elimination of all vertices from $\mathcal{C}_0$ having only one adjacent vertex. Not more than two iterations are needed to obtain the space $\mathcal{C}_0$ in which each vertex has at least two neighbors. Hence the space $\mathcal{C}$ has properties (a) and (b). Property (c) follows from Lemma 5.5.5.

The rank 2 vertex $l \in \mathrm{ske}_0(\mathcal{C}_0)$ has not more than two adjacent vertices l' for which $\mathrm{Conv}(l') \neq \mathrm{Conv}(l)$ (see statements (2) and (3) in Theorem 4.3.3). Property (d) now follows from Lemma 5.5.4. Proposition 5.5.1 is proved.

5.6. Cr-structure $\mathcal{S}_1^*$ on an absolutely convex submanifold $M_1^* \subset M_1$.

5.6.1. According to the condition of Theorem 1.5, there exists an isomorphism $\alpha : \Gamma_0 \to \Gamma_1$ of fundamental groups $\Gamma_i := \pi_1(M_i)$, $i = 0, 1$. This isomorphism induces the isomorphism Φ of the corresponding simplicial spaces of commensurable Abelian groups. We put $\mathcal{C}_1 := \Phi(\mathcal{C}_0)$. In this case, $\mathcal{C}_1$ is a nonempty connected complete Γ_1-invariant subspace in the simplicial space $\operatorname{sp}\operatorname{Cab}$ of the group Γ_1, the properties (a)–(d) of Proposition 5.5.1 remaining valid for $\mathcal{C}_1$.

5.6.2. LEMMA. *Each vertex l in $\mathcal{C}_1$ has as its representative a subgroup in Γ_i formed by hyperbolic isometries.*

PROOF. Let us first assume that $\operatorname{rk}(l) = 1$. In this case, we can take for the required subgroup the group $a_l = 1 \times \mathbf{Z} \subset b_l$ mentioned in property (c) in Proposition 5.5.1. The group a_l lies at the center of the group b_l and is an infinite cyclic group.

If a_l were not formed by hyperbolic isometries, arguments analogous to those presented in §4 of [**7**] would lead to the conclusion that the group b_l acts freely on a certain horosphere in X_1 and is thus a fundamental group of a certain surface. But this is impossible since $b_l = F_k \times \mathbf{Z}$, where F_k is a free group with $k \geq 2$ generators.

Let us now suppose that $\operatorname{rk}(l) = 2$. According to property (d) in Proposition 5.5.1, the vertex l has two different adjoining vertices l_1 and l_2. These are rank 1 vertices, and the subgroups $a_{l_1} \in l_1$, $a_{l_2} \in l_2$ corresponding to them are infinite cyclic groups represented by hyperbolic isometries. Any group $a_l \in l$ is isomorphic to $\mathbf{Z} \oplus \mathbf{Z}$ and contains subgroups commensurable with a_{l_1} as well as a_{l_2}. Since $l_1 \neq l_2$, class l contains a subgroup generated by hyperbolic isometries, and consisting of hyperbolic isometries since we are dealing with an Abelian group. The lemma is proved.

Thus, all conditions of Theorem 4.3.3 are satisfied for the manifold M_1.

5.6.3. LEMMA. *For any vertex $l \in \operatorname{ske}_0(\mathcal{C}_1)$, the group $\operatorname{Stab}(l)$ acts uniformly on the set $\operatorname{Conv}(l)$, and a natural bijection exists between the set of components of the complement $X_1 \setminus \operatorname{Conv}(l)$ and the set L of vertices in $\mathcal{C}_1$ neighboring l.*

PROOF. Let us first assume that $\operatorname{rk}(l) = 2$. In this case, $\operatorname{Conv}(l) = D \times \mathbf{R}^2$, where $D \leq 1$. Since the group $\operatorname{Stab}(l)$ contains a finite index subgroup isomorphic to $\mathbf{Z} \oplus \mathbf{Z}$, and since the set L consists of two elements (property (d) in Proposition 5.5.1), to prove both statements of the lemma it suffices to verify that D is bounded. Let us suppose that this is not so. Since X_1 is not flat (see 5.2.4), $D = [0, \infty)$. Let l_1 and l_2 be two different vertices neighboring l. We make use of property (3) from Theorem 4.3.3, according to which $\operatorname{Conv}(l_1) \cap \operatorname{Conv}(l_2) = \operatorname{Conv}(l)$. Then at least one of the sets $\operatorname{Conv}(l_1)$, $\operatorname{Conv}(l_2)$ coincides with $\operatorname{Conv}(l)$. But this is impossible since the group $F_k \times \mathbf{Z}$ with $k \geq 2$ cannot act discretely and freely through isometries on the Euclidean plane $E = \partial \operatorname{Conv}(l)$. Putting the component of the complement $X_1 \setminus \operatorname{Conv}(l)$ containing points from the set $\operatorname{Conv}(l_i)$, $i = 1, 2$, in correspondence with the vertex l_i, we obtain the bijection required according to the conditions of the lemma. The lemma is thus proved for rank 2 vertices.

Now we assume that $\operatorname{rk}(l) = 1$. We consider the vertex $l' := \Phi^{-1}(l)$ of the simplicial space $\mathcal{C}_0$ and the set L' of vertices in $\mathcal{C}_0$ adjacent to l'. According to Lemma 5.5.4, each element $l'' \in L'$ has a certain component of the boundary $\partial \operatorname{Conv}(l')$ corresponding to it (namely the component lying in $\operatorname{Conv}(l'')$). Since $\operatorname{Conv}(l')$ lies inside $\operatorname{Conv}(\mathcal{C}_0)$ in accordance with Lemma 5.4.3, this correspondence is a bijection between L' and the set of components of the boundary $\partial \operatorname{Conv}(l')$.

The set L' is invariant under the action of the group $\operatorname{Stab}(l')$. In particular, on this set there is an action of the subgroup $b_{l'} = F_k \times \mathbf{Z}$. Since the subgroup

$a_{l'} \subset b_{l'}$ of the type $1 \times \mathbf{Z}$ belongs to the class l', $a_{l'}$ acts trivially on L'. Hence the set L' is acted upon by the free group F_k. It now becomes clear that the boundary components of the compact surface $W_{l'}$ (see Lemma 5.5.5) are in a one-to-one correspondence with the elements of the set L'/F_k, where the manifold $\operatorname{Conv}(l')/b_{l'}$ is diffeomorphic to the product $W_{l'} \times S^1$.

Further arguments are analogous to those in [**7**, Section 4]). Hence we shall describe them only briefly.

The group $b_l = F_k \times \mathbf{Z}$ acts on the set $\operatorname{Conv}(l) = D \times \mathbf{R}$ by isometries of the type (γ', shift); hence the group F_k acts in a definite manner on the convex set D. This action is free (see 4.2 in [**7**]); hence $W_l := D/F_k$ is a surface with nonpositive curvature. As in Lemma 5.5.4, we find that for any vertex $l_0 \in L$, the set $\operatorname{Conv}(l_0)$ does not lie inside $\operatorname{Conv}(l)$, which means that it contains only one component of the boundary $\partial \operatorname{Conv}(l)$ (it should be recalled that all conditions of Theorem 4.3.3 are satisfied for the manifold M_1 and the space $\mathcal{C}_1$). As before, we obtain a natural mapping of L into the set of components of the boundary $\partial \operatorname{Conv}(l)$, which can be characterized at this stage as only injective since we have not yet proved that $\operatorname{Conv}(l)$ lies inside $\operatorname{Conv}(\mathcal{C}_1)$. This mapping is natural in the sense that it is equivariant relative to the action of the group F_k. Hence the number of boundary components of the surface W_l is not smaller than that of $W_{l'}$. Gluing the corresponding holes in W_l and $W_{l'}$ with tori having one hole each, we obtain homotopy equivalent surfaces, the surface obtained from $W_{l'}$ being closed. Hence the surface W_l is also closed. Thus, the group b_l, and the more so the group $\operatorname{Stab}(l)$, acts uniformly on the set $\operatorname{Conv}(l)$, and the above correspondence between the set L and the set of components of the boundary $\partial \operatorname{Conv}(l)$ is a bijection.

The lemma is proved.

5.6.4. COROLLARY. *The set* $\operatorname{Conv}(\mathcal{C}_1) := \{\operatorname{Conv}(l) \colon l \in \operatorname{ske}_0(\mathcal{C}_1)\}$ *is open in* X_1.

Indeed, according to Lemma 5.6.3, for any rank 2 vertex $l \in \operatorname{ske}_0(\mathcal{C}_1)$, the set $\operatorname{Conv}(l)$ lies inside $\operatorname{Conv}(l_1) \cup \operatorname{Conv}(l_2)$, where l_1 and l_2 are vertices neighboring l. Since the boundary $\operatorname{Conv}(l')$ lies in the union $\bigcup \operatorname{Conv}(l)$ for the rank 1 vertex $l' \in \operatorname{ske}_0(\mathcal{C}_1)$, where l runs through all the vertices neighboring l', $\operatorname{Conv}(l')$ lies inside $\operatorname{Conv}(\mathcal{C}_1)$. Hence the set $\operatorname{Conv}(\mathcal{C}_1)$ is open in X_1.

5.7. Completion of the proof of Theorem 1.5. According to Lemmas 5.6.2, 5.6.3 and Corollary 5.6.4, we can apply Theorem 4.3.4 to the manifold $M_1 = X_1/\Gamma_1$ and to the corresponding space $\mathcal{C}_1$. Thus the manifold M_1 contains an absolutely convex open subset $M_1^* = \operatorname{Conv}(\mathcal{C}_1)/\Gamma_1$ on which a Cr-structure $\mathcal{S}_1^*$ exists, the metric on M_1^* induced by the inclusion $M_1^* \subset M_1$ matching with $\mathcal{S}_1^*$.

Our aim is to prove that the manifolds M_0^* and M_1^* are diffeomorphic. This is sufficient for proving Theorem 1.5, since in this case the fundamental group of the manifold M_1^*, hence of the manifold M_1, does not contain nontrivial peripheral elements, and the results in 5.4 can be applied to M_1^* and M_1. Since M_0^* and M_1^* are three-dimensional, it is sufficient to construct the homeomorphism $f \colon M_0^* \to M_1^*$. This will be done in such a way that f induces a given isomorphism $\alpha \colon \Gamma_0 \to \Gamma_1$ of the fundamental groups of M_0 and M_1.

The isomorphism α and the equivariant isomorphism $\Phi\colon \mathcal{C}_0 \to \mathcal{C}_1$ (see 5.6.1) induce the isomorphism $\beta\colon S_0^* \to S_1^*$ of corresponding simplicial spaces $S_i^* = \mathcal{C}_i/\Gamma_i$, $i = 0, 1$, of the Cr-structures S_0^* and S_1^*.

5.7.1. For any vertex $s \in S_0^*$, the isomorphism α induces an isomorphism of fundamental groups of blocks V_s, $V_{\beta(s)}$ of the Cr-structures $\mathcal{S}_0^*$ and $\mathcal{S}_1^*$ respectively.

This follows from the fact that $V_s = \operatorname{Conv}(l)/\operatorname{Stab}(l)$, where s is the orbit $\Gamma_0(l)$ of the vertex $l \in \operatorname{ske}_0(\mathcal{C}_0)$ (see Remark 4.3.5), and the isomorphism α transforms the group $\operatorname{Stab}(l) \subset \Gamma_0$ into the group $\operatorname{Stab}(\Phi(l)) \subset \Gamma_1$.

5.7.2. It should also be recalled (see §3) that for the rank 2 vertex $s \in \operatorname{ske}_0(S_0^*)$, the manifold V_s can be treated as a submanifold in M_0^*. For the case $\operatorname{rk}(s) = 1$, the same can be stated about the closure A_s of the set $V_s \setminus \bigcup_{s'>s} V_{s'}$ (in other words, A_s is obtained from V_s by removing the open collar of the boundary). The same statement also holds for the Cr-structure $\mathcal{S}_1^*$.

5.7.3. PROPOSITION. *There exists a continuous mapping $f_0\colon M_0^* \to M_1^*$ which induces the isomorphism α and possesses the following properties.*

(a) *For any rank 2 vertex $s \in \operatorname{ske}_0(S_0^*)$, the restriction $f|V_s$ is a homeomorphism $V_s \to V_{\beta(s)}$ of the corresponding blocks.*

(b) *For any rank 1 vertex $s \in \operatorname{ske}_0(S_0^*)$, the restriction $f|V_s$ transforms A_s into $A_{\beta(s)}$ and ∂A_s into $\partial A_{\beta(s)}$.*

Before proving this proposition, we shall show that it involves the existence of the required homeomorphism for f.

Indeed, for each rank 1 vertex $s \in \operatorname{ske}_0(S_1^*)$ the mapping $f_0|A_s$ satisfies all conditions of Theorem A in [**12**]. The manifold A_s is homeomorphic to $V_s = \operatorname{Conv}(l)/\operatorname{Stab}(l)$ and, as can be seen easily from Proposition 5.5.1 (c), differs from a linear foliation over a closed surface. According to this theorem, there exists a homotopy of the mapping $f_0|A_s$ (which is constant on ∂A_s) to the homeomorphism $A_s \to A_{\beta(s)}$. Combining these homotopies over all rank 1 vertices, we obtain the homotopy f_τ, $0 \le \tau \le 1$, between f_0 and the required homeomorphism $f_1 = f : M_0^* \to M_1^*$.

Thus, to complete the proof of Theorem 1.5 it remains to prove Proposition 5.7.3.

5.8. Proof of Proposition 5.7.3.

5.8.1. Let $s \in \operatorname{ske}_0(S_0^*)$ be a rank 2 vertex, and $l \in s$ the corresponding vertex in $\mathcal{C}_0$. We shall show that there exists a homeomorphism $V_s \to V_{\beta(s)}$ which induces the homeomorphism $\alpha : \pi_1(V_s) \to \pi_1(V_{\beta(s)})$ mapped by the components of the boundary ∂V_s in concurrence with $\Phi|L(l)$, where $L(l)$ is the set of vertices neighboring l in $\mathcal{C}_0$ (see 5.6.3).

Indeed, the manifolds V_s and $V_{\beta(s)}$ are homeomorphic to linear fibrations over the same closed surface, be it the torus or the Klein bottle. The action of the group $\operatorname{Stab}(l) = \pi_1(V_s)$ on the set $L(l)$ determines whether or not this fibration is trivial. It is found to be nontrivial if the group $\operatorname{Stab}(l)$ transposes the elements of $L(l)$, and trivial otherwise. The same is true for the vertex $\beta(s)$ and the group $\operatorname{Stab}(\Phi(l)) = \alpha(\operatorname{Stab}(l))$. Hence the manifolds V_s and $V_{\beta(s)}$ are homeomorphic and the corresponding homeomorphism can be chosen in the manner described above.

Combining the homeomorphisms constructed above over all rank 2 vertices in S_0^*, we obtain a homeomorphism which we shall denote by h. This is the restriction of the would-be mapping f_0 to the submanifold $\bigsqcup_{\mathrm{rk}(s)=2} V_s \subset M_0^*$.

5.8.2. Let $s \in \mathrm{ske}_0(S_0^*)$ be a rank 1 vertex and $l \in s$ the corresponding vertex in $\mathcal{C}_0$. The isomorphism $\alpha : \pi_1(A_s) \to \pi_1(A_{\beta(s)})$ "respects" the peripheral structure, i.e., for any component B of the boundary ∂A_s there exists a component C of the boundary $\partial A_{\beta(s)}$ such that the image $\alpha(i_*(\pi_1(B)))$ lies in a subgroup of the group $\pi_1(A_{\beta(s)})$ that is conjugate with $i_*(\pi_1(C))$ (here i_* stands for the homeomorphism of inclusion). This follows from the fact that the image $i_*(\pi_1(B))$ coincides (up to a conjugation in $\pi_1(A_s)$) with the subgroup of $\mathrm{Stab}(l')$ of index ≤ 2 that acts on the vertices neighboring l', where l' is a vertex neighboring l and corresponding to the component B.

Hence, there exists a mapping $g : (A_s, \partial A_s) \to (A_{\beta(s)}, \partial A_{\beta(s)})$ which induces $\alpha|\pi_1(A_s)$ (see Lemma 6.3 in **[14]**). In this case, g maps components of the boundary ∂A_s into the same components of the boundary $\partial A_{\beta(s)}$ as the homeomorphism h in 5.7.4. Both g and h induce on ∂A_s the same isomorphism of fundamental groups and are hence homotopic on ∂A_s. Continuing the homotopy of the mapping $g|\partial A_s$ to the entire A_s, we obtain the mapping f_0 required in Proposition 5.7.3. This proves Proposition 5.7.3, and hence Theorem 1.5.

§6. Appendix. Proof of the global compatibility property

In keeping with the promise held out in 2.4.2, we prove in this section that under the conditions of Theorem 4.3.4 the global compatibility property is satisfied (see 2.3.6) for a Cr-structure defined as in 4.3.5.

Let $s \leq s'$ for the vertices s and s' of the simplicial space $S^* := \mathcal{C}/\Gamma$. This means that we can find vertices $l \in s$ and $l' \in s'$ in the simplicial space $\mathcal{C}$ with $l \subset l'$. In this case, $\mathrm{Conv}(l') \subset \mathrm{Conv}(l)$ and, in particular, $\mathrm{Int}(\mathrm{Conv}(l')) \subset \mathrm{Int}(\mathrm{Conv}(l))$. Hence the condition $U_s' \subset U_s$ is satisfied for $U_s := \varphi_s(\mathrm{Int}(V_s))$, i.e., condition (i) in 2.3.6 holds.

Let $\mathrm{Conv}(l') = D \times \mathbf{R}^{k'}$, $k' = \mathrm{rk}(l')$. The inverse image in $\mathrm{Conv}(l)$ of the image $\varphi_{s'}(q')$ of the leaf q' of the fibration $\mathcal{T}_{s'}$ consists of sets of the type $\mathrm{Conv}(l) \cap \gamma(E)$, where $\gamma \in \Gamma$ and E is a Euclidean k'-plane of the type $d' \times \mathbf{R}^{k'} \subset D \times \mathbf{R}^{k'}$. Let $\gamma \in \Gamma$ be an isometry such that $\mathrm{Conv}(l) \cap \gamma(E)$ is not empty. In this case, the intersection $\mathrm{Conv}(l) \cap \mathrm{Conv}(\gamma(l')) = \mathrm{Conv}(l) \cap \gamma(\mathrm{Conv}(l')) := A$ for the vertices l, $\gamma(l') \in \mathcal{C}$ contains the set $\mathrm{Conv}(l) \cap \gamma(E)$. According to property (3) in Theorem 4.3.3, there exists a vertex $l_0 \in \mathcal{C}$ for which l, $\gamma(l') \subset l_0$, and $\mathrm{Conv}(l_0) = A$. Thus the Euclidean planes E_1, $\gamma(E)$, and E_2 pass through the point $x \in \mathrm{Conv}(l) \cap \gamma(E)$, where $\partial_\infty E_1 = l$, $\partial_\infty(\gamma(E)) = \gamma(l')$, $\partial_\infty E_2 = l_0$. Since l, $\gamma(l') \subset l_0$, we have E_1, $\gamma(E) \subset E_2$. Thus E_1, $\gamma(E)$, $E_2 \subset \mathrm{Conv}(l_0) \subset \mathrm{Conv}(l)$.

If the image $\varphi_{s'}(q')$ of the leaf q' of the fibration $\mathcal{T}_{s'}$ touches the set U_s', the Euclidean k'-plane $\gamma(E)$ touches the inside of the convex set $\mathrm{Conv}(l)$ for a suitable isometry $\gamma \in E$. In this case, $\gamma(E) \subset \mathrm{Int}(\mathrm{Conv}(l))$, i.e., $\varphi_{s'}(q') \subset U_s$. This proves property (ii) in 2.3.6.

If the image $\varphi_{s'}(q')$ touches the image $\varphi_s(B)$ of a certain component of the boundary ∂V_s, the Euclidean k'-plane $\gamma(E)$ passes, for an appropriate isometry $\gamma \in \Gamma$, through the boundary point x of the convex set $\mathrm{Conv}(l)$. Then the planes E_1 and E_2 with $\partial_\infty E_1 = l$, $\partial_\infty E_2 = l_0$ also pass through x, where l, $\gamma(l') \subset l_0$

and $\operatorname{Conv}(l_0) = A$. In this case, $E_2 \subset \partial \operatorname{Conv}(l)$ is the boundary leaf. We put $s'' := \Gamma(l_0)$, and take the fiber q'' of the fibration $\mathcal{T}_{s''}$ as the image of the plane E_2. In view of the fact that $E_1 \subset E_2 \subset \partial \operatorname{Conv}(l)$, we have $\varphi_{s'}(q') \subset \varphi_{s''}(q'') \subset \varphi_s(B)$ since it can be assumed that the component of the boundary $\partial \operatorname{Conv}(l)$ containing E_2 is mapped onto $\varphi_s(B)$. Thus, property (iii) in 2.3.6 is also satisfied.

References

1. S. V. Buyalo, *Collapsing manifolds of nonpositive curvature*, Algebra i Analiz **1** (1989), no. 5, 74–94; English transl. in Leningrad Math. J. **1** (1990).
2. J. Cheeger and M. Gromov, *Collapsing Riemannian manifolds while keeping their curvature bounded*, J. Differential Geom. **23** (1986), 309–346.
3. S. V. Buyalo, *Collapsing manifolds of nonpositive curvature*, Algebra i Analiz **1** (1989), no. 6, 70–97; English transl. in Leningrad Math. J. **1** (1990).
4. J. Cheeger and M. Gromov, *Collapsing Riemannian manifolds while keeping their curvature bounded.* II, J. Differential Geom. **32** (1990), 269–298.
5. S. V. Buyalo, *Homotopy invariance of some geometric properties of three-dimensional manifolds of nonpositive curvature*, Algebra i Analiz **3** (1991), no. 4, 93–112; English transl. in St.-Petersburg Math. J. **3** (1992).
6. ———, *Euclidean planes in open 3-manifolds of nonpositive curvature*, Algebra i Analiz **3** (1991), no. 1, 102–117; English transl. in St.-Petersburg Math. J. **3** (1992).
7. V. Schroeder, *Rigidity of nonpositively curved graph-manifolds*, Math. Ann. **274** (1986), 19–26.
8. P. E. Conner and F. Raymond, *Deforming homotopy equivalences to homeomorphisms in aspherical manifolds*, Bull. Amer. Math. Soc. **83** (1977), 36–85.
9. W. Ballmann, M. Gromov, and V. Schroeder, *Manifolds of nonpositive curvatures*, Birkhäuser, Boston, MA., 1985.
10. J. A. Wolf, *Spaces of constant curvature*, Univ. of California Press, Berkeley, CA, 1972.
11. P. Scott, *There are no fake Seifert fiber spaces with infinite π_1*, Ann. of Math. (2) **117** (1983), 35–70.
12. W. Heil, *On P^2-irreducible 3-manifolds*, Bull. Amer. Math. Soc. **75** (1969), 772–775.
13. J. R. Munkres, *Elementary differential topology*, Princeton Univ. Press, Princeton, NJ, 1966.
14. F. Waldhausen, *On irreducible 3-manifolds which are sufficiently large*, Ann. of Math. (2) **87** (1968), 56–88.
15. I. Tamura, *The topology of foliations*, Iwanami Shoten, Tokyo, 1976; English transl., Amer. Math. Soc., Providence, RI, 1992.
16. P. Scott, *The geometries of 3-manifolds*, Bull. London Math. Soc. **15** (1983), 401–487.

30, PR. KOSYGINA, KORP. 2, APT. 148, ST.-PETERSBURG 195298, RUSSIA

Amer. Math. Soc. Transl.
(2) Vol. **176**, 1996

Algebraic Surfaces with an Infinite Set of Skew Symmetry Planes. Mutual Arrangement of Linear Spans of Four Orbits of Symmetry Directions

V. F. Ignatenko

Let F_n be an $(m-1)$-dimensional noncylindrical surface of nth order in a real space E^m, which is invariant under an infinite group G generated by skew reflections with respect to planes, N the set of all symmetry directions defined by vectors, and let the μ_j-planes $\Pi^{\mu_j}(j = 0,\dots,p)$ be linear spans of infinite $G(\mathbf{u})$ orbits of the vectors $\mathbf{u} \in N$. Then Π^{μ_j} are direct sums of γ_j-planes Π^{γ_j} and d_j-planes Π^{d_j} such that the symmetry planes F_n conjugate to the vectors Π^{d_j} are parallel to the former planes [**1**]. The mutual arrangement of the Π^{μ_j} is defined by the γ_j-planes Π^{γ_j}. The case with $p = 2$ was considered by us earlier [**1**]. In the present paper, we shall consider the mutual arrangement of four planes $\Pi^{\gamma_j}(\gamma_0 \geq \gamma_1 \geq \gamma_2 \geq \gamma_3)$. The complete solution of this problem will be obtained for $\Pi^{\gamma_j} \cap \Pi^{\gamma_k} = 0$, $k = 0,\dots,3$; $j \neq k$, $r = \dim(\Pi^{\gamma_0} + \Pi^{\gamma_1} + \Pi^{\gamma_2}) = \gamma_0 + \gamma_1 + \gamma_2$ (see §2), and a partial solution for $r < \gamma_0 + \gamma_1 + \gamma_2$ (see §3). We shall obtain the equations for the special surfaces F_n whose symmetry groups G cannot be expanded, and also single out the relevant sets of symmetry planes. In. §4, the intersection of Π^{γ_j} along a straight line will be considered by using a method which is practically independent of the value of $p > 2$.

1. Definitions and general results

We say that a vector $\mathbf{u}$ defines an asymptotic direction for the surface F_n if the straight line parallel to $\mathbf{u}$ passes through an improper point of F_n. Let the surface F_n be defined in Cartesian coordinates by the equation $\varphi(x_i) = 0$, where $\varphi(x_i)$ is an nth degree polynomial of x_i, $i = 1,\dots,m$). The set of all surfaces F_n invariant under the same group G corresponds to the set of polynomials $\{\varphi(x_i)\}$ and forms a ring K^G. If the symmetry directions are nonasymptotic for F_n, the ring K^G is finitely generated (see Theorem 4 in [**2**]).

1991 *Mathematics Subject Classification*. Primary 51F15.

It is natural to single out two types of asymptotic symmetry directions in the following way. The vector $\mathbf{u}\|\Pi^\lambda \in \{\Pi^{\mu_j}\}$ specifies a direction of the type t (or s) if at least one λ-plane which does not lie on the surface F_n and is parallel to Π^λ does not (or does) intersect F_n along $(\lambda - 1)$-quadrics of the same symmetry (real and imaginary). If the vectors $\mathbf{u}$ of all Π^{μ_j} belong to the type t, the ring K^G is finitely generated (see Theorem 6 in [**2**]).

It should be noted that Theorems 4 and 6 of [**2**], mentioned above, list all the generators of the relevant rings K^G.

However, if the vectors Π^{μ_j} belong to the type s, the linear span of the set N is not a direct sum of Π^{μ_j} in the general case. Here $\dim(\Pi^{\gamma_j} \cap \Pi^{\gamma_k}) \leq 1$, $j \neq k$ [**2**]. Zalesskiĭ[**3**] was the first to construct in E^{11} an example of a group G whose algebra of invariants is not free. His student Veles′ko [**4**] proved the existence of groups G with infinitely generated rings of invariants. The structure of [**4**] is based on the well-known Nagata example (see [**5**]). The results obtained in [**3**] and [**4**] emphasize, among other things, the complexity of groups G corresponding to symmetry directions of the type s. The general canonical equation of the surface F_n invariant with respect to G is given in Theorem 10 of [**1**]. This theorem gives the principle of constructing the complete symmetry group $\overline{G}$ of the surface F_n and hence may serve as the basis for investigating the groups G, $\overline{G}$, and the rings of their invariants.

In the analysis of the geometry of groups G, the description of the mutual arrangement of Π^{γ_k} $(1 < k \leq p)$ and $\Pi^{\gamma_0} + \dots + \Pi^{\gamma_t}$ $(t < k)$ is one of the most important problems. The case $p = 2$ is analyzed completely in the following theorems from [**1**].

THEOREM 1. *If three different spans Π^{μ_α}, Π^{μ_β}, Π^{μ_γ}, $0 \leq \alpha, \beta, \gamma \leq p$, intersect along a straight line, the dimension of their sum is $\mu_\alpha + \mu_\beta + \mu_\gamma - 2$.*

THEOREM 2. *Suppose that the γ_j-planes Π^{γ_j}, $j = 0, 1, 2$ do not intersect along a straight line. Then the mutual arrangement of the Π^{γ_j} can be arbitrary, i.e., for any arrangement of the Π^{γ_j} there exists a surface F_n with the group $\overline{G} = G$.*

We shall use these theorems for investigating the case $p = 3$, which is a complex problem.

2. The case $r = \gamma_0 + \gamma_1 + \gamma_2$

Let us suppose that $\Pi^{\gamma_j} \cap \Pi^{\gamma_k} = 0$ $(j, k = 0, \dots, 3; j \neq k)$, $r = \dim(\Pi^r = \Pi^{\gamma_0} + \Pi^{\gamma_1} + \Pi^{\gamma_2}) = \gamma_0 + \gamma_1 + \gamma_2$. We introduce the following notation: $\gamma_0 = \lambda$, $\gamma_1 = \mu$, $\gamma_2 = \nu$, $\gamma_3 = \sigma$ $(\lambda \geq \mu \geq \nu \geq \sigma)$; $\Pi^{r_1} = \Pi^\lambda \oplus \Pi^\mu$, $\Pi^{r_2} = \Pi^\lambda \oplus \Pi^\nu$, $\Pi^{r_3} = \Pi^\mu \oplus \Pi^\nu$, $\Pi^v = \Pi^\sigma \cap \Pi^r$, $\Pi^{v_t} = \Pi^\sigma \cap \Pi^{r_t}$ $(t = 1, 2, 3)$. The relation $\Pi^v = F\Pi^v$ indicates that the position of the v-plane Π^v can be arbitrary. Without any loss of generality, we put $d_j = 1$ and $\sigma = v$. We shall use the following new notation for the appropriately chosen coordinate frames: $\Pi^{\mu_0} = \Pi^1(y_1) \oplus \Pi^\lambda(z_i)$, $i = 1, \dots, \lambda$; $\Pi^{\mu_1} = \Pi^1(y_2) \oplus \Pi^\mu(z_{\lambda+j})$, $j = 1, \dots, \mu$; $\Pi^{\mu_2} = \Pi^1(y_3) \oplus \Pi^\nu(z_{r_1+k})$, $k = 1, \dots, \nu$; $\Pi^{d_4} = \Pi^1(y_4)$.

We define the surface F_n $(n > 2)$ with the complete symmetry group $\overline{G} = G$ (which does not permit any expansion) by the equation

$$(1)\quad R\left(y_1^2 + \sum_{i=1}^{\lambda} \xi_i z_i\right) + S\sum_{j=1}^{\mu} \zeta_j z_{\lambda+j} + T\left(y_3^2 + \sum_{k=1}^{\nu} \chi_k z_{r_1} + k\right) + Py_4^2 = c,$$

where the polynomials R, S, T, and P and the linear functions depend on the variables x_τ, $\tau = 1, \dots, q$, $q \geq 2$.

The case $\Pi^v \in \Pi^{r_t} (1 \leq t \leq 3)$ is similar to that with $p = 2$.

1°. $v_1 > 0$, $v_2 > 0$, $v_3 > 0$. We write in Π^r the following equations for Π^v:

$$(2)\quad z_{v+\varepsilon} = \sum_{p=1}^{v} a_{\varepsilon p} z_p, \quad \varepsilon = 1, \dots, \lambda - v,$$

$$(3)\quad z_{\lambda+j} = \sum_{p=1}^{v} b_{jp} z_p, \quad j = 1, \dots, \mu,$$

$$(4)\quad z_{r_1+k} = \sum_{p=1}^{v} c_{kp} z_p, \quad k = 1, \dots, \nu;$$

rank $\|b_{jp}\| = v - v_2$, rank $\|c_{kp}\| = v - v_1$, and $\sum_j b_{jp}^2 > 0$, $\sum_k c_{kp}^2 > 0$ for any p.

The v_1-plane Π^{v_1} is defined in Π^{r_1} by (2), (3), and the equation $\sum_{p=1}^{v} c_{kp} z_p = 0$ (see (4)), while Π^{v_2} is defined in Π^{r_2} by equations (2), (4), and $\sum_{p=1}^{v} b_{jp} z_p = 0$. If $z_p = 0$, equations (2)–(4) define the point $O(v_3 = 0)$.

The new coordinate axes Oz'_p will be placed in Π^v. The corresponding formulas of coordinate transformation can be written as follows:

$$(5)\quad \begin{aligned} z_p &= z'_p, \quad p = 1, \dots, v, \\ z_{v+\varepsilon} &= z'_{v+\varepsilon} + \sum_{p=1}^{v} a_{\varepsilon p} z'_p, \quad \varepsilon = 1, \dots, \lambda - v, \\ z_{\lambda+j} &= z'_{\lambda+j} + \sum_{p=1}^{v} b_{jp} z'_p, \quad j = 1, \dots, \mu, \\ z_{r_1+k} &= z'_{r_1+k} + \sum_{p=1}^{v} c_{kp} z'_p, \quad k = 1, \dots, \nu; \end{aligned}$$

while other coordinate axes remain unchanged. Let us consider the following equation:

$$(6)\quad \begin{aligned} &R\left(y_1^2 + \sum_{\varepsilon=1}^{\lambda-v} \xi_{v+\varepsilon} z'_{v+\varepsilon}\right) + S\left(y_2^2 + \sum_{j=1}^{\mu} \zeta_j z'_{\lambda+j}\right) \\ &\qquad + T\left(y_3^2 + \sum_{k=1}^{\nu} \chi_k z'_{r_1+k}\right) + P\left(y_4^2 + \sum_{p=1}^{v} \kappa_p z'_p\right) = c, \end{aligned}$$

where the κ_p are linear functions of the x_τ. Equation (1) can be reduced to (6) by using the transformation (5) if

$$(7)\qquad R\left(\xi_p + \sum_{\varepsilon=1}^{\lambda-v} a_{\varepsilon p}\xi_{v+\varepsilon}\right) + S\sum_{j=1}^{\mu} b_{jp}\zeta_j + T\sum_{k=1}^{\nu} c_{kp}\chi_k = P\kappa_p, \quad p = 1,\dots,v.$$

Consequently, we arrive at

LEMMA 1. *For the surface F_n defined by equation* (1), *the case $v_1 > 0$, $v_2 > 0$, $v_3 > 0$ is possible only under the condition* (7).

Let

$$(8)\qquad P = R + S + T,$$

$$(9)\qquad \kappa_p = \xi_p + \sum_{\varepsilon=1}^{\lambda-v} a_{\varepsilon p}\xi_{v+\varepsilon} = \sum_{j=1}^{\mu} b_{jp}\zeta_j = \sum_{k=1}^{\nu} c_{kp}\chi_k.$$

Then in view of the structure of $\|b_{jp}\|$ and $\|c_{kp}\|$, there exists a linear dependence between the ξ_i. However, this is ruled out since the surface F_n is not cylindrical. Consequently, Lemma 1 leads to

COROLLARY. *Under conditions* (8) *and* (9), *relation* (7) *does not hold.*

Let us write the complete set of data for the special groups G for which condition (7) is observed.

Let us suppose that $v_1 = v_2 = 1$. In this case, we can take κ_l, $l = 1,\dots,v-1$, according to formulas (9), and condition (7) is preserved. We put

$$(10)\qquad \kappa_v = \xi_v + \sum_{\varepsilon=1}^{\lambda-v} a_{\varepsilon v}\xi_{v+\varepsilon},$$

$$(11)\qquad S = S_0\kappa_v, \quad T = T_0\kappa_v, \quad S_0 \neq cT_0.$$

Then relations (7), (8) and (11) lead to

$$(12)\qquad S_0\sum_{j=1}^{\mu} b_{jv}\zeta_j + T_0\sum_{k=1}^{\nu} c_{kv}\chi_k = (S_0 + T_0)\kappa_v.$$

With a special choice of ζ_j, χ_k, we have

$$(13)\qquad \begin{aligned} \sum_{j=1}^{\mu} b_{jv}\zeta_j &= \kappa_v + A_1, \\ \sum_{k=1}^{\nu} c_{kv}\chi_k &= \kappa_v + A_2, \end{aligned}$$

where A_1 and A_2 are linear functions such that $A_1 \neq cA_2$.

It follows from (12) and (13) that $S_0A_1 + T_0A_2 = 0$. Therefore,

$$(14)\qquad S_0 = S_1A_2, \quad T_0 = -S_1A_1.$$

Thus, we arrive at

LEMMA 2. *If* $v_1 = v_2 = 1$ *and* $v_3 = 0$, *then* $\Pi^v = F\Pi^v$. *The polynomials* κ_p $(p = 1, \dots, v)$, S, T, P *can be defined by formulas* (8)–(11), (14).

Statements like Lemmas 2 and 3 are helpful, among other things, in that they impose certain restrictions on the choice of the set of planes with corresponding reflections belonging to the group G.

2°. $v_1 > 0$, $v_2 = 0$, $v_3 > 0$. We define Π^v in Π^r by the following equations:

$$\begin{aligned} z_i &= \sum_{p=1}^{v} a_{ip} z_{\lambda+p}, \quad i = 1, \dots, \lambda, \\ z_{\lambda+x+\varepsilon} &= \sum_{p=1}^{v} b_{\varepsilon p} z_{\lambda+p}, \quad \varepsilon = 1, \dots, \mu - v, \\ z_{r_1+k} &= \sum_{p=1}^{v} c_{kp} z_{\lambda+p}, \quad k = 1, \dots, \nu, \end{aligned} \tag{15}$$

where $\operatorname{rank} \|a_{ip}\| = v - v_3$, $\operatorname{rank} \|c_{kp}\| = v - v_1$ and $\sum_i a_{ip}^2 > 0$, $\sum_k c_{kp}^2 > 0$.

Equations (15) define in Π^r a new system of coordinates,

$$\begin{aligned} z_i &= z'_i + \sum_{p=1}^{v} a_{ip} z'_{\lambda+p}, \quad i = 1, \dots, \lambda, \\ z_{\lambda+p} &= z'_{\lambda+p}, \quad p = 1, \dots, v, \\ z_{\lambda+v+\varepsilon} &= z'_{\lambda+v+\varepsilon} + \sum_{p=1}^{v} b_{\varepsilon p} z'_{\lambda+p}, \quad c = 1, \dots, \mu - v, \\ z_{r_1+k} &= z'_{r_1+k} + \sum_{p=1}^{v} c_{kp} z'_{\lambda+p}, \quad k = 1, \dots, \nu. \end{aligned} \tag{16}$$

Since $\Pi^v = \Pi^v(z'_{\lambda+p})$, in view of (16) equation (1) can be written in the form

$$R\left(y_1^2 + \sum_{i=1}^{\lambda} \xi_i z'_i\right) + S\left(y_2^2 + \sum_{\varepsilon=1}^{\mu-v} \zeta_{v+\varepsilon} z'_{\lambda+v+\varepsilon}\right) + T\left(y_3^2 + \sum_{k=1}^{\nu} \chi_k z'_{r_1+k}\right) + P\left(y_4^2 + \sum_{p=1}^{v} \eta_p z'_{\lambda+p}\right) = c$$

if

$$R\sum_{i=1}^{\lambda} a_{ip}\xi_i + S\left(\zeta_p + \sum_{\varepsilon=1}^{\mu-v} b_{\varepsilon p}\zeta_{v+\varepsilon}\right) + T\sum_{k=1}^{\nu} c_{kp}\chi_k = P\eta_p, \quad p = 1, \dots, v. \tag{17}$$

For $v' = \min(v - v_1, v - v_3)$ the functions η_l, $l = 1, \dots, v'$, can be defined by the formulas

$$\eta_l = \sum_{i=1}^{\lambda} a_{il}\xi_i = \zeta_l + \sum_{\varepsilon=1}^{\mu-v} b_{\varepsilon l}\zeta_{v+\varepsilon} = \sum_{k=1}^{\nu} c_{kl}\chi_k. \tag{18}$$

The functions η_p $(p > v')$ cannot be defined in a similar way, i.e., a statement of the type of Lemma 1 holds (we shall not single out this case).

Condition (17) holds if, for example, $v_1 = v_3 = 1$ (cf. (7) and (17)). In this case, η_l can be found using formulas (18).

3°. $v_1 = 0$, $v_2 > 0$, $v_3 > 0$. Let the v-plane Π^v (in Π^r as before) be defined by the equations

$$(19)\qquad \begin{aligned} z_i &= \sum_{p=1}^{v} a_{ip} z_{r_1+p}, \quad i = 1, \dots, \lambda, \\ z_{\lambda+j} &= \sum_{p=1}^{v} b_{jp} z_{r_1+p}, \quad j = 1, \dots, \mu, \\ z_{r_1+v+\varepsilon} &= \sum_{p=1}^{v} c_{\varepsilon p} z_{r_1+p}, \quad \varepsilon = 1, \dots, \nu - v; \end{aligned}$$

rank $\|a_{ip}\| = v - v_3$, rank $\|c_{jp}\| = v - v_2$ and $\sum_i a_{ip}^2 > 0$, $\sum_j c_{jp}^2 > 0$.

Using these relations, we can derive the formulas of a coordinate transformation of type (16). After this, the line of reasoning used in 2° can be extended to this case practically without any change.

4°. $v_1 > 0$, $v_2 = v_3 = 0$. Equations (2)–(4) define Π^v in Π^r (rank $\|a_{jp}\| = v$). In particular, for $z_{\lambda+j} = 0$ the rank of the matrix of the system of equations (2)–(4) is equal to $\lambda + \nu$, and this system has only the zero solution $(v_2 = 0)$.

In this case statements similar to Lemmas 1 and 2 are true, but the choice of the functions κ_p is wider: we can put

$$\kappa_p = \sum_{j=1}^{\mu} b_{jp} \zeta_j, \quad p = 1, \dots, v.$$

The cases $v_2 > 0$, $v_1 = v_3 = 0$ and $v_3 > 0$, $v_1 = v_2 = 0$ can be described using (15) and (19), and have nothing new to offer.

5°. $v_1 = v_2 = v_3 = 0$. In this case, equations (2)–(4) also specify Π^v, but all the κ_p can be determined from (9). Therefore, the following lemma holds.

LEMMA 3. *If* $v_1 = v_2 = v_3 = 0$, *then* $\Pi^v = F\Pi^v$.

6°. $v_t > 0$ $(t = 1, 2, 3)$. An analysis of this case requires a modification of the method used for proving Lemmas 1 and 3.

Equations (2) and (3) define Π^{v_1} in Π^{r_1} if we replace the index v by v_1 under the condition rank $\|b_{jp}\| = v_1$. Using this substitution, the first three formulas from (5) define in Π^{r_1} a new system of coordinates in which equation (1) of the surface F_n assumes the form (6). We put

$$(20)\qquad \kappa_p = \lambda_0^{-1} \left(\xi_p + \sum_{\varepsilon=1}^{\lambda - v_1} a_{\varepsilon p} \xi_{v+\varepsilon} \right) = \lambda_1^{-1} \sum_{j=1}^{\mu} b_{jp} \zeta_j,$$

where the numbers λ_0 and λ_1 are nonzero. In this case,

$$(21)\qquad P = \lambda_0 R + \lambda_1 S.$$

For certain ζ_j, the polynomials κ_p are linearly independent. According to (20) and (21), conditions of type (7) hold.

Henceforth, we shall use the plane $\Pi^{v_3} \in \Pi^{r_3}$ defined by the second and third formulas from (15) if p and v are replaced by q and v_3 respectively for rank $\|c_{kq}\| = v_3$. As in 2°, we write the equation for F_n in the new coordinate system:

$$\begin{aligned}(22)\qquad & R\left(y_1^2 + \sum_{\varepsilon=1}^{\lambda-v_1} \xi_{v_1+\varepsilon} z'_{v_1+\varepsilon}\right) + S\left(y_2^2 + \sum_{\varepsilon=1}^{\mu-v_3} \zeta_{v_3+\varepsilon} z'_{\lambda+v_3+\varepsilon}\right) \\ & + T\left(y_3^2 + \sum_{k=1}^{\nu} \chi_k z'_{r_1+k}\right) + P\left(y_4^2 + \sum_{p=1}^{v_1} \kappa_p z'_p + \sum_{q=1}^{v_3} \eta_q z'_{\lambda+q}\right) = c,\end{aligned}$$

where

$$(23)\qquad S\left(\zeta_q + \sum_{\varepsilon=1}^{\mu-v_3} b_{\varepsilon q}\zeta_{v_3+\varepsilon}\right) + T\sum_{k=1}^{\nu} c_{kq}\chi_k = P\eta_q, \quad q = 1, \dots, v_3.$$

Let us suppose that

$$(24)\qquad \eta_q = \zeta_q + \sum_{\varepsilon=1}^{\mu-v_3} b_{\varepsilon q}\zeta_{v_3+\varepsilon} = \lambda_2^{-1}\sum_{k=1}^{\nu} c_{kq}\chi_k,$$

where $\lambda_2 \neq 0$. The polynomial P is defined as

$$(25)\qquad P = S + \lambda_2 T.$$

Since $v_1 < \mu$ and $v_3 < \nu$, the functions κ_p and η_q can be chosen to be linearly independent according to formulas (20) and (24). It follows from (21) and (25) that

$$(26)\qquad \lambda_2 T = \lambda_0 R + (\lambda_1 - 1)S, \quad \lambda_1 \neq 1.$$

7°. The method of constructing the equation of the surface F_n used in 6° also makes it possible to analyze the case $v_1 > 0$, $v_2 = 0$, $v_3 > 0$ (see 2°).

Indeed, if $v = v_1 + v_3$, this case is analyzed in 6°. Let $v - v_1 - v_3 = w > 0$, and let $\Pi^w = \Pi^v \ominus (\Pi^{v_1} \oplus \Pi^{v_3})$ be a certain w-plane. We choose Π^w as the coordinate w-plane in the new variables z''_{v_1+t}, $t = 1$, $w \leq \lambda$. For this purpose, we define $\Pi^w \in \Pi^r$ in the coordinate system used in 6° by the following equations:

$$(27)\qquad \begin{aligned} z'_p &= 0, \quad p = 1, \dots, v_1, \\ z'_{v_1} + w + \delta &= \sum_{t=1}^{w} a_{\delta t} z'_{v_1+t}, \quad \delta = 1, \dots, \lambda - v_1 - w, \\ z'_{\lambda+q} &= 0, \quad q = 1, \dots, v_3, \\ z'_{\lambda+v_3+\varepsilon} &= \sum_{t=1}^{w} b_{\varepsilon t} z'_{v_1+t}, \quad \varepsilon = 1, \dots, \mu - v_3, \\ z_{r_1+k} &= \sum_{t=1}^{w} c_{kt} z'_{v_1+t}, \quad k = 1, \dots, \nu, \end{aligned}$$

where rank $\|b_{\varepsilon t}\| = \operatorname{rank} \|c_{kt}\| = w$.

Equations (27) define in Π^r the coordinate axes Oz_i'' $(i = \overline{1,r})$ (see (15) and (16)). In this case, equation (22) assumes the form

$$
\begin{aligned}
&R\left(y_1^2 + \sum_{\delta=1}^{\lambda-v_1-w} \xi_{v_1+w+\delta} z''_{v_1+w+\delta}\right) + S\left(y_2^2 + \sum_{\varepsilon=1}^{\mu-v_3} \zeta_{v_3+\varepsilon} z''_{\lambda+v_3+\varepsilon}\right) \\
&\quad + T\left(y_3^2 + \sum_{k=1}^{\nu} \chi_k z''_{r_1+k}\right) \\
&\quad + P\left(y_4^2 + \sum_{p=1}^{v_1} \kappa_p z''_p + \sum_{t=1}^{w} \kappa_{v_1+t} z''_{v_1+t} + \sum_{q=1}^{v_3} \eta_q z''_{\lambda+q}\right) = c.
\end{aligned} \tag{28}
$$

We can write relations similar to (23)–(26):

$$
\begin{aligned}
R\left(\xi_{v_1+t} + \sum_{\delta=1}^{\lambda-v_1-w} a_{\delta t}\xi_{v_1+w+\delta}\right) + S\sum_{\varepsilon=1}^{\mu-v_3} b_{\varepsilon t}\zeta_{v_3+\varepsilon} + T\sum_{k=1}^{\nu} c_{kt}\chi_k& \\
= P\eta_{v_1+t}, \quad t = 1, \dots, w;&
\end{aligned} \tag{29}
$$

$$
\begin{aligned}
\eta_{v_1+t} = \xi_{v_1+t} + \sum_{\delta=1}^{\lambda-v_1-w} a_{\delta t}\xi_{v_1+w+\delta} &= \lambda_3^{-1}\sum_{\varepsilon=1}^{\mu-v_3} b_{\varepsilon t}\zeta_{v_3+\varepsilon} \\
&= \lambda_4^{-1}\sum_{k=1}^{\nu} c_{kt}\chi_k, \qquad v_1 + v_3 < \nu;
\end{aligned} \tag{30}
$$

$$P = R + \lambda_3 S + \lambda_4 T. \tag{31}$$

From (21), (26), and (31), we find that

$$
\begin{aligned}
&\lambda_0(1 - \lambda_2^{-1}\lambda_4) = 1 \quad (\lambda_0 \neq 1), \\
&\lambda_3 = \lambda_1 - \lambda_2^{-1}\lambda_4(\lambda_2 - 1).
\end{aligned} \tag{32}
$$

The set of symmetry planes of the surface F_n in the symmetry directions $\mathbf{u}\|\Pi^{\mu_3}$ consists of (certain) diametral planes of a quadric defined as

$$y_4^2 + \sum_{p=1}^{v_1} \kappa_p z''_p + \sum_{t=1}^{w} \kappa_{v_1+t} z''_{v_1+t} + \sum_{q=1}^{v_3} \eta_q z''_{\lambda+q} = c. \tag{33}$$

The linear functions κ_p, η_q, κ_{v_1+t} can be determined using formulas (20), (24), and (30) respectively.

Thus, we have the following result.

LEMMA 4. *For $v_1 > 0$, $v_2 = 0$, $v_3 > 0$, the v-plane is defined as $\Pi^v = F\Pi^v$. For fixed Π^λ, Π^μ, and Π^ν, each arrangement of Π^v corresponds to a certain group G for which the symmetry planes conjugate to the vectors Π^μ are diametral planes of the quadric defined by equation* (33). *In equation* (28) *of the surface F_n, the polynomials R, S, T, and P are constructed so that relations* (26), (31), *and* (32) *hold.*

Let us now consider the cases $v_1 > 0$, $v_2 > 0$, $v_3 = 0$ and $v_1 = 0$, $v_2 > 0$, $v_3 > 0$ (see 1° and 3°).

8°. For $v_1 > 0$ in $\Pi^{r_2-v_1}(z'_{v_1+\varepsilon}, z'_{r_1+k})$, where $\varepsilon = 1, \lambda - v_1$, the h-plane $\Pi^h = \Pi^v \cap \Pi^{r_2-v_1}$ $(h > 0)$ will be defined as

$$(34)\qquad \begin{aligned} z'_{\psi+\delta} &= \sum_{q+1}^{h} a_{\delta q} z'_{v_1+q}, \qquad \delta = 1, \dots, \lambda - \psi, \\ z'_{r_1+k} &= \sum_{q=1}^{h} c_{kq} z'_{v_1+q}, \qquad k = 1, \dots, \nu, \end{aligned}$$

where $\psi = v_1 + h$ and rank $\|c_{kq}\| = h$, $\sum_k c_{kq}^2 > 0$. As in 7°, equations (34) define in $\Pi^{r_2-v_1}$ new coordinate axes $Oz''_{v_1+\varepsilon}$ and Oz''_{r_1+k}. We shall write equations (28) in the form

$$(35)\qquad \begin{aligned} &R\left(y_1^2 + \sum_{\delta=1}^{\lambda-\psi} \xi_{\psi+\delta} z''_{\psi+\delta}\right) + S\left(y_2^2 + \sum_{j=1}^{\mu} \zeta_j z'_{\lambda+j}\right) \\ &+ T\left(y_3^2 + \sum_{k=1}^{\nu} \chi_k z''_{r_1+k}\right) + P\left(y_4^2 + \sum_{p=1}^{v_1} \kappa_p z'_p + \sum_{q=1}^{h} \kappa_{v_1+q} z''_{v_1+q}\right) = c, \end{aligned}$$

where the functions κ_p can be found using formulas (20).

According to (35), relations similar to (29) have a simple form

$$(36)\qquad R\left(\xi_{v_1+q} + \sum_{\delta=1}^{\lambda-\psi} a_{\delta q} \xi_{\psi+\delta}\right) + T \sum_{k=1}^{\nu} c_{kq} \chi_k = P \kappa_{v_1+q}, \qquad q = 1, \dots, h.$$

We put

$$(37)\qquad \kappa_{v_1+q} = \rho_0^{-1}\left(\xi_{v_1+q} + \sum_{\delta=1}^{\lambda-\psi} a_{\delta q} \xi_{\psi+\delta}\right) = \rho_1^{-1} \sum_{k=1}^{\nu} c_{kq} \chi_k,$$

where ρ_0 and ρ_1 are certain numbers.

According to (36) and (37), the polynomial P can be written in the form

$$(38)\qquad P = \rho_0 R + \rho_1 T.$$

Since each of formulas (21) and (38) is the equation of the surface F_n, we have

$$(39)\qquad (\lambda_0 - \rho_0) R = \rho_1 T - \lambda_1 S, \quad \lambda_0 \neq \rho_0.$$

Let $w = v - v_1 - h > 0$ ($v_3 = 0$). We shall define a certain w-plane $\Pi^w = \Pi^v \ominus (\Pi^{v_1} \oplus \Pi^h)$ in Π^r by the following equations:

$$
(40)\qquad
\begin{aligned}
z'_p &= 0, \quad p = 1, \dots, v_1, \\
z''_{v_1+q} &= 0, \quad q = 1, \dots, h, \\
z''_{\psi+w+\beta} &= \sum_{t=1}^{w} a'_{\beta t} z''_{\psi+t}, \quad \beta = 1, \dots, \lambda - \psi - w, \\
z'_{\lambda+j} &= \sum_{t=1}^{w} b_{jt} z''_{\psi+t}, \quad j = 1, \dots, \mu, \\
z''_{r_1+k} &= \sum_{t=1}^{w} c'_{kt} z''_{\psi+t}, \quad k = 1, \dots, \nu; \\
\operatorname{rank} \|b_{jt}\| &= \operatorname{rank} \|c'_{kt}\| = w.
\end{aligned}
$$

It should be noted that equations (40) define a special (and not arbitrary) Π^w.

We choose Π^w as the coordinate w-plane and $O\tilde{z}_i$ ($i = 1, \dots, r$) as new axes in Π^r. The coordinate transformation formulas of type (16) follow from (40). In the new coordinate system, the surface F_n is defined as

$$
(41)\qquad
\begin{aligned}
&R\left(y_1^2 + \sum_{\beta=1}^{\lambda-\psi-w} \xi_{\psi+w+\beta} \tilde{z}_{\psi+w+\beta}\right) \\
&+ S\left(y_2^2 + \sum_{j=1}^{\mu} \zeta_j \tilde{z}_{\lambda+j}\right) + T\left(y_3^2 + \sum_{k=1}^{\nu} \chi_k \tilde{z}_{r_1+k}\right) \\
&+ P\left(y_4^2 + \sum_{p=1}^{v_1} \kappa_p \tilde{z}_p + \sum_{q=1}^{h} \kappa_{v_1+q} \tilde{z}_{v_1+q} + \sum_{t=1}^{w} \kappa_{\psi+t} \tilde{z}_{\psi+t}\right) = c.
\end{aligned}
$$

Let us suppose that in this equation the functions are defined as

$$
(42)\qquad
\begin{aligned}
\kappa_{\psi+t} &= \lambda_5^{-1} \left(\xi_{\psi+t} + \sum_{\beta=1}^{\lambda-\psi-w} a'_{\beta t} \xi_{\psi+w+\beta}\right) \\
&= \lambda_6^{-1} \sum_{j=1}^{\mu} b_{jt} \zeta_j = \lambda_7^{-1} \sum_{k=1}^{\nu} c'_{kt} \chi_k, \qquad t = 1, \dots, w.
\end{aligned}
$$

Expressions (41) and (42) lead to

$$
(43)\qquad P = \lambda_5 R + \lambda_6 S + \lambda_7 T.
$$

Formulas (21), (38), and (43) along with (39) give the following results:

$$
(44)\qquad
\begin{aligned}
&\frac{\lambda_1}{\rho_0 - \lambda_0} = \frac{\lambda_6 - \lambda_1}{\lambda_0 - \lambda_5} = \frac{\lambda_6}{\rho_0 - \lambda_5}, \\
&\frac{\rho_1}{\lambda_0 - \rho_0} = \frac{\lambda_7}{\lambda_0 - \lambda_5} = \frac{\lambda_7 - \rho_1}{\rho_0 - \lambda_5}; \\
&\lambda_0 \neq \lambda_5, \quad \lambda_0 \neq \rho_0 \neq \lambda_5, \quad \lambda_1 \neq \lambda_6, \quad \rho_1 \neq \rho_7.
\end{aligned}
$$

For $h = 0$, we also use the coordinate transformation defined by formulas (40). Thus, we have the following result.

LEMMA 5. *If* $v_1 > 0$, $v_2 > 0$, *and* $v_3 = 0$, *then* $\Pi^v = F\Pi^v$. *The surface* F_n *is defined by equation* (41) *with relations* (20), (21), (37), (38), (42)–(44).

Similarly, the following lemma can be formulated.

LEMMA 6. *For* $v_1 = 0$, $v_2 > 0$, *and* $v_3 > 0$, *the* v*-plane* $\Pi^v = F\Pi^v$.

9°. Let us now continue the analysis of the case $v_t > 0$ $(t = 1, 2, 3)$ started in 6°.

Applying the coordinate transformation to (22) following from formulas (34), we obtain the equation of the surface F_n:

$$\begin{aligned} &R\left(y_1^2 + \sum_{\delta=1}^{\lambda-\psi} \xi_{\psi+\delta} z''_{\psi+\delta}\right) + S\left(y_2^2 + \sum_{\varepsilon=1}^{\mu-v_3} \zeta_{v_3+\varepsilon} z'_{\lambda+v_3+\varepsilon}\right) \\ &\quad + T\left(y_3^2 + \sum_{k=1}^{\nu} \chi_k z''_{r_1+k}\right) \\ &\quad + P\left(y_4^2 + \sum_{p=1}^{v_1} \kappa_p z'_p + \sum_{q=1}^{h} \kappa_{v_1+q} z'_{v_1+q} + \sum_{q'=1}^{v_3} \eta_{q'} z'_{\lambda+q'}\right) = c, \end{aligned} \tag{45}$$

where formulas (20), (37), and (24) are valid for the linear functions κ_p, κ_{v_1+q}, $\eta_{q'}$ respectively. Therefore, the numerical parameters ρ_0 and ρ_1 in (38) are defined by

$$\rho_0 = \frac{\lambda_0}{1-\lambda_1}, \quad \rho_1 = \frac{\lambda_1\lambda_2}{\lambda_1 - 1}. \tag{46}$$

It should be noted that formulas (24) do not contradict (37) since $v_3 + h < \nu$.

Let us suppose that $w = v - v_3 - \psi > 0$ $(\psi = v_1 + h)$. In equations (27), we replace v_1 by the index ψ and z'_{v_1+t}, $z'_{v_1+w+\delta}$, z'_{r_1+k} by the variables $z''_{\psi+t}$, $z''_{\psi+w+\delta}$, z''_{r_1+k}. These equations define in Π^r the w-plane $\Pi^w = \Pi^v \ominus (\Pi^{v_1} \oplus \Pi^{v_3} \oplus \Pi^h)$. Moreover, they define new coordinate axes $O\tilde{z}_i$ $(i = 1, \dots, r)$ such that $O\tilde{z}_{\psi+t} \in \Pi^w$ $(t = 1, \dots, w)$. The corresponding formulas of coordinate transformation are similar to (16). In this case, equation (45) assumes the form

$$\begin{aligned} &R\left(y_1^2 + \sum_{\delta=1}^{\lambda-\psi-w} \xi_{\psi+w+\rho} \tilde{z}_{\psi+w+\rho}\right) + S\left(y_2^2 + \sum_{\varepsilon=1}^{\mu-v_3} \zeta_{v_3+\varepsilon} \tilde{z}_{\lambda+v_3+\varepsilon}\right) \\ &\quad + T\left(y_3^2 + \sum_{k=1}^{\nu} \chi_k \tilde{z}_{r_1+k}\right) \\ &+ P\left(y_4^2 + \sum_{p=1}^{v_1} \kappa_p \tilde{z}_p + \sum_{q=1}^{h} \kappa_{v_1+h} \tilde{z}_{v_1+h} + \sum_{t=1}^{w} \kappa_{\psi+t} \tilde{z}_{\psi+t} + \sum_{q'=1}^{v_3} \eta_{q'} \tilde{z}_{\lambda+q'}\right) = c. \end{aligned} \tag{47}$$

For this equation, we single out relations (29)–(31) for the substitution of variables (and index v_1) mentioned above. Along with (46), we find that the parameters

λ_0 and λ_1 in (21) can be found from the formulas

$$\lambda_0 = \frac{\lambda_2}{\lambda_2 - \lambda_4}, \quad \lambda_1 = \frac{\lambda_2\lambda_3 - \lambda_4}{\lambda_2 - \lambda_4}. \tag{48}$$

Since (46) and (48) do not lead to a contradiction, the surface F_n can be defined by equation (47) for any arrangement of the v-plane Π^v.

Consequently, we arrive at

LEMMA 7. *If $v_t > 0$ $(t = 1, 2, 3)$, then $\Pi^v = F\Pi^v$. Equation (47) defines the surface F_n which is invariant relative to the group G.*

According to Lemmas 3–7, the following theorem holds [**6**].

THEOREM 1. *Let any two γ_j-planes Π^{γ_j} $(j = 0, \dots, 3)$ intersect at only one point, and let $\gamma_0 \geq \gamma_1 \geq \gamma_2 \geq \gamma_3$ and $\dim(\Pi^{\gamma_0} + \Pi^{\gamma_1} + \Pi^{\gamma_2}) = \gamma_0 + \gamma_1 + \gamma_2$. Then the arrangement of Π^{γ_j} can be arbitrary, namely for any arrangement of Π^{γ_j} there exists a surface F_n with a certain group G which is its complete symmetry group.*

3. The case $r < \gamma_0 + \gamma_1 + \gamma_2$

1°. Let us assume that the following condition is imposed on the γ_j-planes Π^{γ_j} considered in §2: two planes from this set (and only these planes) intersect along a straight line. If (for definiteness) $\Pi^{\gamma_0} \cap \Pi^{\gamma_1} = Oz_1$, then in equation (1) we have

$$R = R_0\zeta_1, \quad \zeta = R_0\xi_1, \tag{49}$$

where $j = 2, \dots, \mu$ [**2**]. Relations (49) essentially do not affect the arguments in §2. The condition $v_1 < \mu = \gamma_1$ is not related to the choice of Π^v. Consequently, the v-plane $\Pi^v = F\Pi^v$, i.e., the result described in Theorem 1 is also valid for the case considered here.

If two more of the planes Π^{γ_j} intersect along a straight line, additional analysis associated with a special structure of the linear functions appearing in the equation of the surface F_n is required.

2°. Now let us consider an arrangement of the Π^{γ_j} for which any two of the planes intersect only at the origin O. In addition to the notation introduced at the beginning of §2, we put $\Pi^g = \Pi^\nu \cap \Pi^{r_1}$ $(\rho = \nu - g)$. Then the surface F_n with the symmetry group $\overline{G} = G$ is defined by equation (1) if we replace ν by ρ.

In the r_1-plane Π^{r_1}, the g-plane Π^g will be defined by the following equations:

$$\begin{aligned} z_{g+\varepsilon} &= \sum_{s=1}^{g} A_{\varepsilon s} z_s, \quad \varepsilon = 1, \dots, \lambda - g, \\ z_{\lambda+j} &= \sum_{s=1}^{g} B_{js} z_s, \quad j = 1, \dots, \mu, \end{aligned} \tag{50}$$

where $\operatorname{rank} \|B_{js}\| = g$ and $\sum_j B_{js}^2 > 0$ for any s.

Let us specify the conditions which must be satisfied by equation (1). In other words, we shall construct the surface F_n with a special group G. For this purpose, we shall use the method applied in §2. In Π^g we introduce new coordinate axes Oz'_s. Equations (50) lead to the following formulas for the coordinate transformation:

$$
\begin{aligned}
&z_s = z'_s, \quad s = 1, \dots, g, \\
&z_{g+\varepsilon} = z'_{g+\varepsilon} + \sum_{s=1}^{g} A_{\varepsilon s} z'_s, \quad \varepsilon = 1, \dots, \lambda - g, \\
&z_{\lambda+j} = z'_{\lambda+j} + \sum_{s=1}^{g} B_{js} z'_s, \quad j = 1, \dots, \mu.
\end{aligned} \tag{51}
$$

Taking these expressions into account, we can write equation (1) in the form

$$
\begin{aligned}
&R\left(y_1^2 + \sum_{\varepsilon=1}^{\lambda-g} \xi_{g+\varepsilon} z'_{g+\varepsilon}\right) + S\left(y_2^2 + \sum_{j=1}^{\mu} \zeta_j z'_{\lambda+j}\right) \\
&\qquad + T\left(y_3^2 + \sum_{s=1}^{g} H_s z'_s + \sum_{k=1}^{\rho} \chi_k z_{r_1+k}\right) + P y_4^2 = c.
\end{aligned} \tag{52}
$$

We put

$$
H_s = h_0^{-1}\left(\xi_s + \sum_{\varepsilon=1}^{\lambda-g} A_{\varepsilon s} \xi_{g+\varepsilon}\right) = h_1^{-1} \sum_{j=1}^{\mu} B_{js} \zeta_j, \quad s = 1, \dots, g, \tag{53}
$$

where h_0 and h_1 are numerical parameters. In this case,

$$
T = h_0 R + h_1 S. \tag{54}
$$

3°. $v = v_1$ ($\Pi^v \in \Pi^{r_1}$). In Π^{r_1}, equations (2) and (3) define Π^v. Equation (6) defines F_n if $k = 1, \dots, \rho$. We assume that the linear functions κ_p $(p = 1, \dots, v)$ in (6) can be determined using (20). Since the symmetry groups G of the surface F_n cannot be expanded, we have $T \neq cP$. According to (21) and (54),

$$
h_0 \neq c\lambda_0, \quad h_1 \neq c\lambda_1.
$$

Let us suppose, for definiteness, that $v \leq g$. Then (20) and (53) lead to

$$
\xi_p = h_0 h_1^{-1} \sum_{j=1}^{\mu} B_{jp} \zeta_j - \sum_{\varepsilon=1}^{\lambda-g} A_{\varepsilon p} \xi_{g+\varepsilon} = \lambda_0 \lambda_1^{-1} \sum_{j=1}^{\mu} b_{jp} \zeta_j - \sum_{\varepsilon=1}^{\lambda-v} a_{\varepsilon p} \bar{\xi}_{v+\varepsilon} \tag{55}
$$

In view of the independence of ζ_j, relations (55) for $v = g = \lambda$ give

$$
h_0 h_1^{-1} B_{jp} - \lambda_0 \lambda_1^{-1} b_{jp} = 0, \quad p = 1, \dots, v. \tag{56}
$$

Therefore, the arrangement of Π^v depends on the choice of Π^g $(g = v)$, and the coefficients of their equations satisfy formula (56).

Therefore, the following lemma holds.

LEMMA 8. *If the linear functions κ_p $(p = 1, \dots, v)$ and H_s $(s = 1, \dots, g)$ in equations* (6) *and* (52) *of the surface F_n can be found by formulas* (20) *and* (53) *respectively, then in general $\Pi^v \neq F\Pi^v$, $v_1 = v$, for a given Π^g.*

In the case when $\lambda = \mu = v = 1$, the straight line $\Pi^1 = F\Pi^1$, which follows from (56).

4°. $v = v_2$ $(\Pi^v \in \Pi^{r_2}, \Pi^v \notin \Pi^{r_1})$. Let the v'-plane $\Pi^{v'}$ be given by $\Pi^{v'} = \Pi^v \cap \Pi^{r_1}$. Further, let $\Pi^{v_0} = \Pi^v \cap (\Pi^\lambda \oplus \Pi^\rho)$, $\Pi^w \ominus = \Pi^v \ominus (\Pi^{v'} \oplus \Pi^{v_0})$. Since $\Pi^{v'} = \Pi^v \cap (\Pi^\lambda \oplus \Pi^g)$, we have $v' \leq g$. Therefore, we can write relations of type (55) if $\Pi^{v'}$ is defined by equations of types (2) and (3). This means that for $\Pi^{v'}$ a statement similar to Lemma 8 is valid, i.e., $\Pi^{v'} \neq F\Pi^{v'}$ in the general case.

Let $v = v_0$ $(v \leq \rho)$. In $\Pi^\lambda \oplus \Pi^\rho$, we define the v-plane Π^v by equations (2) and (4) for rank $\|c_{kp}\| = v$, $\nu = \rho$. The equation of the surface F_n has the form (6), where $k = 1, \dots, \rho$. The functions κ_p are defined on F_n by the formulas

$$(57)\qquad \kappa_p = h_3^{-1}\left(\xi_p + \sum_{\varepsilon=1}^{\lambda-v} a_{\varepsilon p}\xi_{v+\varepsilon}\right) = h_4^{-1}\sum_{k=1}^{\rho} c_{kp}\chi_k, \quad p = 1, \dots, v.$$

The polynomial P is defined as

$$(58)\qquad P = h_3 R + h_4 T.$$

For certain ζ_j, χ_k, relations (53) and (57) hold. Consequently, we have the following result.

LEMMA 9. *In the case $v = v_0 = v_2$, the v-plane $\Pi^v = F\Pi^v$. In equation* (1) *of the surface F_n, the linear functions ξ_s $(s = 1, \dots, g)$, ξ_p $(p = 1, \dots, v)$, ζ_j $(j = 1, \dots, \mu)$, χ_k $(k = 1, \dots, \rho)$ and the polynomials R, S, T, and P satisfy formulas* (53), (57), *and* (54), (58).

We define the plane Π^w in Π^r by equations (2)–(4), replacing the index v with w for rank $\|b_{jp}\| =$ rank $\|c_{kp}\| = w$. This gives relations similar to (55). For example, the coefficients b_{jp} are functions of B_{jp}. Consequently, the following lemma holds.

LEMMA 10. *For $v = v_2$ and $w \neq 0$, we have $\Pi^v \neq F\Pi^v$ in the general case.*

5°. $v = v_3$ $(\Pi^v \in \Pi^{r_3}, \Pi^v \notin \Pi^{r_1})$. As in 4°, choose $\Pi^{v'}$, $\Pi^h = \Pi^v \cap (\Pi^\mu \oplus \Pi^\rho)$ and Π^w, $\Pi^{v_1} = \Pi^v \cap (\Pi^\mu \oplus \Pi^g)$, $v' \leq g$. Repeating the reasoning of 4°, we find that the following lemma holds.

LEMMA 11. *In each case when $v' \neq 0$ and $w \neq 0$ $(v = v_3)$, we have $\Pi^v \neq F\Pi^v$ in the general case.*

In Lemmas 10 and 11, we assume that the general form of the linear functions determining the relevant sets of the symmetry planes of F_n is the same as in Lemma 8.

Let us assume that $v = h$. We define $\Pi^\mu \oplus \Pi^\rho$ in Π^v by the second and third formulas from (15) for rank $\|c_{kp}\| = v$, $\nu = \rho$. We choose coordinate axes Oz'_i $(i = \lambda + 1, r)$ such that $Oz'_{\lambda+p} \in \Pi^v$. The relevant coordinate transformations of type (51) define the equations of Π^v. This gives the following equation of the surface F_n:

$$
\begin{aligned}
(59)\qquad & R\left(y_1^2+\sum_{i=1}^{\lambda}\xi_i z_i\right)+S\left(y_2^2+\sum_{\varepsilon=1}^{\mu-v}\zeta_{v+\varepsilon}z'_{\lambda+v+\varepsilon}\right)\\
&\qquad +T\left(y_3^2+\sum_{k=1}^{\rho}\chi_k z'_{r_1+k}\right)+P\left(y_4^2+\sum_{p=1}^{v}\eta_p z'_{\lambda+p}\right)=c.
\end{aligned}
$$

If

$$
(60)\qquad \eta_p=h_5^{-1}\left(\zeta_p+\sum_{\varepsilon=1}^{\mu-v}a_{\varepsilon p}\zeta_{v+\varepsilon}\right)=h_6^{-1}\sum_{k=1}^{\rho}c_{kp}\chi_k,
$$

the polynomial P is given by

$$
(61)\qquad P=h_sS+h_6T.
$$

According to (53), (59)–(61), the following lemma, similar to Lemma 9, holds.

LEMMA 12. *If $v=v_3=h$, then $\Pi^v=F\Pi^v$. There exists a surface F_n with the linear functions H_s, $s=1,\dots,g$, η_p, $p=1,\dots,v$, in the coresponding equations* (52) *and* (59) *determined by formulas* (53) *and* (60) *respectively.*

6°. $v_1=v_2=v_3=0$. The number $v\le\rho=\nu-g$. Equations (2)–(4) with the index $\nu=\rho$ define Π^v in Π^r. The surface F_n is defined by equation (6). Let us make relations (8) and (9) more stringent. We choose

$$
(62)\qquad \kappa_p=h_7^{-1}\left(\xi_p+\sum_{\varepsilon=1}^{\lambda-v}a_{\varepsilon p}\xi_{v+\varepsilon}\right)=h_8^{-1}\sum_{j=1}^{\mu}b_{jp}\zeta_j=h_9^{-1}\sum_{k=1}^{\rho}c_{kp}\chi_k;
$$

$$
(63)\qquad P=h_7R+h_8S+h_9T.
$$

Formulas (54) and (63) lead to

$$
(64)\qquad P=(h_7+h_0h_9)R+(h_8+h_1h_9)S.
$$

The nonvanishing of the coefficients of R and S in (64) is attained by an appropriate choice of the numerical parameters. Consequently, the existence of the surface F_n defined by equation (6), which is invariant with respect to G, depends on formulas (53) and (62). Consequently, the following lemma holds.

LEMMA 13. *Suppose that $v_1=v_2=v_3=0$ and H_s, $s=1,\dots,g$, κ_p, $p=1,\dots,v$, are determined by formulas* (53) *and* (62). *Then $\Pi^v\ne F\Pi^v$ in the general case.*

It should be recalled that the structure of the set of the symmetry planes of F_n depends on H_s and κ_p. The corresponding change in κ_p makes it possible to construct the surface F_n for any arrangement of Π^v. Indeed, if

$$
(65)\qquad t=g+v\le\mu,
$$

then the v-plane Π^v can be defined in Π^r by equations (2)–(4) if we replace z_p, $z_{v+\varepsilon}$, $z_{\lambda+j}$ by the variables z'_{g+p}, $z'_{t+\varepsilon}$, $z'_{\lambda+j}$, respectively ($\varepsilon=1$, $\lambda-t$, $\nu=\rho$). In the new coordinate system $O\tilde{z}$ $(i = 1, \dots, r)$, the axis $O\tilde{z}_{g+p} \in \Pi^v$. Applying the coordinate transformation to equation (52), we obtain

$$
\begin{aligned}
&R\left(y_1^2 + \sum_{\varepsilon=1}^{\lambda-t} \xi_{t+\varepsilon}\tilde{z}_{t+\varepsilon}\right) + S\left(y_2^2 + \sum_{j=1}^{\mu} \zeta_j \tilde{z}_{\lambda+j}\right) \\
&\quad + T\left(y_3^2 + \sum_{s=1}^{g} H_s \tilde{z}_s + \sum_{k=1}^{\rho} \chi_k \tilde{z}_{r_1+k}\right) + P\left(y_4^2 + \sum_{p=1}^{v} \kappa_{g+p}\tilde{z}_{g+p}\right) = c.
\end{aligned} \tag{66}
$$

In this equation, we choose the linear functions κ_{g+p} in the form

$$
\kappa_{g+p} = q_0^{-1}\left(\xi_{g+p} + \sum_{\varepsilon=1}^{\lambda-t} a_{\varepsilon p}\xi_{t+\varepsilon}\right) = q_1^{-1}\sum_{j=1}^{\mu} b_{jp}\zeta_j = q_2^{-1}\sum_{k=1}^{\rho} c_{kp}\chi_k, \quad p = 1, \dots, v, \tag{67}
$$

where q_0, q_1, q_2 are numerical parameters.

The linear functions ξ_s, ξ_{g+p}, and ζ_j satisfying formulas (53) and (67) exist. Since $v \le \nu - g$, the inequality $v > \mu - g$ $(\nu \le \mu)$ is ruled out. Consequently, we have the following result.

LEMMA 14. *If* $v_1 = v_2 = v_3 = 0$, *then* $v \le \nu - g$ *and* $\Pi^v = F\Pi^v$.

7°. Let us consider in more detail some conditions imposed on the arrangement of the Π^{γ_j} $(j = 0, 3)$ and on the choice of the numbers γ_j for which there exists a surface F_n with the symmetry group $\overline{G} = G$.

Let relation (65) be satisfied in the case $v = v_1$ (see 3°). We take the new axes $O\tilde{z}_{g+p}$ in Π^v and take for κ_{g+p} the linear functions in (67), which contain the factors q_0^{-1} and q_1^{-1}. Since $\Pi^v \cap \Pi^g = 0$, the rank of the $\mu \times t$ matrix with the rows B_{js} and b_{jp} is equal to t. Therefore, H_s and κ_{g+p} can be determined by using the formulas indicated above. Consequently, the following lemma holds.

LEMMA 15. *For* $g + v \le \mu$ $(v = v_1)$, *the* v*-plane* $\Pi^v = F\Pi^v$.

If $t > \mu$ $(v = v_1)$, we put

$$
d = t - \mu. \tag{68}
$$

Since $v \le \mu$, we have $d \le g$. We present Π^v in the form $\Pi^d \oplus \Pi^{\mu-g}$, where $\Pi^g \cap \Pi^{\mu-g} = 0$. We choose in Π^{r_1} the coordinate axes Oz_i' $(i = 1, \dots, r)$ so that $Oz_{g+\delta}' \in \Pi^{\mu-g}$ $(\delta = 1, \dots, \mu - g)$. Let us write the equation for the surface F_n:

$$
\begin{aligned}
&R\left(y_1^2 + \sum_{\varepsilon=1}^{\lambda-\mu} \xi_{\mu+\varepsilon} z_{\mu+\varepsilon}'\right) + S\left(y_2^2 + \sum_{j=1}^{\mu} \zeta_j z_{\lambda+j}'\right) \\
&\quad + T\left(y_3^2 + \sum_{s=1}^{g} H_s z_s' + \sum_{k=1}^{\rho} \chi_k z_{r_1+k}\right) + P\left(y_4^2 + \sum_{\delta=1}^{\mu-g} \kappa_{g+\delta} z_{g+\delta}'\right) = c.
\end{aligned} \tag{69}
$$

The linear functions $\kappa_{g+\delta}$ will be defined by formulas of type (67):

$$\kappa_{g+\delta} = q_3^{-1}\left(\xi_{g+\delta} + \sum_{\varepsilon=1}^{\lambda-\mu} a_{\varepsilon\delta}\xi_{\mu+\varepsilon}\right) = q_4^{-1}\sum_{j=1}^{\mu} b_{j\delta}\zeta_j, \quad \delta = 1,\dots,\mu-g; \tag{70}$$
$$\operatorname{rank}\|b_{j\delta}\| = \mu - g.$$

According to (69) and (70), the polynomial P is given by

$$P = q_3R + q_4S. \tag{71}$$

If $g + d \le \lambda$, we consider along with (69) the equation for F_n in new coordinate axes Oz_i'' $(i = 1,\dots,r_1)$; $Oz_{g+l}'' \in \Pi^d$, $l = 1,\dots,d$. In order to derive this equation of F_n, we define Π^d in Π^{r_1} as follows:

$$\begin{aligned} z_s &= 0, \quad s = 1,\dots,q,\\ z_{g+d+\varepsilon} &= \sum_{l=1}^{d} a_{\varepsilon l}' z_{g+l}, \quad \varepsilon = 1,\dots,\lambda-g-d,\\ z_{\lambda+j} &= \sum_{l=1}^{d} b_{jl}' z_{g+l}, \quad j = 1,\dots,\mu;\\ &\operatorname{rank}\|b_{jl}'\| = d. \end{aligned} \tag{72}$$

Equations (72) specify the coordinate transformation of the form

$$\begin{aligned} z_{g+l} &= z_{g+l}'', \quad l = 1,\dots,d,\\ z_{g+l+\varepsilon} &= z_{g+l+\varepsilon}'' + \sum_{l=1}^{d} a_{\varepsilon l}' z_{g+l}'', \quad \varepsilon = 1,\dots,\lambda-g-d,\\ z_{\lambda+j} &= z_{\lambda+j}'' + \sum_{l=1}^{d} b_{jl}' z_{g+l}''. \end{aligned} \tag{73}$$

Applying the transformation (73) to equation (1) for $\nu = \rho$, we obtain

$$\begin{aligned} &R\left(y_1^2 + \sum_{s=1}^{g}\xi_s z_s + \sum_{\varepsilon=1}^{\lambda-g-d}\xi_{g+d+\varepsilon} z_{g+d+\varepsilon}''\right) + S\left(y_2^2 + \sum_{j=1}^{\mu}\zeta_j z_{\lambda+j}''\right)\\ &\qquad + T\left(y_3^2 + \sum_{k=1}^{\rho}\chi_k z_{r_1+k}\right) + P\left(y_4^2 + \sum_{l=1}^{d} C_l z_{g+l}''\right) = c. \end{aligned} \tag{74}$$

Here

$$R\left(\xi_{g+l} + \sum_{\varepsilon=1}^{\lambda-g-d} a_{\varepsilon l}'\xi_{g+d+\varepsilon}\right) + S\sum_{j=1}^{\mu} b_{jl}'\zeta_j = PC_l, \quad l = 1,\dots,d. \tag{75}$$

Since ξ_i and ζ_j are connected through relations (53) and (70), the functions C_l cannot be determined with the help of the same formulas for an arbitrary arrangement of Π^d.

We rewrite (75) in the form

$$RA_l + SB_l = PC_l. \tag{76}$$

The form of the linear functions A_l and B_l is clear from (75) and (76).

According to (71) and (76), we have

$$\frac{R}{S} = \frac{q_4C_l - B_l}{A_l - q_3C_l}, \qquad l = 1, \dots, d, \quad \mu > g. \tag{77}$$

Consequently,

$$R = R_0(q_4C_l - B_l), \quad S = R_0(A_l - q_3C_l)$$

or

$$\begin{aligned} A_1 - q_3C_1 &= \dots = A_d - q_3C_d, \\ q_4C_1 - B_1 &= \dots = q_4C_d - B_d. \end{aligned} \tag{78}$$

For $d > 1$, these relations imply

$$\begin{aligned} q_3(C_{l_1} - C_{l_2}) &= A_{l_1} - A_{l_2}, \\ q_4(C_{l_1} - C_{l_2}) &= B_{l_1} - B_{l_2}, \end{aligned} \tag{79}$$

where l_1, $l_2 = 1, \dots, d$, $l_1 < l_2$. According to (79),

$$q_3^{-1}(A_{l_1} - A_{l_2}) = q_4^{-1}(B_{l_1} - B_{l_2}). \tag{80}$$

Formulas (53) and (70) lead to

$$\zeta_j = H_j(\xi_i), \quad j = 1, \dots, \mu, \tag{81}$$

where the H_j (ξ_i) are linear functions in ξ_i, $i = 1, \dots, \lambda$. Substituting into (80) the values of ζ_j determined using (81), we obtain a linear dependence between ξ_i if the arrangement of Π^v is not special.

For $d = 1$, we put

$$C_1 = A_1, \quad S = S_0A_1. \tag{82}$$

According to (77) and (82), we have

$$\frac{R}{S_0A_1} = \frac{B_1 - q_4A_1}{(q_3 - 1)A_1}.$$

This means that $(q_3 - 1)R = (B_1 - q_4A_1)S_0$, and the surface F_n $(n > 2)$ with the constructed group G does exist.

Thus, we have proved the following lemma.

LEMMA 16. *If $g + v > \mu$ $(v = v_1)$ and $d > 1$, $g < \mu$, $g + d \leq \lambda$ and the linear functions H_s $(s = 1, \dots, g)$, $\kappa_{g+\delta}$ $(\delta = 1, \dots, \mu - g)$ can be obtained from formulas (53) and (70), then the v-plane $\Pi^v \neq F\Pi^v$. In this case, $\Pi^v = F\Pi^v$ if $d = 1$.*

In the case $g = \mu$, $g + d \leq \lambda$, we consider equation (74) of the surface F_n. We put in (75)

$$C_1 = A_1, \quad C_2 = B_2, \quad R = R_0B_2, \quad S = S_0A_1. \tag{83}$$

This gives

$$\frac{R_0}{S_0} = \frac{A_1 - B_1}{B_2 - A_2}. \tag{84}$$

Consequently,

$$R_0 = R_0'(A_1 - B_1), \quad S_0 = R_0'(B_2 - A_2), \tag{85}$$

and the value $d \leq 2$ is realized.

The condition $g + d \leq \lambda$ can be dropped. Indeed, we choose $\Pi^d = \Pi^d(z_l')$. Formula (75) takes the form

$$R\left(\xi_l + \sum_{\varepsilon=1}^{\lambda-d} a'_{\varepsilon l}\xi_{d+\varepsilon}\right) + S\sum_{j=1}^{\mu} b'_{jl}\zeta_j = PC_l, \quad l = 1, \ldots, d. \tag{86}$$

We assume that (76) is an alternative form of (86). Then (83)–(85) are preserved; they give no additional relations between ξ_i and ζ_j. If necessary, we can take $C_1 = B_1$, $R = R_0B_1$ instead of formulas (82), $d = 1$.

The first statement of Lemma 16 ($\Pi^v \neq F\Pi^v$) is preserved if we drop the condition $g + d \leq \lambda$. Consequently, we obtain the following modification of Lemma 16.

LEMMA 17. *If $g + v > \mu$ ($v = v_1$) and $d > 1$, $g < \mu$, and the linear functions H_s ($s = 1, \ldots, g$), $\kappa_{g+\delta}$ ($\delta = 1, \ldots, \mu - g$) can be determined from formulas (53) and (70), then the v-plane $\Pi^v \neq F\Pi^v$. If, however, $d = 1$ ($g < \mu$) or $d \leq 2$ for $g = \mu$, then $\Pi^v = F\Pi^v$.*

8°. Let us suppose that $v > v_1 > 0$, $v_2 = v_3 = 0$; $w = v - v_1$. First we consider the case (65) for $t \leq \mu$. According to Lemma 15, the arrangement of Π^{v_1} can be arbitrary. This is also valid for Π^v, $t < \mu$. The number $w \leq \rho = \nu - g$ is used in an equation of form (66). For $d = t - \mu > 0$, instead of (70) we find formulas similar to (62), and the polynomial P is defined by (63) and (64). Since P depends only on R, S, and the relevant parameters, the statement of Lemma 17 is essentially preserved. This means that the following lemma holds.

LEMMA 18. *For $v > v_1 > 0$ and $v_2 = v_3 = 0$, the v-plane $\Pi^v = F\Pi^v$ if*

$$g + v \leq \mu + 1. \tag{87}$$

9°. If $v > v_2 > 0$, $v_1 = v_3 = 0$, then $\Pi^v \cap \Pi^{r_1} = 0$. Therefore, $v_2 \leq \rho$. Moreover, $v \leq \rho$. We replace v in (57) and (62) respectively by v_2 and $v - v_2$, retaining $\xi_{v+\varepsilon}$ in (62). Further, instead of κ_p, ξ_p in (62) we write κ_{v_2+p}, ξ_{v_2+p}. Relations (58) and (63) hold. As a result, we obtain the formulas for linear functions that define the structure of the set of symmetry planes of F_n. Consequently, we have the following lemma.

LEMMA 19. *If $v > v_2 > 0$ and $v_1 = v_3 = 0$, then $\Pi^v = F\Pi^v$.*

In the case $v > v_3 > 0$, $v_1 = v_2 = 0$, we also have $v \leq \rho$. Therefore, a statement similar to Lemma 19 holds, and $\Pi^v = F\Pi^v$.

10°. Let us suppose that $v_1 > 0$, $v_3 > 0$, $v_2 = 0$ ($\Pi^v \notin \Pi^{r_1}$), and the h-plane $\Pi^h = \Pi^v \cap (\Pi^\mu \oplus \Pi^\rho)$, $0 < h \leq \rho$. Essentially, we have the case $v_1 > 0$, $v_2 = 0$, $h > 0$. Since the functions ξ_i do not appear in formulas (60), $t - h \leq \mu$ if Π^v is chosen arbitrarily. For $v = v_1 + h$, we confine ourselves to an analysis of formulas (53), (60), and Lemma 17, and the h-plane $\Pi^h = F\Pi^h$. Thus, the following lemma holds.

LEMMA 20. *For $v_1 > 0$, $h > 0$, $v \geq v_1 + h$, we have $\Pi^v = F\Pi^v$ provided*

$$g + v \leq \mu + h + 1. \tag{88}$$

Inequalities (87) and (88) indicate that the presence of Π^h increases the upper estimate for $g + v$.

11°. Let $v_1 > 0$, $v_2 > 0$, $v_3 = 0$; $\Pi^{v_0} = \Pi^v \cap (\Pi^\lambda \oplus \Pi^\rho)$, $v_0 > 0$. In the case $v = v_0 + v_1$, we use (57) along with (53). Here $g + v_1 \leq \mu$ if Π^v is arranged arbitrarily. In other words, Lemma 20 can be extended to the case when h is replaced by v_0.

For $v > v_0 + v_1$, we must also take into account formulas of type (67). This gives $v - v_1 \leq \rho$, $g + v - v_0 \leq \mu$ (for relevant symmetry planes of F_n). However, the first inequality holds in view of the choice of linear spans. Therefore, Lemma 20 is valid in this case also.

Let us suppose that $v_2 > 0$, $v_3 > 0$, $v_1 = 0$. Then $v_2 \geq v_0$, $v_3 \geq h$ (see 10° and 11°), and $v \leq \rho$. Based on (53), (57), (60), and (67), we arrive at the case considered in 9°.

LEMMA 21. *If $v_1 = 0$, $v_2 > 0$, $v_3 > 0$, then $\Pi^v = F\Pi^v$.*

12°. In the case $v_t > 0$ $(t = 1, 2, 3)$, we introduce the same numbers v_0 and h as in 10° and 11°. Since $v_0 + h \leq \rho$, inequality (88) becomes more stringent:

$$g + v \leq \mu + v_0 + h + 1. \tag{89}$$

This inequality holds for $g < \mu$. If $g = \mu$ $(v_0 = h = 0)$, the upper estimate of v in Lemma 17 increases.

Thus, Lemmas 9, 12, 14, 18–21 and 9°, 11°, and 12° lead to the following result (see [**7**]).

THEOREM 2. *Let any two of the γ_j-planes Π^{γ_j} $(j = 0, \dots, 3)$ intersect only at a single point, and $\dim(\Pi^{\gamma_0} + \Pi^{\gamma_1} + \Pi^{\gamma_2}) < \gamma_0 + \gamma_1 + \gamma_2$ $(\gamma_0 \geq \gamma_1 \geq \gamma_2 \geq \gamma_3)$. Then the arrangement of the v-plane Π^v can be arbitrary if the number $v_1 = 0$, or one of the following conditions holds for $v_1 > 0$:*

$$g + v \leq \mu + v_0 + h + 1; \quad g = \mu, \quad v \leq 2.$$

As in 11°, we note that the result given in Theorem 2 can be extended (perhaps, with a refinement of conditions) to the case when only two of the planes Π^{γ_j} intersect along a straight line.

As an example we consider the general case when Π^σ and Π^ν intersect along the straight line d.

13°. If $d \notin \Pi^{r_1}$, we choose $Oz_{r_1+1} \| d$. Then in equation (1) of the surface F_n we have

$$T = T_0\omega, \quad P = T_0\chi_1, \tag{90}$$

where $\omega \neq c\chi_1$ is a linear function of x_τ [**2**].

Let us suppose that $v_1 > 0$ $(v > v_1)$. From (54) and (90), we find that

$$T_0\omega = h_0 R + h_1 S. \tag{91}$$

Relation (21) holds for $v = v_1$ as well. Therefore,

(92) $$T_0\chi_1 = \lambda_0 R + \lambda_1 S.$$

14°. Let us suppose that $v_0 = h = 1$ (see 12°) and $v = v_1 + 2$. We take

(93) $$R = T_0\omega_1, \quad S = T_0\omega_2, \quad \omega_1 \neq c\omega_2.$$

Formulas (91)–(93) give

(94) $$\omega = h_0\omega_1 + h_1\omega_2, \quad \chi_1 = \lambda_0\omega_1 + \lambda_1\omega_2,$$

i.e., the surface F_n exists and is defined by equation (1), $\nu = \rho$, under conditions (90)–(94).

15°. $v_0 = 1$, $h > 1$, $v = v_1 + h + 1$. According to (61) and (90), we have

(95) $$T_0\chi_1 = h_5 S + h_6 T_0\omega.$$

Under conditions (94) and (95), we have

(96) $$\chi_1 = h_5\omega_2 + h_6\omega.$$

A comparison of (94) and (96) gives

(97) $$\lambda_0 = h_0 h_6, \quad \lambda_1 = h_s + h_1 h_6.$$

Thus, the case under investigation is possible for polynomials R and S defined by formulas (93). Limitations (97) imposed on corresponding parameters affect the choice of the symmetry planes of F_n.

16°. $v_0 > 1$, $h > 1$, $v = v_0 + v_1 + h$. Since formula (58) is valid in the case $\Pi^v \in \Pi^{r_2}$ as well, in view of (90) and (93) the function χ_1 is defined as follows:

(98) $$\chi_1 = h_3\omega_1 + h_4\omega.$$

Substituting into this expression the values of ω and χ_1 determined from formulas (94), we obtain

(99) $$\lambda_0 = h_3 + h_0 h_4, \quad \lambda_1 = h_1 h_4.$$

This case takes place since conditions (99) do not contradict (97).

17°. $v_0 > 1$, $h > 1$, $v > v_0 + v_1 + h$. Here, a relation of form (63) appears, i.e.,

(100) $$\chi_1 = h_7\omega_1 + h_8\omega_2 + h_9\omega.$$

On the basis of (94) and (100), we obtain

(101) $$\lambda_0 = h_7 + h_0 h_9, \quad \lambda_1 = h_8 + h_1 h_9.$$

Consequently, there exists a surface F_n invariant under the group G. Equalities (101) impose certain restrictions on the choice of G.

Other cases can be realized in a similar way.

Let $v_1 = 0$. Then the limitations imposed on the structure of the polynomials R, S, T, and P become less stringent since relation (21) is ruled out. Therefore, the value $v_1 = 0$ cannot be realized.

18°. If $d \in \Pi^{r_1}$, we choose $Oz_1 \| d$. In equation (52) of the surface F_n, we have

$$T = T_0\omega', \quad P = T_0 H_1,$$

where the linear function H_1 can be determined from formula (53). In other words, the structure of the polynomials T and P is the same as for $Oz_{r_1+1}\|d$.

Thus, Theorem 2 leads to the following

COROLLARY. *Theorem 2 is valid when one of the conditions is altered, i.e.,* Π^{γ_2} *and* Π^{γ_3} *intersect along a straight line.*

It should be noted that Theorems 1 and 2 can be used, for example, for an analysis of the problem of classification of quadratic forms whose coefficients depend on a point of a manifold (see the paper by Gel′fand and Mishchenko [**8**]).

4. Intersection of Π^{μ_j} along a straight line

Let μ_j-planes Π^{μ_j} $(j = 0, \ldots, p)$ intersect along a straight line coinciding with the axis Oz_1. Then the equation of the surface F_n invariant relative to the group G can be written in one of the following forms [**1**]:

$$\sum_{t=0}^{s} R_t A^{s-t} = 0, \tag{102}$$

$$\sum_{t=0}^{s} S_{jt} \left(\zeta_j(x_\tau) z_1 + B_j\right)^{s-t} = 0, \quad j \geq 1, \tag{103}$$

where the quadratic forms are given by

$$A = y_1^2 + \sum_{i=1}^{\gamma_0} \xi_i(x_\tau) z_i,$$

$$B_j = y_{j+1}^2 + \sum_{k=1}^{\rho_j} \xi_{jk}(x_\tau) z_{\alpha_j+k},$$

and the polynomials R_t and S_{jt} depend on x_τ $(\tau = 1, \ldots, q \geq 2)$ and other variables which do not appear in A or in the corresponding B_j.

In formulas (102) and (103), $d_j = 1$, $\gamma_0 \geq \gamma_1 \geq \cdots \geq \gamma_p$; $\alpha_1 = \gamma_0$, $\rho_1 = \gamma_1 - 1$, $\alpha_2 = \gamma_0 + \gamma_1 - 1$, $\rho_2 = \gamma_2 - 1$ (see §1); $\alpha_j = \sum_{k=1}^{j-1} \alpha_k + \rho_{j-1}$, $(j > 2)$ and $\rho_j = \gamma_j - v_j$, if v_j is the dimension of $\Pi^{v_j} = \Pi^{\gamma_j} \cap (\Pi^{\gamma_0} + \cdots + \Pi^{\gamma_{j-1}})$. Further,

$$R_0(x_\tau) = R \prod_{j=1}^{p} \zeta_j^s, \quad S_{j0}(x_\tau) = R \left(\xi_1 \zeta_1 \ldots \hat{\zeta}_j \ldots \zeta_p\right)^s; \tag{104}$$

and $\hat{\zeta}_j$ indicates the absence of ζ_j.

LEMMA 22. *If* μ_j*-planes* Π^{μ_j} $(j = 0, \ldots, p \geq 1)$ *intersect along the axis* Oz_1*, the equation of the surface* F_n *has the form*

$$\sum_{t=0}^{s} Q_t \left(A \prod_{j=1}^{p} \zeta_j + \xi_1 \sum_{j=1}^{p} B_j \zeta_1 \ldots \hat{\zeta}_j \ldots \zeta_p\right)^{s-t} = 0, \tag{105}$$

where the Q_t *are polynomials in* x_τ *and other variables do not appear in* A *and* B_j*.*

PROOF. The case $p \leq 2$ is considered in [7]. We assume that the lemma is valid for $pG(\mathbf{u})$ orbits. Then we can write

$$\sum_{t=0}^{s} C_t \left(\sum_{j=1}^{p} B_j \zeta_1 \dots \hat{\zeta}_j \dots \zeta_p \right)^{s-t} = 0, \tag{106}$$

$$\sum_{t=0}^{s} D_{kt} \left(A\zeta_1 \dots \hat{\zeta}_k \dots \zeta_p + \xi_1 \sum_{j=1}^{p} a_k B_j \zeta_1 \dots \hat{\zeta}_j \dots \hat{\zeta}_k \dots \zeta_p \right)^{s-t} = 0, \tag{107}$$
$$k = 1, \dots, p.$$

Here $a_k = 0$ for $j = k$ and $a_k = 1$ for $j \neq k$.

These equations lead to

$$C_t = C'_t \xi_1^{s-t}, \quad D_{kt} = D'_{kt} \zeta_k^{s-t}. \tag{108}$$

In order to obtain this relation, it is sufficient to use the functional coefficients of B_j^{s-t}. Consequently, equations (106) and (107) of the surface F_n assume the form

$$\sum_{t=0}^{s} C'_t \left(\xi_1 \sum_{j=1}^{p} B_j \zeta_1 \dots \hat{\zeta}_j \dots \zeta_p \right)^{s-t} = 0 \tag{109}$$

or

$$\sum_{t=0}^{s} D'_{kt} \left(A\zeta_1 \dots \zeta_p + \xi_1 \sum_{j=1}^{p} a_k B_j \zeta_1 \dots \hat{\zeta}_j \dots \zeta_p \right)^{s-t} = 0. \tag{110}$$

The left-hand sides of formulas (109) and (110) can be presented as polynomials in $\xi_1 \sum_{j=1}^{p} B_j \zeta_1 \dots \zeta_j \dots \zeta_p$ $(j \neq k)$. Therefore, applying the method used in the proof of Theorem 8 in [2] and taking (104) into account, we obtain (105). The lemma is proved.

If $p+1$ is the maximum number of infinite $G(\mathbf{u})$ orbits, then $Q_t = Q_t(x_\tau)$, and equation (105) is a canonical equation of F_n.

THEOREM 3. *If different μ_j-planes Π^{μ_j}, $j = 0, \dots, p$, intersect along a straight line, the dimensions of their sum are*

$$\sum_{j=1}^{p} \mu_j - p.$$

PROOF. For $p = 2$, the theorem is proved in [9]. Let us suppose that it is valid for p linear spans of infinite $G(\mathbf{u})$ orbits. In the r-plane $\Pi^r = \Pi^{\gamma_0} + \dots + \Pi^{\gamma_{p-1}}$, we define a certain $(v-1)$-plane $\Pi^{v-1} = \Pi^v \ominus \Pi^1\ (z_\lambda)$, where $\lambda = \gamma_0$, $Oz_\lambda \| d$ and $v = v_p > 1$, by the equations

$$
\begin{aligned}
z_{v+\varepsilon} &= \sum_{l=1}^{v-1} a_{\varepsilon l} z_l, \quad \varepsilon = 0, \ldots, \lambda - v, \\
z_{\alpha_j + k} &= \sum_{l=1}^{v-1} b_{jkl} z_l, \quad k = 1, \ldots, \rho_j,
\end{aligned} \tag{111}
$$

on the basis of the assumption that the number $\rho_j = \gamma_j - 1$ $(j = 3, \ldots, p-1)$.

We take new coordinate axes Oz'_l in Π^{v-1} (other axes remaining unchanged). According to (111), the coordinate transformation is defined by the formulas

$$
\begin{aligned}
z_l &= z'_l \quad (l = 1, \ldots, v-1), \\
z_{v+\varepsilon} &= z'_{v+\varepsilon} + \sum_{l=1}^{v-1} a_{\varepsilon l} z'_l, \\
z_{\alpha_j+k} &= z'_{\alpha_j+k} + \sum_{l=1}^{v-1} b_{jkl} z'_l.
\end{aligned} \tag{112}
$$

Substituting ξ_λ for ξ_1 in (105), we obtain the following equation of the surface F_n (in the new system of coordinates):

$$
\begin{aligned}
\sum_{t=0}^{s} Q_t \Bigg[\Bigg(y_1^2 + \sum_{\varepsilon=0}^{\lambda-v} \xi_{v+\varepsilon} z'_{v+\varepsilon} \Bigg) \prod_{j=1}^{p} \zeta_j \\
+ \xi_\lambda \sum_{j=1}^{p-1} \Bigg(y_{j+1}^2 + \sum_{k=1}^{\gamma_j - 1} \xi_{jk} z'_{\alpha_j+k} \Bigg) \zeta_1 \cdots \hat{\zeta}_j \cdots \zeta_p \\
+ \xi_\lambda \Bigg(y_{p+1}^2 + \sum_{l=1}^{v-1} \chi_l z'_l + \sum_{k=1}^{\rho_p} \xi_{pk} z_{r+k} \Bigg) \zeta_1 \cdots \zeta_{p-1} \Bigg]^{s-t} = 0,
\end{aligned} \tag{113}
$$

where the linear functions χ_l depend on the variables x_τ. On the other hand, (113) can be obtained from an equation of F_n of form (105) by using transformation (112). Therefore

$$
\begin{aligned}
\Bigg(\xi_l + \sum_{\varepsilon=0}^{\lambda-v} a_{\varepsilon l} \xi_{v+\varepsilon} \Bigg) \prod_{j=1}^{p} \zeta_j + \xi_\lambda \sum_{j=1}^{p-1} \sum_{k=1}^{\gamma_j-1} b_{jkl} \xi_{jk} \zeta_1 \cdots \hat{\zeta}_j \cdots \zeta_p \\
= \chi_l \xi_\lambda \zeta_1 \cdots \zeta_{p-1}, \qquad l = 1, \ldots, v-1.
\end{aligned} \tag{114}
$$

Since the surface F_n differs from the cylindrical surface, the functions ξ_i $(i = 1, \ldots, \lambda)$ are linearly independent; moreover, $\xi_\lambda \neq c\zeta_j$ $(1 \leq j \leq p)$ [**2**]. Consequently, equalities (114) do not hold, and the theorem is proved.

Lemma 22 and Theorem 3 give a complete description of the group G and its ring of invariants K^G.

REMARK. Recently, the following results have been obtained.

1. Complete systems of the groups G have been established for three $G(\mathbf{u})$ orbits ($p = 2$).

2. As expected in 1° in §3, the case of the intersection of γ_j-planes Π^{γ_j} ($j = 0, \dots, 3$) along two straight lines required considerable additional investigations. Together with Issam Diban, we obtained geometrical inequalities for the dimensions γ_j, under which the mutual arrangement of the Π^{γ_j} can be arbitrary. The relevant sets of skew symmetry planes were singled out.

References

1. V. F. Ignatenko, *On the geometric theory of group invariants generated by reflections*, Problemy Geometrii, vol. 21, VINITI, Moscow, 1989, pp. 155–208; English transl. in J. Soviet Math. **55** (1991), no. 6.
2. ______, *Algebraic surfaces with an infinite set of skew symmetry planes*, Ukrain. Geometr. Sb. **32** (1989), 47–60; English transl. in J. Soviet Math. **59** (1992), no. 2.
3. A. E. Zalesskiĭ, *The fixed algebra of a group generated by reflections is not always free*, Arch. Math. (Basel) **41** (1983), 434–437.
4. A. E. Veles′ko, *The existence of groups generated by reflections with infinitely generated invariant rings*, Dokl. Akad. Nauk BSSR **30** (1986), 105–107. (Russian)
5. J. Dieudonné and J. Carrell, *Invariant theory, old and new*, Academic Press, 1971.
6. V. F. Ignatenko, *Infinite groups generated by skew reflections. Mutual arrangement of linear spans of four symmetry direction orbits,* Part I, Simferopol Univ., Simferopol, 1988; Deposited at UrkNIINTI, Manuscript no. 2373-Uk89. (Russian)
7. ______, *Infinite groups generated by skew reflections. Mutual arrangement of linear spans of four symmetry direction orbits,* Part II, Simferopol Univ., Simferopol, 1989; Deposited at UrkNIINTI, Manuscript no. 244-Uk90. (Russian)
8. I. M. Gel′fand and A. S. Mishchenko, *Quadratic forms over commutative group rings and K-theory*, Funktsional. Anal. i Prilozhen. **3** (1969), no. 4, 28–33; English transl. in Functional Anal. Appl. **3** (1969).
9. V. F. Ignatenko, *On special algebraic surfaces with infinite set of skew symmetry planes,* Ukrain. Geometr. Sb. **33** (1990), 52–55; English transl. in J. Soviet Math. **53** (1991), no. 5.

10, 60-LETIYA OKTJABRJA, APT. 14, SIMFEROPOL' 333044, UKRAINE

Amer. Math. Soc. Transl.
(2) Vol. **176**, 1996

Curves and Discontinua with Paradoxical Geometric Properties

A. V. Kuz′minykh

In this paper, we prove the existence of curves and discontinua having unusual geometrical and analytical properties in Euclidean space.

It is well known (see, for example, [**10**]) that there exists a simple arc (i.e., a set homeomorphic to a closed segment) in $\mathbf{R}^3$ whose projection (throughout this paper, by projection we always mean the orthogonal projection) on the plane $z = 0$ is a square (this simple arc is obtained by "unwinding" Peano's curve). One can naturally ask if there exists a curve homeomorphic to a straight line in $\mathbf{R}^n$ (i.e., a subset of the space $\mathbf{R}^n$ homeomorphic to the straight line or, in other words, an image of the straight line embedded into $\mathbf{R}^n$; such a curve is also called an elementary curve), whose projection on each hyperplane is the entire hyperplane. This means that the curve (which, being homeomorphic to a straight line, is, in particular, not dense anywhere in $\mathbf{R}^n$) intersects with each straight line $\lambda \subset \mathbf{R}^n$.

The main purpose of this paper is to provide a positive answer to this and other related questions. A more general question can be posed in this connection in the following form: what geometrical and analytical properties can be possessed by the topological imbeddings of simple topological objects in $\mathbf{R}^n$?

As an example, we prove the existence of a homeomorphism $F : \mathbf{R}^n \to \mathbf{R}^n$, $n \geq 2$, such that for each straight line $l \subset \mathbf{R}^n$ the projection of the curve $F(l)$ onto any hyperplane $P \subset \mathbf{R}^n$ is the entire hyperplane P (moreover, the intersection $\lambda \cap F(l)$ has the cardinality of a continuum for each straight line $\lambda \subset \mathbf{R}^n$.

There exists a curve $\mathfrak{M} \subset \mathbf{R}^n$ homeomorphic to a straight line and differentiable almost anywhere such that for each unbounded regular (doubly differentiable continuously) elementary curve $\gamma \subset \mathbf{R}^n$ of limited curvature, the intersection $\gamma \cap \mathfrak{M}$ has infinite Lebesgue measure defined on the curve γ in the standard manner, and for each plane $Q \subset \mathbf{R}^n$ ($1 \leq \dim Q \leq n$) the intersection $Q \cap \mathfrak{M}$ has infinite Lebesgue measure defined on Q in the standard manner.

The existence of such curves is associated with the existence of discontinua (it should be recalled that a discontinuum is a zero-dimensional perfect compact set;

1991 *Mathematics Subject Classification.* Primary 54F50.

each discontinuum is homeomorphic to the Cantor perfect set) that are "well imbedded" into $\mathbf{R}^n$ in the topological sense but have unexpected geometrical properties. For example, there exists a discontinuum $\mathfrak{F} \subset \mathbf{R}^n$ such that for each hyperplane $P \subset \mathbf{R}^n$ the projection of $\mathfrak{F}$ on P contains an $(n-1)$-dimensional disk of radius 1 such that on each point of the disk a continuum of points from $\mathfrak{F}$ is projected, and the discontinuum $\mathfrak{F}$ has n-dimensional Lebesgue measure zero.

Now we proceed with more complete formulations.

Below $\mathbf{R}$ always denotes the set of real numbers, card $\mathcal{M}$ the cardinality of the set $\mathcal{M}$, $\mathfrak{c}$ the cardinality of a continuum, and m_n the n-dimensional Lebesgue measure in $\mathbf{R}^n$.

THEOREM 1. *There exists a homeomorphism $F\colon \mathbf{R}^n \to \mathbf{R}^n$, $n \geq 2$, with the following properties.*

(1) *For each straight line $l \subset \mathbf{R}^n$, the projection of the curve $F(l)$ on any hyperplane $P \subset \mathbf{R}^n$ is the entire hyperplane P.*

(2) *Each straight line $l \subset \mathbf{R}^n$ bifurcates into a continuum of pairwise disjoint and pairwise homeomorphic subsets $\mathfrak{M}^l_\alpha \subset l$ $(\alpha \in \mathbf{R})$ such that the set $\mathfrak{M}^l_\alpha$ has the following properties for each $\alpha \in \mathbf{R}$:*

(a) *$\mathfrak{M}^l_\alpha$ has s-dimensional Hausdorff measure zero for each $s > 0$; and*

(b) *the following equality holds for each straight line $\lambda \subset \mathbf{R}^n$:*

$$\operatorname{card}(F(\mathfrak{M}^l_\alpha) \cap \lambda) = \mathfrak{c}.$$

(3) *There exist an isotopy $F_t : \mathbf{R}^n \to \mathbf{R}^n$, $0 \leq t \leq 1$, and a closed set $\mathfrak{W} \subset \mathbf{R}^n$ such that F_0 is an identical mapping, $F_1 = F$, $m_n(\mathfrak{W}) < 1$, and the restriction $F_t|_{R^n \setminus \mathfrak{W}}$ is identical for each t, $0 \leq t \leq 1$.*

(4) *There exists a closed set $\Omega \subset R^n$ such that $m_n(\Omega) = 0$, and each point $X \in \mathbf{R}^n \setminus \Omega$ has a neighborhood $U_x \subset \mathbf{R}^n \setminus \Omega$ (open in $\mathbf{R}^n$) such that the restriction $F_t|_{Ux}$ is affine for each t, $0 \leq t \leq 1$, i.e., coincides with the restriction of some affine transformation of the space $\mathbf{R}^n$ to U_x. In particular, the mapping F_t is differentiable almost everywhere for each t, $0 \leq t \leq 1$.*

Throughout this paper, by a curve we mean the image of a straight line under a continuous mapping in $\mathbf{R}^n$, $n \geq 2$.

A simple arc $\Gamma \subset \mathbf{R}^n$ is called a Lipschitz arc if there exist a Cartesian rectangular orthonormal system of coordinates in $\mathbf{R}^n$ and positive numbers M and a such that the arc Γ can be defined in this coordinate system by the equations $x_1(t) = t$, $x_2(t), \dots, x_n(t)$ $(t \in [0, a])$, where the functions $x_i(t)$ $(i = 2, \dots, n)$ are such that for each $t_1, t_2 \in [0, a]$ the following inequality holds:

$$\left(\sum_{i=2}^{n} (x_i(t_1) - x_i(t_2))^2\right)^{\frac{1}{2}} \leq M|t_1 - t_2|$$

(such an arc is also called an M-Lipschitz arc; note that the functions $x_i(t)$ are not assumed to be differentiable).

The curve $\gamma \subset \mathbf{R}^n$ is called a piecewise Lipschitz curve if it is the union of a countable set of simple Lipschitz arcs.

The curve $\gamma \subset \mathbf{R}^n$ is called admissible if there exists an unbounded set $M_\gamma \subset \mathbf{R}^n$ such that $\rho(\gamma, \mu_\gamma) > 0$, where ρ is a Euclidean metric in $\mathbf{R}^n$. (Thus, for example, if there exists an unbounded convex body $V_\gamma \subset \mathbf{R}^n$ such that $\gamma \cap V_\gamma = \emptyset$, then the curve γ is admissible.)

The curve $\gamma \subset \mathbf{R}^n$ is called a curve with limited curvature if γ is a regular (doubly differentiable continuously) elementary curve and there exists a number $M(\gamma) > 0$ such that the curvature of γ at each point is smaller than $M(\gamma)$.

If $\Gamma \subset \mathbf{R}^n$ is a rectifiable simple arc, then in view of the fact that Γ with its intrinsic metric is isometric to a rectilinear segment, we can define on Γ (in the same way as on this segment) a (linear) Lebesgue measure m_Γ (such that $m_\Gamma(\Gamma)$ is equal to the length of Γ).

The measure m_γ on the curve γ with a limited curvature is defined in an identical manner (in particular, for each simple arc $\Gamma \subset \gamma$, $m_\gamma(\Gamma)$ is equal to the length of Γ).

We denote by $\mathfrak{K}$ the Cantor perfect set constructed on the segment $[0,1]$ by using the standard technique, by $\mathbf{N}$ the set of natural numbers, by $\mathbf{Z}$ the set of integers, by $\mathbf{Q}$ the set of rational numbers, by $\operatorname{int} M$, $\operatorname{cl} M$, and ∂M the interior, closure, and boundary of the set M, by $\mathbf{A}(M)$ and $\operatorname{diam} M$ the affine shell and diameter of the set M respectively, by (M, ε) the ε-neighborhood of the set M, and by m_1 the (linear) Lebesgue measure on $\mathbf{R}$.

The open interval $]0,1[\subset \mathbf{R}$ is denoted by J. The mapping $f : J \to \mathbf{R}^n$ is called an isometry at the point $x \in J$ if there exists an open interval J_x such that $x \in J_x \subset J$ and the restriction of f to J_x is an isometry (thus, in particular, $f(J_x)$ is a rectilinear interval).

THEOREM 2. *There exists a curve $\mathfrak{M} \subset \mathbf{R}^n$, $n \geq 2$, homeomorphic to a straight line and possessing the following properties.*

1. *For each hyperplane $P \subset \mathbf{R}^n$, the projection of $\mathfrak{M}$ on P is the entire hyperplane P.*

2. *For each $\alpha \in \mathbf{R}^n$, there exists a set $\mathfrak{N}_\alpha \subset \mathfrak{M}$ homeomorphic to $\mathfrak{K} \setminus \{0\}$ such that the following equality holds for each unbounded admissible piecewise Lipschitz curve $\gamma \subset \mathbf{R}^n$:*

$$\operatorname{card}(\gamma \cap \mathfrak{N}_\alpha) = \mathfrak{c},$$

and $\mathfrak{N}_{\alpha_1} \cap \mathfrak{N}_{\alpha_2} = \emptyset$ for $\alpha_1 \neq \alpha_2$ $(\alpha_1, \alpha_2 \in \mathbf{R})$.

3. *For each $i \in \mathbf{N}$, there exists a set $\mathfrak{S}_i \subset \mathfrak{M}$ homeomorphic to $\mathfrak{K} \setminus \{0\}$ such that for each unbounded admissible piecewise Lipschitz curve $\gamma \subset \mathbf{R}^n$ there exists a simple Lipschitz arc $\Gamma \subset \gamma$ such that*

$$m_\Gamma(\Gamma \cap \mathfrak{M}) \geq m_\Gamma(\Gamma \cap \mathfrak{S}_i) > 0,$$

and $\mathfrak{S}_{i_1} \cap \mathfrak{S}_{i_2} = \emptyset$ for $i_1 \neq i_2$ $(i_1, i_2 \in \mathbf{N})$.

4. *For each $i \in \mathbf{N}$ there exists a set $\mathfrak{T}_i \subset \mathfrak{M}$ homeomorphic to $\mathfrak{K} \setminus \{0\}$ such that the following equalities hold for each unbounded curve $\gamma \subset \mathbf{R}^n$ with limited curvature:*

$$m_\gamma(\gamma \cap \mathfrak{M}) = m_\gamma(\gamma \cap \mathfrak{T}_i) = \infty,$$

and $\mathfrak{T}_{i_1} \cap \mathfrak{T}_{i_2} = \emptyset$ for $i_1 \neq i_2$ $(i_1, i_2 \in \mathbf{N})$.

5. *There exists a set $\mathfrak{U} \subset \mathfrak{M}$ homeomorphic to $\mathfrak{K} \setminus \{0\}$ and possessing the following properties:*

(a) *the following inequalities are valid for each plane $P \subset \mathbf{R}^n$ $(1 \leq \dim P \leq n)$:*

$$m_P(P \setminus \mathfrak{M}) \leq m_P(P \setminus \mathfrak{U}) < 1,$$

where m_P is the standard ($\dim P$*-dimensional) Lebesgue measure in* P;

(b) *for each finite nonempty set* $M \subset \mathbf{R}^n$ *and for each nonzero vector* $v \in \mathbf{R}^n$ *there exists a set* $T(M, v) \subset \mathbf{R}$ *such that* $m_1(T(M, v)) = \infty$ *and for each* $t \in T(M, v)$ *the image of the set* M *obtained as a result of a parallel displacement by the vector* tv *is contained in* $\mathfrak{U}$ (*in particular, it is possible to inscribe into the curve* $\mathfrak{M}$ *a continuum of sets, each of which is congruent to* M).

6. *There exists a homeomorphism* $\varphi : J \to \mathfrak{M}$ *that is an isometry almost everywhere* (*in other words, there exists a set* $\Omega \subset J$ *with a linear measure* 0 *such that the mapping* φ *is an isometry at each point* $x \in J \setminus \Omega$.

COROLLARY 1. *There exists a curve* $\mathfrak{M} \subset \mathbf{R}^n$, $n \geq 2$, *homeomorphic to a straight line and such that for each regular* (*continuously differentiable*) *unbounded curve* $\gamma \subset \mathbf{R}^n$ *whose projection on at least one hyperplane is not dense everywhere in this hyperplane, the following equality holds*:

$$\operatorname{card}(\gamma \cap \mathfrak{M}) = \mathfrak{c}.$$

REMARK 1. It follows from property 5 (in Theorem 2) of the curve $\mathfrak{M}$ that $m_n(\mathfrak{M}) = \infty$. However, there exists a curve $\mathfrak{M}_0 \subset \mathbf{R}^n$, $n \geq 2$, homeomorphic to a straight line such that $m_n(\mathfrak{M}_0) = 0$ and the curve $\mathfrak{M}_0$ possesses property 2 and hence property 1 (in whose formulation we must replace $\mathfrak{M}$ by $\mathfrak{M}_0$) as well as property 6.

Moreover, there exists for each $\varepsilon > 0$ a curve $\mathfrak{M}_\varepsilon \subset \mathbf{R}^n$, $n \geq 2$, homeomorphic to a straight line such that $m_n(\mathfrak{M}_\varepsilon) < \varepsilon$ and the curve $\mathfrak{M}_\varepsilon$ possesses properties 1–3 and 6 above, as well as the following properties.

4′. For each number $i \in \mathbf{N}$ there exists a set $\mathfrak{T}_i^\varepsilon \subset \mathfrak{M}_\varepsilon$ homeomorphic to $\mathfrak{K} \setminus \{0\}$ such that the following inequalities hold for each unbounded curve $\gamma \subset \mathbf{R}^n$ with a bounded curvature:

$$m_\gamma(\gamma \cap \mathfrak{M}_\varepsilon) \geq m_\gamma(\gamma \cap \mathfrak{T}_i^\varepsilon) > 0$$

and $\mathfrak{T}_{i_1}^\varepsilon \cap \mathfrak{T}_{i_2}^\varepsilon = \emptyset$ for $i_1 \neq i_2$ ($i_1, i_2 \in \mathbf{N}$).

5′. There exists a set $\mathfrak{U}_\varepsilon \subset \mathfrak{M}_\varepsilon$ homeomorphic to $\mathfrak{K} \setminus \{0\}$ and such that the following inequalities hold for each plane $P \subset R^n$, $1 \leq \dim P \leq n$:

$$m_P(P \cap \mathfrak{M}_\varepsilon) \geq m_P(P \cap \mathfrak{U}_\varepsilon) > 0.$$

REMARK 2. Obviously, property 1 of the curve $\mathfrak{M}$ (see Theorem 2) is preserved for "quite good deformations" of $\mathfrak{M}$. To be more precise, a homeomorphism $F\colon \mathbf{R}^n \to \mathbf{R}^n$ is called canonical if $F^{-1}(l)$ is a piecewise Lipschitz unbounded admissible curve for each straight line $l \subset \mathbf{R}^n$. The curve $\mathfrak{M}$ has the following property: for each canonical homeomorphism $F : \mathbf{R}^n \to \mathbf{R}^n$ and for each hyperplane $P \subset \mathbf{R}^n$ the projection of $F(\mathfrak{M})$ on P is the entire hyperplane P.

The closed ball and the sphere (in $\mathbf{R}^n$) of radius r and with center at X are denoted by $D^n(X, r)$ and $S^{n-1}(X, r)$ respectively.

If r_1 and r_2 are positive numbers such that $r_1 < r_2$, we denote by $V_n(r_1, r_2)$ the open layer (in $\mathbf{R}^n$) between concentric spheres of radii r_1 and r_2 (with their centers at the origin of coordinates O) respectively, i.e.,

$$V_n(r_1, r_2) \stackrel{\text{def}}{=} (\operatorname{int} D^n(0, r_2)) \setminus D^n(0, r_1).$$

If $\delta \subset \mathbf{R}^n$ is a simple arc or a curve with limited curvature, we say that δ canonically intersects the layer $V_n(r_1, r_2)$ if $\delta \cap S^{n-1}(0, r_i) \neq \emptyset$ for $i = 1, 2$.

A compact set $A \subset R^n$ is called (see [**3**]) cell-decomposed in $\mathbf{R}^n$ if we can inscribe into any neighborhood V of A a neighborhood U whose closure $\operatorname{cl} U$ is a union of a finite number of pairwise disjoint n-dimensional topological balls, $\operatorname{cl} U \subset V$.

THEOREM 3. *Let ε, M, r_1, and r_2 be positive numbers ($r_1 < r_2$) and let n be a natural number. There exist a cell-decomposition in R^n, $n \geq 2$, and discontinua $\mathfrak{D} \subset V_n(r_1, r_2)$ and $\mathfrak{F} \subset \mathfrak{D}$ with the following properties.*

1. *The discontinuum $\mathfrak{F}$ has n-dimensional Lebesgue measure zero: for each $\alpha \in \mathbf{R}$, there exists a discontinuum $\mathfrak{C}_\alpha \subset \mathfrak{F}$ such that the following equality holds for each simple arc $\Gamma \subset \mathbf{R}^n$ that intersects $V_n(r_1, r_2)$ canonically and is a union of simple arcs each of which is an M-Lipschitz arc:*

$$\operatorname{card}(\Gamma \cap \mathfrak{C}_\alpha) = \mathfrak{c}$$

$\mathfrak{C}_{\alpha_1} \cap \mathfrak{C}_{\alpha_2} = \emptyset$ for $\alpha_1 \neq \alpha_2$ ($\alpha_1, \alpha_2 \in \mathbf{R}$) (in particular, the projection of the discontinuum $\mathfrak{C}_\alpha$ onto the hyperplane P contains, for each $\alpha \in \mathbf{R}$ and for each hyperplane $P \subset \mathbf{R}^n$, an $(n-1)$-dimensional disk of radius r_1 onto each of whose points a continuum of points from $\mathfrak{C}_\alpha$ is projected).

2. *For each simple arc $\Gamma \subset \mathbf{R}^n$ which intersects $V_n(r_1, r_2)$ canonically and is a union of N simple, M-Lipschitz arcs, the following inequality holds:*

$$m_\Gamma(\Gamma \cap \mathfrak{D}) > 0.$$

3. *For each curve $\gamma \subset \mathbf{R}^n$ of limited curvature smaller than M at each point such that γ canonically intersects $V_n(r_1, r_2)$, the following inequality is valid:*

$$m_\gamma(\gamma \cap \mathfrak{D}) > r_2 - r_1 - \varepsilon.$$

4. *For each plane $P \subset \mathbf{R}^n$, $1 \leq \dim P \leq n$, the following inequality holds:*

$$m_P(P \cap \mathfrak{D}) > m_P\left(P \cap V_n(r_1, r_2)\right) - \varepsilon.$$

COROLLARY 2. *Since the discontinuum $\mathfrak{D}$ is cell-decomposed in $\mathbf{R}^n$, L. V. Keldysh's theorem (see Theorem* 1 *in* [3], *and also* [4]) *leads to the existence of a simple arc $\Gamma(\mathfrak{D})$ such that*

$$\mathfrak{D} \subset \Gamma(\mathfrak{D}) \subset V_n(r_1, r_2)$$

and there exists an isotopy $F_t : \mathbf{R}^n \to \mathbf{R}^n$, $0 \leq t \leq 1$, of the identical mapping F_0 such that $F_1(\Gamma(\mathfrak{D}))$ is a rectilinear segment.

COROLLARY 3. *Let r_1 and r_2 be positive numbers, $r_1 < r_2$. There exists a curve $\mathfrak{L} \subset V_n(r_1, r_2)$, $n \geq 2$, homeomorphic to a straight line and such that the projection of $\mathfrak{L}$ onto each hyperplane $P \subset \mathbf{R}^n$ is an open (in P) $(n-1)$-dimensional ball of radius r_2 (with its center at the projection of the point O).*

The results presented in this paper (as well as other similar results) were announced in [**5–8**].

Now we begin to prove the statements formulated above, assuming that $n \geq 2$.

By a block we mean an n-dimensional parallelepiped with mutually orthogonal edges, whose base is an $(n-1)$-dimensional cube, and whose height (i.e., the length of the edge orthogonal to the base; we call it the vertical edge) is less than the length of the side of the base. For each block K, we put

$$\omega(K) \overset{\text{def}}{=} \frac{a(K)}{8h(K)},$$

where $a(K)$ is the length of the side of the base, and $h(K)$ is the height of the block K.

An oriented block is a pair (K, K^1), where K is a block and K^1 its base (the second base of the block K will be denoted by K^2). Thus, an oriented block is a block with a selected ("first") base. Without introducing any superfluous notation, we denote the oriented block (K, K^1) and the block K by the same letter K whenever it does not lead to confusion.

Let $M > 0$ be a number and $p \subset \mathbf{R}^n$ a (closed) segment. A simple arc $\Gamma \subset \mathbf{R}^n$ is callled an M-Lipschitz arc with base p if there is a rectangular Cartesian orthonormal coordinate system (in $\mathbf{R}^n$) such that the segment p belongs to the first coordinate axis, and the arc Γ is described by the equations $x_1(t) = t$, $x_2(t), \dots, x_n(t)$, $t \in p$, where the functions $x_i(t)$, $i = 2, \dots, n$, are such that the following inequality holds for each $t_1, t_2 \in p$:

$$\left(\sum_{i=2}^{n} (x_i(t_1) - x_i(t_2))^2\right)^{\frac{1}{2}} \leq M|t_1 - t_2|$$

(such a coordinate system is called (M,p)-canonical for Γ).

For each block K, we denote by $\mathcal{H}(K)$ the set of all $\omega(K)$-Lipschitz simple arcs each of which has (any) vertical edge of the block K as its base.

For each $(n-1)$-dimensional cube $T \subset \mathbf{R}^n$, denote by T_λ (where $\lambda > 0$ is a certain number) the image of the cube T obtained by homothety with coefficient λ and center of homothety at the center of the cube T.

We say that a simple arc $\Gamma \subset \mathbf{R}^n$ canonically intersects an oriented block K if $\Gamma \cap K^1_{1/4} \neq \emptyset$. Note that for each simple arc $\Gamma \subset \mathcal{H}(K)$ which intersects an oriented block K canonically, we have $\Gamma \cap K^2 \subset K^2_{1/2}$.

For each straight line $L \subset \mathbf{R}^n$, denote by κ_l the orthogonal projection $\kappa_L : \mathbf{R}^n \to L$ of the space $\mathbf{R}^n$ on the straight line L.

For each oriented block K, denote by $O^i(K)$, $i = 1, 2$, the center of the base K^i of the block K, by $L(K)$ the straight line joining the points $O^1(K)$ and $O^2(K)$, and by $I(K)$ the segment $[O^1(K), O^2(K)]$.

The oriented block Q is called standard for an oriented block K if $Q \subset K$, the edges of Q and K are pairwise parallel (the vertical edges of Q and K are parallel), and the first base (Q^1) of Q is the base that is closer to K^1.

The finite set W of blocks standard for an oriented block K is called ε-canonical for K (where $\varepsilon > 0$ is a number) if it possesses the following properties

1. For each block $Q \in W$, we have $Q \subset \operatorname{int} K$, $\operatorname{diam} Q < \varepsilon$, $\rho(Q, K^1) < \varepsilon$, $\omega(Q) > \omega(K)$

2. For each arc $\Gamma \in \mathcal{H}(K)$ canonically intersecting K, there exists a block $Q(\Gamma) \in W$ such that Γ canonically intersects $Q(\Gamma)$.

3. The projections $\kappa_{L(K)}(Q)$ of blocks $Q \in W$ do not intersect pairwise.

We will need the following lemmas.

LEMMA 1. *For each oriented block K and for each $\varepsilon > 0$, there exists a set of blocks ε-canonical for K.*

PROOF. Let K be an oriented block and $\varepsilon > 0$ a number. We denote by $\mathcal{T}$ a finite set of $(n-1)$-dimensional cubes whose edges are parallel to the corresponding edges of the cube K^1, such that for each cube $T \in \mathcal{T}$ we have $\operatorname{diam} T < \frac{\varepsilon}{2}$ and $T \subset \operatorname{int} K^1$ (interior in the topology $A(K^1)$, where $A(\mathcal{M})$ is the affine span of the set $\mathcal{M}$), and the following inclusion is valid:

$$K^1_{1/4} \subset \bigcup_{T \in \mathcal{T}} T_{1/8}.$$

We define $a_0 \stackrel{\text{def}}{=} \min\{a(T) : T \in \mathcal{T}\}$ (where $a(T)$ is the length of an edge of the cube T), and $\varepsilon_0 \stackrel{\text{def}}{=} \min\{\varepsilon/2, h(K)/2, a_0 \cdot (16 \cdot \omega(K))^{-1}\}$.

For each cube $T \in \mathcal{T}$, we denote by T' the $(n-1)$-dimensional cube obtained by parallel displacement of T by a vector of length $< \varepsilon_0$ orthogonal to K^1, such that $T' \subset \operatorname{int} K$ and the distances $\rho(T', K^1)$ $(T \in \mathcal{T})$ are pairwise distinct.

It can easily be seen that for each cube $T \in \mathcal{T}$ there exists a block $Q_T \subset \operatorname{int} K$ which is standard for K and for which $\operatorname{diam} Q_T < \varepsilon$, $\rho(Q_T, K^1) < \varepsilon$, $\omega(Q_T) > \omega(K)$, the base Q'_T of the block Q_T coincides with T', and the projections $\kappa_{L(K)}(Q_T)$ of blocks Q_T $(T \in \mathcal{T})$ are pairwise disjoint.

Let $\Gamma \in \mathcal{H}(K)$ be an arc intersecting K canonically, A a point such that $\{A\} = \Gamma \cap K^1_{1/4}$, and $\tilde{T} \in \mathcal{T}$ a cube such that $A \in \tilde{T}_{1/8}$. In view of the conditions satisfied by $\tilde{T}'$, and in the light of the fact that $\Gamma \in \mathcal{H}(K)$, we find that $\Gamma \cap \tilde{T}'_{1/4} \neq \emptyset$, i.e., the arc Γ canonically intersects the block $Q_{\tilde{T}}$. We put $W \stackrel{\text{def}}{=} \{Q_T : T \in \mathcal{T}\}$. Obviously, the set W is ε-canonical for K. Lemma 1 is proved.

LEMMA 2. *For each oriented block K and for each $\varepsilon > 0$, there exist disjoint finite sets W_0 and W_1 of standard blocks for K, such that the set $W_0 \cup W_1$ is ε-canonical for K, and for $i = 0, 1$ and each arc $\Gamma \in \mathcal{H}(K)$ intersecting K canonically, there exist (different) blocks Q_1, $Q_2 \in W_i$ such that Γ intersects Q_1 and Q_2 canonically.*

PROOF. By Lemma 1, there exists a set of blocks ε-canonical for K (we denote this set by W^1).

For each finite set $\mathfrak{W}$ of blocks, we put

$$\varepsilon(\mathfrak{W}) \stackrel{\text{def}}{=} \min\left\{\rho(Q, K^1) : Q \in \mathfrak{W}\right\}.$$

If W^j, $j \in \mathbf{N}$, is a set of blocks that is ε'-canonical for K (for a certain $\varepsilon' > 0$), we denote by W^{j+1} a set of blocks which is $\frac{1}{2}\varepsilon(W^j)$-canonical for K (the existence of this set follows from Lemma 1).

Thus, a finite set of blocks W^j is defined by induction for each $j \in \mathbf{N}$. It can be easily seen that for each $j \in \mathbf{N}$

$$\bigcup_{Q \in W^{j+1}} Q \subset (K^1, \varepsilon(W^j)),$$

where $(\mathcal{M}, \delta)$ is the δ-neighborhood of the set $\mathcal{M}$. Consequently, the projections $\kappa_{L(K)}(Q)$ $(Q \in \bigcup_{j=1}^{\infty} W^j)$ are pairwise disjoint.

It can be verified directly that the sets $W_0 \stackrel{\text{def}}{=} W^1 \cup W^2$ and $W_1 \stackrel{\text{def}}{=} W^3 \cup W^4$ are the required ones. Lemma 2 is proved.

LEMMA 3. *For each oriented block K there exist discontinua $\mathcal{A}(K) \subset \operatorname{int} K$, $\mathcal{R}(K) \subset I(K)$ and a simple arc $\mathfrak{M}(K) \subset K$ having the following properties.*

1. *The discontinuum $\mathcal{A}(K)$ is cell-decomposed in $\mathbf{R}^n$.*
2. *The inclusion $\mathcal{A}(K) \subset \mathfrak{M}(K)$ is valid.*
3. *The simple arc $\mathfrak{M}(K)$ is projected bijectively on the segment $I(K)$.*
4. *The ends of the arc $\mathfrak{M}(K)$ coincide with the points $O^1(K)$, $O^2(K)$.*
5. *$\mathfrak{M}(K) \setminus \{O^1(K), O^2(K)\} \subset \operatorname{int} K$.*
6. *$\kappa_{L(K)}(\mathcal{A}(K)) \subset \mathcal{R}(K)$, and $\{O^1(K), O^2(K)\} \subset \mathcal{R}(K)$.*
7. *For each $s > 0$, the s-dimensional Hausdorff measure of the discontinuum $\mathcal{R}(K)$ equals 0.*
8. *$m_n(\mathfrak{M}(K)) = 0$.*
9. *If the point $X \in \mathfrak{M}(K)$ is such that $\kappa_{L(K)}(X) \notin \mathcal{R}(K)$, then there exists a (rectilinear) open interval $I_X \subset \mathfrak{M}(K)$ containing X and parallel to $I(K)$.*
10. *For each $\alpha \in \mathbf{R}$, there exists a discontinuum $\mathfrak{N}_\alpha(K) \subset \mathcal{A}(K)$ such that for each simple arc $\Gamma \in \mathcal{H}(K)$ canonically intersecting K, the following equality holds:*

$$\operatorname{card}(\Gamma \cap \mathfrak{N}_\alpha(K)) = \mathfrak{c}_0,$$

and $\mathfrak{N}_{\alpha_1}(K) \cap \mathfrak{N}_{\alpha_2}(K) = \emptyset$ for $\alpha_1, \alpha_2 \in \mathbf{R}, \alpha_1 \neq \alpha_2$.

PROOF. Let us fix an oriented block K. For each block $Q \subset K$ that is standard for K and for each $\varepsilon > 0$, we denote by $W_0(Q, \varepsilon)$ and $W_1(Q, \varepsilon)$ disjoint finite sets of blocks standard for K such that the set $W(Q, \varepsilon) \stackrel{\text{def}}{=} W_0(Q, \varepsilon) \cup W_1(Q, \varepsilon)$ is ε-canonical for Q, for each block $Q' \in W(Q, \varepsilon)$ we have $\operatorname{diam} Q' < \frac{1}{2} \operatorname{diam} Q$, while for $i = 0, 1$ and for each arc $\Gamma \in \mathcal{H}(K)$ intersecting Q canonically, there exist blocks Q_1, $Q_2 \in W_i(Q, \varepsilon)$ such that Γ intersects Q_1 and Q_2 canonically (the existence of $W_i(Q, \varepsilon)$, $i = 0, 1$, follows from Lemma 2).

Thus, for each block $Q \subset K$ standard for K and each $\varepsilon > 0$ we have chosen the sets $W_i(Q, \varepsilon)$, $i = 0, 1,$.

For each $j \in \mathbf{N}$, we define a finite set V_j of blocks that are standard for K as follows. Let

$$V_1 \stackrel{\text{def}}{=} \bigcup_{i=0}^{1} W_i(K, 1).$$

Suppose that a finite (nonempty) set V_j of blocks standard for K has been defined for some $j \in \mathbf{N}$. We set

$$\delta_j \stackrel{\text{def}}{=} \frac{1}{2} \cdot \left((j+1) \operatorname{card} V_j\right)^{-j},$$

$$V_{j+1} \stackrel{\text{def}}{=} \bigcup_{k=0}^{1} \bigcup_{Q \in V_j} W_k(Q, \delta_j).$$

Thus, the set V_j of blocks has been defined by induction for each $j \in \mathbf{N}$. For each $j \in \mathbf{N}$, we put

$$\tilde{V}_j \stackrel{\text{def}}{=} \bigcup_{Q \in V_j} Q.$$

and set

$$\mathcal{A}(K) \stackrel{\text{def}}{=} \bigcap_{j=1}^{\infty} \tilde{V}_j.$$

Obviously, the set $\mathcal{A}(K)$ is injectively projected into $I(K)$.

For each $j \in \mathbf{N}$ and each block $Q \in V_j$, denote by $\tilde{W}(Q, \delta_j)$ the union of all blocks that are elements of the set

$$\bigcup_{k=0}^{1} W_k(Q, \delta_j).$$

Obviously, for each $j \in \mathbf{N}$ and for each block $Q \in V_j$, the set $\kappa_{L(K)}(\tilde{W}(Q, \delta_j))$ is contained in the segment of length $2\delta_j$ (since $\tilde{W}(Q, \delta_j))$ is contained in the $2\delta_j$-neighborhood of the base Q^1 of the block Q). Hence, for each $j \in \mathbf{N}$, the set $\tilde{V}_{j+1}$ possesses the following property: there exist segments $I_1, \ldots, I_{m(j)} \subset I(K)$ (where $m(j) \stackrel{\text{def}}{=} \operatorname{card} V_j$) such that for each k ($k = 1, \ldots, m(j)$), $|I_k| < 1/j$ (where $|I|$ is the length of the segment I), the relations

$$\kappa_{L(K)}(\tilde{V}_j) \subset \bigcup_{k=1}^{m(j)} I_k, \quad \sum_{k=1}^{m(j)} |I_k|^{1/j} < 1/j$$

hold. Consequently, the set $\kappa_{L(K)}(\mathcal{A}(K))$ has s-dimensional Hausdorff measure zero for each $s > 0$. Obviously, $\mathcal{A}(K)$ is a cell-decomposed discontinuum in $\mathbf{R}^n$.

By a binary sequence we mean a sequence $\{b_j\}_{j=1}^{\infty}$ such that $b_j \in \{0, 1\}$ for each j. Let $B \stackrel{\text{def}}{=} \{b_j\}_{j=1}^{\infty}$ be a binary sequence. We define the sets $\mathcal{U}_j(B)$, $j \in \mathbf{N}$, of blocks standard for K in the following manner. Put $\mathcal{U}_1(B) \stackrel{\text{def}}{=} W_{b_1}(K, 1)$ and assume that for some $j \in \mathbf{N}$ a finite nonempty set $\mathcal{U}_j(B)$ of blocks standard for K is defined. Further, we put

$$\mathcal{U}_{j+1}(B) \stackrel{\text{def}}{=} \bigcup_{Q \in \mathcal{U}_j(B)} W_{b_{j+1}}(Q, \delta_j).$$

Thus, the set $\mathcal{U}_j(B)$ of blocks is defined by induction for each $j \in \mathbf{N}$. For each $j \in \mathbf{N}$, let us define

$$\tilde{\mathcal{U}}_j(B) \stackrel{\text{def}}{=} \bigcup_{Q \in \mathcal{U}_j(B)} Q$$

and put

$$\mathcal{A}(B) \stackrel{\text{def}}{=} \bigcap_{j=1}^{\infty} \tilde{\mathcal{U}}_j(B).$$

It can be easily seen that $\mathcal{A}(B)$ is a discontinuum with the following properties.

1. $\mathcal{A}(B) \subset \mathcal{A}(K)$.

2. For each arc $\Gamma \in \mathcal{H}(K)$ intersecting K canonically, the set $\Gamma \cap \mathcal{A}(B)$ contains a discontinuum (this follows directly from the properties of $W_i(Q, \varepsilon)$, $i \in \{0, 1\}$; see above). Consequently, $\operatorname{card}(\Gamma \cap \mathcal{A}(B)) = \mathfrak{c}$.

We denote by $\mathcal{B}$ the set of all binary sequences. For two different sequences B', $B'' \in \mathcal{B}$, we have $\mathcal{A}(B') \cap \mathcal{A}(B'') = \emptyset$.

Let $\chi : \mathbf{R} \to \mathcal{B}$ be a bijective mapping (which exists in view of the fact that $\operatorname{card}\mathcal{B} = \mathfrak{c}$). For each $\alpha \in \mathbf{R}$, we put $\mathfrak{N}_\alpha(K) \stackrel{\text{def}}{=} \mathcal{A}(\chi(\alpha))$.

Denote by M the set of maximum open intervals contained in

$$\mathcal{Y} \stackrel{\text{def}}{=} I(K) \setminus \left(\kappa_{l(K)}(\mathcal{A}(K)) \cup \{O^1(K), O^2(K)\}\right)$$

(in other words, M is a set of connectivity components of the set $\mathcal{Y}$).

Let $I \in M$ be a certain interval and Y_1 and Y_2 be its ends.

It can be proved that there exists a continuous mapping $\varphi_I : \operatorname{cl} I \to K$ having the following properties.

1. For each point $X \in \operatorname{cl} I$, the vector $\overrightarrow{X\varphi_I(X)}$ is parallel to the base of the block K (in particular, the points X and $\varphi_I(X)$ may coincide).

2. If $Y \in \{Y_1, Y_2\}$ (and hence $Y \in I(K) \setminus \mathcal{Y}$), then $\varphi_I(Y) = \kappa^{-1}_{L(K)}(Y) \cap \mathcal{A}(K)$ for $Y \in \kappa_{L(K)}(\mathcal{A}(K))$ and $\varphi_I(Y) = Y$ for $Y \in \{O^1(K), O^2(K)\}$.

3. $\varphi_I(\operatorname{cl} I)$ is either a segment parallel to $I(K)$ or is contained in a (two-dimensional) rectangle with two sides parallel to $I(K)$ and two other sides parallel to the base K^1 of block K; $\varphi_I(Y_1)$ and $\varphi_I(Y_2)$ are the opposite vertices of this rectangle.

4. There exists a discontinuum $\mathfrak{D}_I \subset \operatorname{cl} I$ containing the ends of the segment $\operatorname{cl} I$ and such that $\mathfrak{D}_I$ has s-dimensional Hausdorff measure zero for each $s > 0$, and for each open interval $p \subset (\operatorname{cl} I) \setminus \mathfrak{D}_I$ the restriction of the mapping φ_I to p is a parallel displacement along the base K^1 (to be more precise, the restriction of the mapping φ_I to p coincides with the restriction of a parallel displacement of the space $\mathbf{R}^n$ by a vector parallel to $A(K^1)$).

Indeed, such a mapping φ_I is obtained using the Cantor function constructed (see, for example, [**2**]) for the discontinuum $\mathfrak{D}_I \subset \operatorname{cl} I$, containing the ends of the interval I and having s-dimensional Hausdorff measure zero for each $s > 0$ (if the segment $[\varphi_I(Y_1), \varphi_I(Y_2)]$ is parallel to $I(K)$, then naturally such a Cantor function degenerates into a constant function). In particular, we fix for each $I \in M$ a discontinuum $\mathfrak{D}_I \subset \operatorname{cl} I$.

Let us now define the mapping $\varphi : I(K) \to K$ as follows: if the point $X \in I(K)$ belongs to $\operatorname{cl} I$ (where $I \in M$), we put $\varphi(X) \stackrel{\text{def}}{=} \varphi_I(X)$. If, however, $X \in I(K) \setminus \bigcup_{I \in M} \operatorname{cl} I$, we put $\varphi(X) \stackrel{\text{def}}{=} \kappa^{-1}_{L(K)}(X) \cap \mathcal{A}(K)$. Obviously, the mapping φ is continuous and injective, and $\mathcal{A}(K) \subset \varphi(I(K))$, $\varphi(O^1(K)) = O^1(K)$, $\varphi(O^2(K)) = O^2(K)$, $\varphi(I(K)) \setminus \{O^1(K), O^2(K)\} \subset \operatorname{int} K$.

We put $\mathcal{R}(K) \stackrel{\text{def}}{=} \kappa_{L(K)}(\mathcal{A}(K)) \cup \bigcup_{I \in M} \mathfrak{D}_I$. Obviously, $\mathcal{R}(K)$ is a discontinuum of s-dimensional Hausdorff measure zero for each $s > 0$. We put $\mathfrak{M}(K) \stackrel{\text{def}}{=} \varphi(I(K))$.

It can be verified directly that the discontinua $\mathcal{A}(K)$, $\mathcal{R}(K)$, $\mathfrak{N}_\alpha(K)$, $\alpha \in \mathbf{R}$, and the simple arc $\mathfrak{M}(K)$ have the properties required by Lemma 3.

Lemma 3 is proved.

PROOF OF THEOREM 1. Let $\tilde{\mathcal{Z}} \subset S^{n-1}(0,1)$ be a countable set which is dense in the sphere $S^{n-1}(0,1)$ (0 is the origin). We enumerate the set $\tilde{\mathcal{Z}}$ as $\tilde{\mathcal{Z}} = \{Z_1, Z_2, \dots\}$.

For each oriented block K, denote by $C(K)$ an oriented block whose first base $(C(K))^1$ is obtained from the base K^1 of K by homothety with coefficient 2 and with center at the center of the base K^1. The height $h(C(K))$ of $C(K)$ is equal to the height $h(K)$ of K, $C(K) \supset K$.

Suppose that the oriented blocks $T_k \subset \mathbf{R}^n$, $k \in \mathbf{N}$, satisfy the following conditions.

1. For each k, the bundle $l(0, Z_k)$ (by $l(X, Y)$ we shall denote a closed bundle with origin X passing through the point Y, $Y \neq X$) is orthogonal to the base of block T_k and passes through its center, the base T_k^1 of block T_k lying closer to 0 than its other base T_k^2

2. For each k, the following inequalities hold (recall that $a(T_k)$ is the length of the side of the base of block T_k):

$$a(T_k) > (\rho(0, T_k))^2 > k^2.$$

3. For each k, the following relation holds:

$$m_n(C(T_k)) < 2^{-k}.$$

4. For each $k > 1$, we have

$$\rho(0, C(T_k)) > \max\left\{\rho(0, X) : X \in \bigcup_{i=1}^{k-1} C(T_i)\right\}.$$

It can easily be seen that blocks T_k ($k \in \mathbf{N}$) satisfying these conditions exist.

Obviously, $\lim_{k\to\infty} h(T_k) = 0$, $\lim_{k\to\infty} \omega(T_k) = \infty$. Now let us define the mappings $F_t\colon \mathbf{R}^n \to \mathbf{R}^n$, $0 \leq t \leq 1$, as follows. For each $k \in \mathbf{N}$, we fix discontinua $\mathcal{A}(T_k)$, $\mathcal{R}(T_k)$, $\mathfrak{N}_\alpha(T_k)$ ($\alpha \in \mathbf{R}$) and a simple arc $\mathfrak{M}(T_k)$, whose existence follows from Lemma 3. Let t, $0 \leq t \leq 1$, be a number and k a natural number. For each hyperplane $P \subset \mathbf{R}^n$ parallel to the bases of T_k and intersecting the segment $I(T_k)$, let the restriction of the mapping F_t to $P \cap T_k$ be a parallel displacement of the $(n-1)$-dimensional cube $P \cap T_k$ by the vector $t \cdot \overrightarrow{X_P Y_P}$, where $\{X_P\} \stackrel{\text{def}}{=} P \cap I(T_k)$, $\{Y_P\} \stackrel{\text{def}}{=} P \cap \mathfrak{M}(T_k)$. For each $(n-2)$-dimensional face Q of the $(n-1)$-dimensional cube $P \cap C(T_k)$, denote by Q' the $(n-2)$-dimensional face of the $(n-1)$-dimensional cube $P \cap C(T_k)$ parallel to Q (for $n > 2$) and closest to Q (simply the closest for $n = 2$). We define the restriction of the mapping F_t to $\operatorname{conv}(Q \cup Q')$ (where $\operatorname{conv} \mathcal{M}$ is the convex hull of the set $\mathcal{M}$) as the restriction to $\operatorname{conv}(Q \cup Q')$ of the affine mapping $\mu : P \to P$ such that $\mu|_{Q'}$ is the identity mapping and $\mu|_Q = F_t|_Q$ (note that $F_t|_Q$ was defined earlier since $Q \subset P \cap T_k$).

Thus, for each t, $0 \leq t \leq 1$, and for each $k \in \mathbf{N}$, the restriction of the mapping F_t to $C(T_k)$ has been defined.

We denote by H_k ($k \in \mathbf{N}$) the union of all hyperplanes parallel to the base block T_k and intersecting the set $\mathcal{R}(T_k)$. We denote by $\tilde{G}$ the set of $(n-1)$-dimensional lateral faces of T_k (i.e., the faces other than the bases). For each face $G \in \tilde{G}$, we denote by G' the $(n-1)$-dimensional lateral face of $C(T_k)$ parallel to the face G and closest to it. We put

$$H_k^0 \stackrel{\text{def}}{=} \bigcup_{G \in \tilde{G}} \partial\left(\operatorname{conv}(G \cup G')\right).$$

It can be verified directly that each point $X \in (\operatorname{int} C(T_k)) \setminus (H_k \cup H_k^0)$ has a neighborhood V such that $X \in V \subset (\operatorname{int} C(T_k)) \setminus (H_k \cup H_k^0)$ and for each t, $0 \leq t \leq 1$, the restriction $F_t|_V$ is affine, i.e., is a restriction of some affine transformation of the space $\mathbf{R}^n$ to V.

We put

$$\Omega_k \stackrel{\text{def}}{=} (C(T_k) \cap H_k) \cup H_k^0$$

and

$$\mathcal{W} \stackrel{\text{def}}{=} \bigcup_{k=1}^{\infty} C(T_k).$$

Since $m_n(C(T_k)) < 2^{-k}$ for each $k \in \mathbf{N}$, we find that $m_n(\mathcal{W}) < 1$. Obviously, the set $\mathcal{W}$ is closed.

Thus, we have defined the restriction $F_t|_{\mathcal{W}}$. We redefine F_t by letting $F_t|_{\mathbf{R}^n \setminus \mathcal{W}}$ to be the identity mapping, and put $F \stackrel{\text{def}}{=} F_1$.

It can be verified by a standard technique that $F_t \colon \mathbf{R}^n \to \mathbf{R}^n$, $0 \le t \le 1$, is a homeomorphism and F_t, $0 \le t \le 1$, is an isotropy, with F_0 being an identical mapping.

We put $\Omega \stackrel{\text{def}}{=} \bigcup_{k=1}^{\infty} \Omega_k$. It can easily be seen that the set Ω has property 4 in the formulation of Theorem 1.

Let $l_0 \subset \mathbf{R}^n$ and $\lambda_0 \subset \mathbf{R}^n$ be bundles which are not antiparallel (i.e., if l_1 and λ_1 are bundles obtained by parallel displacement from l_0 and λ_0 by making their origins coincide with the point 0, the angle between the bundles l_1 and λ_1 is smaller than π. In the forthcoming analysis we denote by ξ a bundle with origin O, which is the bisector of this angle). Note that no assumptions are made about the coincidence of the origins of the bundles l_0 and λ_0.

By the conditions satisfied by the blocks T_k, $k \in \mathbf{N}$, and in view of the fact that the set $\widetilde{\mathcal{Z}}$ is dense in $S^{n-1}(0,1)$, we find that there exists a number $k(l_0, \lambda_0) \in \mathbf{N}$ such that the block $T \stackrel{\text{def}}{=} T_{k(l_0,\lambda_0)}$ has the following properties.

1. The segments $l_0' \stackrel{\text{def}}{=} l_0 \cap T$ and $\lambda_0' \stackrel{\text{def}}{=} \lambda_0 \cap T$ belong to $\mathcal{H}(T)$ and canonically intersect T.

2. Denote by $\tau^i : A(T^i) \to A(T^i)$ $(i = 1, 2)$ the parallel displacement which translates the point $l_0 \cap T^i$ to the point $O^i(T)$ (recall that T^i is the ith base of the block T). Then the segment λ_0' with ends $\tau^1(\lambda_0 \cap T^1)$ and $\tau^2(\lambda_0 \cap T^2)$ belongs to $\mathcal{H}(T)$ and canonically intersects T.

(For T we can select a block far removed from the point O such that the bundle $l(0, Z_k)$ is close to the bisector ξ.)

By $\eta \colon \mathbf{R}^n \to \mathbf{R}^n$ we denote an affine mapping such that $\eta|_{A(T^i)} = \tau^i$, $i = 1, 2$. Obviously, such an affine mapping exists since τ^i is a parallel displacement. Thus, $\eta(l_0') = \mathbf{I}(T)$, $\eta(\lambda_0') = \tilde{\lambda}_0'$, and for each hyperplane $P \subset \mathbf{R}^n$ parallel to $A(T^1)$ the restriction $\eta|_P$ is a parallel displacement $\eta(P) = P$.

Now we assume that $\alpha \in \mathbf{R}$ is some number. Since the segment $\tilde{\lambda}_0'$ belongs to $\mathcal{H}(T)$ and intersects T canonically, we have

$$\operatorname{card}\left(\tilde{\lambda}_0' \cap \mathfrak{N}_\alpha(T)\right) = \mathfrak{c}.$$

Let $X \in \tilde{\lambda}_0' \cap \mathfrak{N}_\alpha(T)$ be a point. We denote by P_X a hyperplane containing X and parallel to $A(T^1)$, and by S the point $P_X \cap I(T)$. Since $X \in \mathfrak{N}_\alpha(T) \subset \mathfrak{M}(T)$, we obtain from the definition of the mapping F that $F|_{P_X \cap T}$ is a parallel displacement by the vector $\overrightarrow{SX}$.

We put $X' \overset{\text{def}}{=} \eta^{-1}(X)$. Obviously, $X' \in \lambda_0'$. Putting $S' \overset{\text{def}}{=} \eta^{-1}(S)$, we find that $S' \in l_0'$. Since $\overrightarrow{S'X'} = \overrightarrow{SX}$, we obtain $X' \in F(l_0')$. Thus, the discontinuum

$$\mathfrak{N}_\alpha(T, l_0) \overset{\text{def}}{=} \eta^{-1}\left(\kappa_{L(T)}\left(\mathfrak{N}_\alpha(T)\right)\right)$$

has the following properties:

$$\mathfrak{N}_\alpha(T, l_0) \subset l_0, \quad \operatorname{card}\left(\lambda_0 \cap F\left(\mathfrak{N}_\alpha(T, l_0)\right)\right) = \mathfrak{c}.$$

For each straight line $l \subset \mathbf{R}^n$, we choose a bundle $l_0 \subset l$ and denote by $T(l)$ the set of all blocks $K \in \{T_1, T_2, \dots\}$ such that $l_0 \cap K$ is the segment canonically intersecting K and belonging to $\mathcal{H}(K)$. For each $\alpha \in \mathbf{R}$, we put

$$\mathfrak{N}_\alpha^l \overset{\text{def}}{=} \bigcup_{K \in T(l)} \left(l_0 \cap \left(\kappa_{L(K)}^{-1}\left(\kappa_{L(K)}\left(\mathfrak{N}_\alpha(K)\right)\right)\right)\right).$$

Let $\Psi : \mathbf{R} \to l \setminus \left(\bigcup_{\alpha \in \mathbf{R}} \mathfrak{N}_\alpha^l\right)$ be a bijective mapping. (Obviously, such a mapping exists.) We put

$$\mathfrak{M}_\alpha^l \overset{\text{def}}{=} \mathfrak{N}_\alpha^l \bigcup \{\Psi(\alpha)\}.$$

In the light of what has been stated above, we find that, for each bundle $\lambda_0 \subset \mathbf{R}^n$ that is not antiparallel to l_0 and for each $\alpha \in \mathbf{R}$,

$$\operatorname{card}\left(F(\mathfrak{M}_\alpha^l) \cap \lambda_0\right) = \mathfrak{c}.$$

Hence for each straight line $\lambda \subset \mathbf{R}^n$,

$$\operatorname{card}\left(F(\mathfrak{M}_\alpha^l) \cap \lambda\right) = \mathfrak{c},$$

since on λ there always exists a bundle that is not antiparallel to l_0.

Note that for each $\alpha \in \mathbf{R}$, the set $\mathfrak{N}_\alpha^l$ is closed and homeomorphic to $\mathcal{K} \setminus \{0\}$ (recall that $\mathcal{K}$ is the standard Cantor perfect set). Hence for each $\alpha \in \mathbf{R}$ the set $\mathfrak{M}_\alpha^l$ is homeomorphic to $(\mathcal{K} \setminus \{0\}) \cup \{2\}$. It can easily be seen that for each straight line $l \subset \mathbf{R}^n$ and for each $\alpha \in \mathbf{R}$, the set $\mathfrak{M}_\alpha^l$ has s-dimensional Hausdorff measure zero for each $s > 0$. Property (1) in the formulation of Theorem 1 is a direct consequence of the properties of the sets $\mathfrak{M}_\alpha^l$.

Theorem 1 is proved.

We have also proved that statement 2 in the formulation of Theorem 1 can be supplemented by the following statement:

(2′) Each bundle $\mathcal{L} \subset \mathbf{R}^n$ has a partition into a continuum of pairwise disjoint and pairwise homeomorphic subsets $\mathfrak{M}_\alpha^{\mathcal{L}} \subset \mathcal{L}$ ($\alpha \in \mathbf{R}$) such that the set $\mathfrak{M}_\alpha^{\mathcal{L}}$ has the following properties for each $\alpha \in \mathbf{R}$:

(a) $\mathfrak{M}_\alpha^{\mathcal{L}}$ has s-dimensional Hausdorff measure zero for each $s > 0$;

(b) for each bundle $\sigma \subset \mathbf{R}^n$ not antiparallel to $\mathcal{L}$, the intersection $F(\mathfrak{M}_\alpha^{\mathcal{L}}) \cap \sigma$ has the cardinality of a continuum.

(In order to verify this, it is sufficient to change the mapping Ψ in the proof appropriately.)

In addition to Lemma 3, the following lemmas are required for proving Theorems 2 and 3.

LEMMA 4. *For each oriented block K and for each $\varepsilon > 0$, there exist a discontinuum $\mathcal{A}(K,\varepsilon)$ and a simple arc $\mathfrak{M}(K,\varepsilon)$ having the following properties.*

1. *The discontinuum $\mathcal{A}(K,\varepsilon)$ is cell-decomposed in $\mathbf{R}^n$.*
2. *The ends $B_1(K,\varepsilon)$, $B_2(K,\varepsilon)$ of the arc $\mathfrak{M}(K,\varepsilon)$ belong to ∂K.*
3. $\mathcal{A}(K,\varepsilon) \subset \mathfrak{M}(K,\varepsilon) \setminus \{B_1(K,\varepsilon), B_2(K,\varepsilon)\} \subset \operatorname{int} K$.
4. *For each simple arc $\Gamma \in \mathcal{H}(K)$ intersecting K canonically,*

$$m_\Gamma\left(\Gamma \setminus \mathcal{A}(K,\varepsilon)\right) < \varepsilon.$$

PROOF. We consider the system of coordinates $(\omega(K), I(K))$ which is canonical for each simple arc $\Gamma \in \mathcal{H}(K)$. We put

$$C \stackrel{\text{def}}{=} \left\{(y_1, y_2, \dots, y_n) : y_1^2 \geq (\omega(K))^{-2}\,(y_2^2 + \cdots + y_n^2)\right\},$$

where y_i, $i = 1, \dots, n$, are the coordinates of a point in this coordinate system (thus, C is a double closed solid circular cone whose axis coincides with $L(K)$). For each point $X \subset \mathbf{R}^n$, we denote by C_X the image of the cone C obtained through a parallel displacement which translates the vertex O' of C into the point X. Obviously, $\Gamma \subset C_X$ for each simple arc $\Gamma \in \mathcal{H}(K)$ and for each point $X \in \Gamma$.

Now we consider another system of coordinates, namely, a skew system in $\mathbf{R}^n$ (with origin at the point O'), whose coordinate hyperplanes $\mathcal{P}_i$ $(i = 1, \dots, n)$ have the following property: $\mathcal{P}_i \cap C = \{O'\}$ for each i. We denote this system of coordinates by σ.

Recall that $\mathbf{Q}$ denotes the set of all rational numbers. For each pair (i,r), $i \in \{1, \dots, n\}$, $r \in \mathbf{Q}$, we denote by $P(i,r)$ the hyperplane which is defined in the coordinate system σ by the equation $x_i = r$.

We put $\widetilde{\mathcal{P}}^1 \stackrel{\text{def}}{=} \{P(i,r) : i \in \{1, \dots, n\}, r \in Q\}$, $\widetilde{\mathcal{P}} \stackrel{\text{def}}{=} \widetilde{\mathcal{P}}^1 \cup \{A(K^1), A(K^2)\}$ (recall that K^1 and K^2 are the bases of the block K). Obviously, the set $\mathcal{P}$ is countable. We enumerate it, $\widetilde{\mathcal{P}} = \{P_1, P_2, \dots\}$.

For each $k \in \mathbf{N}$, there exists $\delta_k > 0$ such that for each simple arc $\Gamma \in \mathcal{H}(K)$ the following inequality holds:

$$m_\Gamma\left(\Gamma \cap (P_k, \delta_k)\right) < 2^{-(k+1)} \min\{\varepsilon, h(K)\}$$

(recall that $(\mathcal{M}, \delta)$ is the δ-neighborhood of the set $\mathcal{M}$). This follows directly from the analysis of (orthogonal) projections of the sets $C_X \cap (P_k, \eta)$, $X \in P_k$, $\eta > 0$, on the straight line $L(K)$ and from the fact that for each simple arc $\Gamma \in \mathcal{H}(K)$ and for each simple arc $\Gamma' \subset \Gamma$ we have

$$m_\Gamma(\Gamma') \leq (\omega(K) + 1)\,\mathcal{L}(\Gamma'),$$

where $\mathcal{L}(\Gamma')$ is the length of the projection of Γ' on $L(K)$.

We put

$$\widetilde{\mathfrak{B}} \stackrel{\text{def}}{=} K \setminus \bigcup_{k=1}^{\infty} (P_k, \delta_k).$$

Let $\varepsilon_0 > 0$ be such that the ε_0-neighborhood of the set $(\partial K) \setminus (K^1 \cup K^2)$ (we denote this ε_0-neighborhood by W) does not intersect any simple arc $\Gamma \in \mathcal{H}(K)$

that intersects K canonically. We put $\mathfrak{B} \stackrel{\text{def}}{=} \widetilde{\mathfrak{B}} \setminus W$. Obviously, $\mathfrak{B}$ is a compact, $\mathfrak{B} \subset \operatorname{int} K$. By the definition of δ_k $(k \in \mathbf{N})$, we have

$$m_{L(K)}\left(I(K) \cap \bigcup_{k=1}^{\infty}(P_k, \delta_k)\right) < \frac{h(K)}{2}$$

($m_{L(K)}$ is linear Lebesgue measure on $L(K)$). Consequently, $m_{L(K)}(I(K) \cap \mathfrak{B}) > 0$. Hence, $\operatorname{card} \mathfrak{B} = \mathfrak{c}$.

We denote by $A(K, \varepsilon)$ the set of all points of condensation of the compact $\mathfrak{B}$. It is well known (see, for example, [**9**]) that $\mathcal{A}(K, \varepsilon)$ is a perfect compact set, and the set $\mathfrak{B} \setminus \mathcal{A}(K, \varepsilon)$ is at most countable. Obviously, the compact set $\mathcal{A}(K, \varepsilon)$ is zero-dimensional, hence a discontinuum. It is also obvious that $\mathcal{A}(K, \varepsilon)$ is cell-decomposed in $\mathbf{R}^n$.

For each simple arc $\Gamma \in \mathcal{H}(K)$ intersecting K canonically, we have

$$m_\Gamma\left(\Gamma \setminus \mathcal{A}(K, \varepsilon)\right) = m_\Gamma(\Gamma \setminus \mathfrak{B}) = m_\Gamma\left(\Gamma \cap \bigcup_{k=1}^{\infty}(P_k, \delta_k)\right) < \frac{\varepsilon}{2} < \varepsilon.$$

The existence of a simple arc $\mathfrak{M}(K, \varepsilon)$ with the properties listed in the formulation of Lemma 4 is essentially a classical fact. Following Luzin [**10**], we shall briefly describe the technique used for proving the existence of such a simple arc.

A parallelepiped $T \subset \mathbf{R}^n$ is called standard if its $(n-1)$-dimensional faces are parallel to the coordinate hyperplanes of the coordinate system σ.

We fix two points B_1, $B_2 \in \partial K$ and define the parallelepipeds $T_1^k, \dots, T_{i_k}^k$ and mappings $f_k : [0,1] \to K$, $k \in \mathbf{N}$, by induction as follows. Obviously, there exist pairwise disjoint standard parallelepipeds $T_1^k, \dots, T_{i_k}^k$ and a mapping $f_k \colon [0,1] \to K$ with the following properties.

1. $\mathcal{A}(K, \varepsilon) \subset \bigcup_{j=1}^{i_1} \operatorname{int} T_j^1$, $\bigcup_{j=1}^{i_1} T_j^1 \subset \operatorname{int} K$.
2. For each $j = 1, \dots, i_1$ we have $\operatorname{diam} T_j < 1$.
3. f_1 is the homeomorphism of the segment $[0,1]$ on $f_1([0,1]), f_1(0) = B_1$, $f_1(1) = B$.
4. For each $j = 1, \dots, i_1$, $f_1([0,1]) \cap T_j^1$ is a simple arc lying completely (except its ends) in T_j^1.
5. $f_1([0,1])$ is a simple arc lying completely (except its ends) in $\operatorname{int} K$.

We assume that for some $k \in \mathbf{N}$ we have defined disjoint standard parallelepipeds and the mapping $f_k : [0,1] \to K$ having, among others, the following properties.

1. $\mathcal{A}(K, \varepsilon) \subset \bigcup_{j=1}^{i_k} \operatorname{int} T_j^k$, $\bigcup_{j=1}^{i_k} T_j^k \subset \operatorname{int} K$.
2. f_k is the homeomorphism of the segment $[0,1]$ on $f_k([0,1])$.
3. For each $j = 1, \dots, i_k$, $f_k([0,1]) \cap T_j^k$ is a simple arc lying completely (except for its ends) inside $\operatorname{int} T_j^k$.

It can easily be seen now that there exist pairwise disjoint standard parallelepipeds $T_1^{k+1}, \dots, T_{i_{k+1}}^{k+1}$ and a mapping $f_{k+1} : [0,1] \to K$ with the following properties:

1. $\mathcal{A}(K, \varepsilon) \subset \bigcup_{q=1}^{i_{k+1}} \operatorname{int} T_q^{k+1}$, $\bigcup_{q=1}^{i_{k+1}} T_q^{k+1} \subset \bigcup_{j=1}^{i_k} \operatorname{int} T_j^k$.
2. For each $q = 1, \dots, i_{k+1}$ we have $\operatorname{diam} T_q^{k+1} < \frac{1}{k+1}$.
3. f_{k+1} is the homeomorphism of the segment $[0,1]$ on $f_{k+1}([0,1])$.

4. For each $q = 1, \dots, i_{k+1}$, $f_{k+1}([0,1]) \cap T_q^{k+1}$ is a simple arc lying completely (except its ends) in $\operatorname{int} T_q^{k+1}$.

5. $\operatorname{diam} f_{k+1}^{-1}(f_{k+1}([0,1]) \cap T_q^{k+1}) < \frac{1}{k+1}$.

6. For each $j = 1, \dots, i_k$, $f_{k+1}([0,1]) \cap T_j^k$ is a simple arc lying completely (except its ends) in $\operatorname{int} T_j^k$.

7. For each point $x \in [0,1] \setminus f_k^{-1}(f_k([0,1]) \cap \bigcup_{j=1}^{i_k} \operatorname{int} T_j^k)$ we have $f_{k+1}(x) = f_k(x)$.

Thus, standard parallelepipeds $T_1^k, \dots, T_{i_k}^k$ and the mapping $f_k : [0,1] \to K$ have been defined by induction for each $k \in \mathbf{N}$. Now let us define $\mathcal{F} : [0,1] \to K$ as follows. Let $x \in [0,1]$ be a number. If there exists a $k \in \mathbf{N}$ such that $x \notin f_k^{-1}(f_k([0,1]) \cap \bigcup_{j=1}^{i_k} T_j^k)$, we put $\mathcal{F}(x) \stackrel{\text{def}}{=} f_k(x)$. If, however, for each $k \in \mathbf{N}$ there exists j_k, $j_k = 1, \dots, i_k$, such that $f_k(x) \in T_{j_k}^k$, we put $\mathcal{F}(x) \stackrel{\text{def}}{=} \bigcap_{k=1}^{\infty} T_{j_k}^k$.

It can be easily seen that the mapping $\mathcal{F}$ is well-defined and is a homeomorphism of the segment $[0,1]$ on $\mathcal{F}([0,1])$.

We put $\mathfrak{M}(K,\varepsilon) \stackrel{\text{def}}{=} \mathcal{F}([0,1])$, $B_1(K,\varepsilon \stackrel{\text{def}}{=} B_1)$, $B_2(K,\varepsilon) \stackrel{\text{def}}{=} B_2$. Obviously, $\mathfrak{M}(K,\varepsilon)$ is a simple arc having the properties described in Lemma 4.

Note that in view of the cellular decomposition of the discontinuum $\mathcal{A}(K,\varepsilon)$ in $\mathbf{R}^n$, the existence of the simple arc $\mathfrak{M}(K,\varepsilon)$ could be proved by using Keldysh's well-known theorem (see Theorem 1 in [**3**]). However, to simplify the task for the reader, we decided to give here a simple version of the classical proof.

Lemma 4 is proved.

We say that a simple arc $\Gamma \subset \mathbf{R}^n$ has curvature smaller than M (where $M > 0$) if Γ lies in a curve which has limited curvature less than M at each point.

LEMMA 5. *For all positive numbers M and ε, there exists a number $\delta(M,\varepsilon) > 0$ with the following property: if $\Gamma \subset \mathbf{R}^n$ is a simple arc of curvature smaller than M and of length smaller than $(2M)^{-1}$, then the following inequality holds for each sphere $S \subset \mathbf{R}^n$ with radius smaller than $(4M)^{-1}$:*

$$m_\Gamma\left(\Gamma \cap (S, \delta(M,\varepsilon))\right) < \varepsilon.$$

PROOF. Let M and ε be positive numbers, $\Gamma \subset \mathbf{R}^n$ a simple arc of curvature smaller than M and of length smaller than $(2M)^{-1}$.

In the two-dimensional plane $\mathbf{R}^2$, we consider the disk

$$D_0 \stackrel{\text{def}}{=} \left\{ (x,y) : x^2 + \left(y - \frac{1}{4M}\right)^2 \le (4M)^{-2} \right\}$$

and the set

$$G_0 \stackrel{\text{def}}{=} \left\{ (x,y) : |y| \le 2Mx^2, \quad x \ge \frac{\varepsilon}{8} \right\}.$$

It can easily be seen that $\rho(D_0, G_0) > 0$. We define $\delta(M,\varepsilon) \stackrel{\text{def}}{=} \frac{1}{3}\rho(D_0, G_0)$ and assume that there exists a sphere $S \subset \mathbf{R}^n$ of radius $r < (4M)^{-1}$ such that

$$m_\Gamma(\Gamma \cap (S, \delta(M,\varepsilon))) \ge \varepsilon.$$

There exist points $B_i \in \Gamma \cap (S, \delta(M, \varepsilon))$, $i = 0, 1, 2$, such that B_0 lies between B_1 and B_2 (on the arc Γ), and for $j = 1, 2$ the length of the arc Γ with ends B_0 and B_j is larger than $\varepsilon/4$. Without loss of generality we can assume that the point B_0 coincides with the origin O (in $\mathbf{R}^n$) and the axis Ox_1 is tangential to Γ at the point $O = B_0$.

For each point $X \in \mathbf{R}^n$, we denote by $x_i(X)$ the i-th coordinate of the point X ($i = 1, \dots, n$), and put $Y(X) \overset{\text{def}}{=} (\sum_{i=2}^n (x_i(X))^2)^{1/2}$. For $X \in \Gamma \setminus \{O\}$, we denote by $s(X)$ the length of the segment of the arc Γ with the ends X and O.

It can be verified by standard techniques (integrating the curvature and imposing a restriction on the length of the arc Γ) that the following relations are observed for each point $X \in \Gamma \setminus \{O\}$:

$$s(X) < 2|x_1(X)|, \qquad Y(X) < \frac{1}{2} M (s(X))^2 < 2M(x_1(X))^2.$$

Let $B_i' \subset S$, $i = 0, 1, 2$, be a point such that $\rho(B_i, B_i') < \delta(M, \varepsilon)$, and let $\tau : \mathbf{R}^n \to \mathbf{R}^n$ be a parallel translation by the vector $\overrightarrow{B_0' B_0}$. We denote by A a point such that the ball $P^h(A, r)$ touches the axis Ox_1 at the point O, while the points $\tau(B_1')$ and A lie in a two-dimensional closed half-plane bounded by the axis Ox_1.

Using the definition of the number $\delta(M, \varepsilon)$ and the inequalities $|x_1(B_1)| > \frac{1}{2} s(B_1) > \frac{\varepsilon}{8}$, $Y(B_1) < 2M(x_1(B_1))^2$, we find that $\rho(\tau(B_1'), A) > r$. Without loss of generality, it can be assumed that $x_1(\tau(B_1')) > 0$. In this case, it is easy to show that if the center A' of the sphere $\tau(S)$ belongs to the half-space $\{(x_1, \dots, x_n) : x_1 \leq 0\}$, then $\rho(\tau(B_1'), A') \geq \rho(\tau(B_1'), A) > r$, a contradiction. Hence the point A' belongs to the open half-space $\{(x_1, \dots, x_n) : x_1 > 0\}$.

Using similar arguments for $\tau(B_2')$, we find that $A' \in \{(x_1, \dots, x_n) : x_1 < 0\}$, a contradiction.

Lemma 5 is proved.

LEMMA 6. *Let M and ε be positive numbers, $\mathcal{V} \overset{\text{def}}{=} V_n(R_1, R_2)$ an open layer between the concentric spheres (with center at O) of radii R_1 and R_2, $0 < R_1 < R_2$, $R_2 - R_1 < (2M)^{-1}$. There exist a discontinuum $\mathfrak{B}(v)$ (for the sake of brevity, we denote by v the ordered triple $(M, \varepsilon, \mathfrak{V})$), a simple arc $\mathfrak{M}(v)$ and a polygonal line $L(v)$ (which is a simple arc) which have the following properties.*

1. *The discontinuum $\mathfrak{B}(v)$ is cell-decomposed in $\mathbf{R}^n$,*

2. *$\mathfrak{B}(v) \subset \mathfrak{M}(v) \setminus \{C_1(v), C_2(v)\} \subset \mathcal{V}$, $\{C_1(v), C_2(v)\} \subset \partial\mathcal{V}$, where $C_i(v)$ are the ends of the arc $\mathfrak{M}(v)$ $(i = 1, 2)$.*

3. *$L(v) \cap \mathfrak{M}(v) = \emptyset$, $L(v) \setminus \{C_1'(v), C_2'(v)\} \subset \mathcal{V}$, where $C_i'(v)$ are the ends of the polygonal line $L(v)$, and*

$$C_i'(v) \in S^{n-1}(0, R_i) \quad (i = 1, 2).$$

4. *Each simple arc Γ having curvature smaller than M, length $R_2 - R_1$, and midpoint (in the sense of the intrinsic metric on Γ) lying on the sphere $S^{n-1}(0, (R_1 + R_2)/2)$ satisfies the inequality $m_\Gamma(\Gamma \setminus \mathfrak{B}(v)) < \varepsilon$.*

5. *For each plane $P \subset \mathbf{R}^n$ $(1 \leq \dim P \leq n)$, we have $m_P(P \cap (\mathcal{V} \setminus \mathfrak{B}(v))) < \varepsilon$.*

PROOF. It can be verified by standard techniques that for any positive numbers r and δ there exists a number $\eta(r,\delta) > 0$ such that the following inequality holds for each plane $P \subset \mathbf{R}^n$ $(1 \leq \dim P \leq h)$ and for each $(n-1)$-dimensional sphere $s \subset \mathbf{R}^n$ of radius r:

$$m_P\left(P \cap (S, \eta(r,\delta))\right) < \delta$$

(here and below, we do not treat a point as a plane).

We denote by $\mathbf{Q}^n$ the set of all rational points of the space $\mathbf{R}^n$ (i.e., points whose coordinates are all rational). We put

$$W(M) \stackrel{\text{def}}{=} \left\{x : x \in \mathbf{Q}, 0 < x < (4M)^{-1}\right\},$$
$$\Omega \stackrel{\text{def}}{=} \left\{S^{n-1}(X,q) : X \in \mathbf{Q}^n, q \in W(M)\right\}.$$

Obviously, the set Ω is countable, and can be enumerated as $\Omega = \{S_1, S_2, \dots\}$.

In view of Lemma 5 and the remark concerning $\eta(r,\delta)$ there exists for each $k \in \mathbf{N}$ a number $\xi_k > 0$ possessing the following properties.

1. If $\Gamma \subset \mathbf{R}^n$ is a simple arc of curvature smaller than M and of length smaller than $(2M)^{-1}$, then

$$m_\Gamma\left(\Gamma \cap (S_k, \xi_k)\right) < 2^{-(k+1)} \min\{\varepsilon, R_2 - R_1\}.$$

2. For each plane $P \subset \mathbf{R}^n$ $(1 \leq \dim P \leq n)$ the following inequality holds:

$$m_P\left(P \cap (S_k, \xi_k)\right) < \varepsilon \cdot 2^{-(k+1)}.$$

Moreover, for $i = 1, 2$, there is an η_i, $0 < \eta_i < \frac{1}{4}\min\{\varepsilon, R_2 - R_1\}$, such that for each plane $P \subset \mathbf{R}^n$ $(1 \leq \dim P \leq n)$, the following inequality holds:

$$m_P\left(P \cap \left(S^{n-1}(0, R_i), \eta_i\right)\right) < \frac{\varepsilon}{4}.$$

We put

$$\mathcal{W}_1 \stackrel{\text{def}}{=} \bigcup_{k=1}^{\infty} (S_k, \xi_k),$$
$$\mathcal{W} \stackrel{\text{def}}{=} \mathcal{W}_1 \cup \bigcup_{i=1}^{2} \left(S^{n-1}(O, R_i), \eta_i\right),$$
$$\mathfrak{B} \stackrel{\text{def}}{=} \mathcal{V} \setminus \mathcal{W}.$$

Obviously, $\mathfrak{B}$ is a compact set, $\mathfrak{B} \subset \mathcal{V}$, and $\mathfrak{B}$ is obtained by deleting from $\mathcal{V}$ a countable set of open "spherical layers".

We denote by $\mathcal{L}$ a (rectilinear) segment of length $R_2 - R_1$ one of whose ends belongs to the sphere $S^{n-1}(O, R_1)$ and the other to the sphere $S^{n-1}(O, R_2)$. By the definition of ξ_k, $k \in \mathbf{N}$, and η_i, $i = 1, 2$, we have $m_{\mathcal{L}}(\mathcal{L} \cap \mathcal{W}) < R_2 - R_1$. Consequently, $m_{\mathcal{L}}(\mathcal{L} \cap \mathfrak{B}) > 0$, whence $\operatorname{card} \mathfrak{B} = \mathfrak{c}$.

We denote by $\mathfrak{B}(v)$ the set of all limit points of the compact set $\mathfrak{B}$. It is well known that $\mathfrak{B}(v)$ is a perfect compact set, while the set $\mathfrak{B} \setminus \mathfrak{B}(v)$ is not more than countable. Obviously, the set $\mathfrak{B}(v)$ is zero-dimensional, hence a discontinuum.

Let $\Gamma \subset \mathbf{R}^n$ be a simple arc of curvature smaller than M whose length is $R_2 - R_1$ and whose midpoint lies on the sphere $S^{n-1}(O, (R_1 + R_2)/2)$. Obviously,

$$m_\Gamma \left(\Gamma \cap \bigcup_{i=1}^{2} (S^{n-1}(O, R_i), \eta_i) \right) < \frac{\varepsilon}{2}, \quad m_\Gamma(\Gamma \cap \mathcal{W}_1) < \frac{\varepsilon}{2}.$$

It follows that $m_\Gamma(\Gamma \setminus \mathfrak{B}(v)) = m_\Gamma(\Gamma \setminus \mathfrak{B}) < \varepsilon$.

By the definition of ξ_k, $k \in \mathbf{N}$, and η_i, $i = 1, 2$, the inequality $m_P(P \cap \mathcal{W}) < \varepsilon$ holds for each plane $P \subset \mathbf{R}^n$, $1 \le \dim P \le n$. Consequently, $m_P(P \cap (\mathcal{V} \setminus \mathfrak{B}(v))) = m_P(P \cap (\mathcal{V} \setminus \mathfrak{B})) < \varepsilon$.

We show that the discontinuum $\mathfrak{B}(v)$ is cell-decomposed in $\mathbf{R}^4$.

By a spherical segment we shall mean the intersection of a sphere with a closed half-space whose boundary hyperplane intersects the open ball bounded by this sphere. Let $r > 0$ be a number. We define a compact set $H(r)$ as follows. Let ξ denote a vector all whose coordinates are equal to 1, and let l_ξ be a straight line parallel to it, $O \in l_\xi$. For each $i = 1, \dots, n$ and for each real number a, we denote by $A_i(a)$ a point such that $x_i(A_i(a)) = a$, $x_j(A_i(a)) = 0$ $(j \ne i)$, where $x(X)$ is the ith coordinate of the point $X \in \mathbf{R}^n$.

We assume that $\delta > 0$ is such that $\delta < r$ and for each i and each point of the spherical segment

$$T^i \stackrel{\text{def}}{=} S^{n-1}(A_i(r), r) \cap \{(x_1, \dots, x_n) : x_i \le \delta\}$$

the normal to the sphere $S^{n-1}(A_i(r), r)$ at this point is not orthogonal to ξ. Note that $\bigcap_{i=1}^n T^i = \{O\}$ (indeed, if this intersection contains a certain point $X \ne O$, then the hyperplane orthogonal to the segment $[O, X]$ and passing through its midpoint passes through each $A_i(r)$, $i = 1, \dots, n$, i.e., $X \in l_\xi$). This contradicts the assumption about the normals. For each i and each real number q, we denote by $T^i(a)$ the image of the segment T^i obtained by parallel displacement by a vector $\overrightarrow{OA_i(a)}$.

Suppose that the positive number d satisfies the following two conditions.

1. For each $i = 1, \dots, n$ we denote by H_i the union of all segments $T^i(a)$, $0 \le a \le d$. Let $H(r) \stackrel{\text{def}}{=} \bigcap_{i=1}^n H_i$. We denote by b_i the (closed) segment $[A_i(r), A_i(r+d)]$. Then for any (different) points $X_1, X_2 \in H(r)$, the hyperplane orthogonal to the segment $[X_1, X_2]$ and passing through its midpoint cannot intersect each of the segments b_i.

2. For each i $(i = 1, \dots, n)$, $l_\xi \cap T^i(d)$ is a single-point intersection and

$$\rho\left(O, l_\xi \cap T^i(d)\right) < \left(r^2 - (r - \delta)^2\right)^{1/2}.$$

It can easily be seen that such a number d exists.

We denote by $G(r)$ the cube $\{(x_1, \dots, x_n) : 0 \le x_i \le d, i = 1, \dots, n\}$ and define the mapping $\varphi_r : H(r) \to G(r)$ as follows: for each point $X \in H(r)$, we denote by $\varphi_r(X)$ a point such that $X \in T^i(x_i(\varphi_r(X)))$ for each i. The mapping φ_r is injective (in view of condition 1) and continuous.

We show that the mapping φ_r is surjective. Let $X \in G(r)$ be a point. We show that $\bigcap_{i=1}^n T^i(x_i(X)) \ne \emptyset$. For each i, denote by θ^i an unbounded convex body which is a union of all $T^i(a)$ $(a \ge x_i(X))$, and put $\theta \stackrel{\text{def}}{=} \bigcap_{i=1}^n \theta^i$. Obviously,

θ is a compact set and $\theta \neq \emptyset$ (in view of condition 2). Let $X_0 \in \theta$ be the point closest to 0. In view of condition 2, we have $X_0 \in T^i(x_i(X))$ or $X_0 \in \operatorname{int}\theta^i$ for each i. For definiteness, we assume that $X_0 \notin T^1(x_1(X))$, i.e., $X_0 \in \operatorname{int}\theta^1$. In this case, for each $j = 2, \dots, n$ there exists a point $X_0^j \in \theta^j \setminus \{X_0\}$ such that the segment $[X_0, X_0^j]$ is parallel to the axis Ox_1, $0 < x_1(X_0^j) < x_1(X_0)$. Consequently, there exists a point $X_0' \in \theta$ for which $\rho(O, X_0') < \rho(O, X_0)$. This is a contradiction. Thus, the mapping φ_r is a homeomorphism of the compact set $H(r)$ on the cube $G(r)$.

Let r_0 be a rational number, $0 < r_0 < (4M)^{-1}$. We consider the compact set $H(r_0)$. Let $\lambda_0 > 0$ be a number such that $H(r_0)$ contains a ball of radius $2\lambda_0$.

Since the discontinuum $\mathfrak{B}(v)$ is zero-dimensional, there exist (see, for example, [**1**]) $p \in \mathbf{N}$ and pairwise disjoint closed (nonempty) subsets $\mathfrak{B}_1, \dots, \mathfrak{B}_P$ of the discontinuum $\mathfrak{B}(v)$, such that $\mathfrak{B}(v) = \bigcup_{k=1}^{P} \mathfrak{B}_k$ and $\operatorname{diam}\mathfrak{B}_k < \lambda_0$ for each k $(k = 1, \dots, p)$. Since $\mathfrak{B}(v)$ is a discontinuum, it is obvious that all these sets $\mathfrak{B}_k$ $(k = 1, \dots, p)$ are also discontinua.

Let $V \subset \mathbf{R}^n$ be a neighborhood of $\mathfrak{B}(v)$. For each $k = 1, \dots, p$ we denote by V_k a neighborhood of $\mathfrak{B}_k$ such that $\operatorname{diam} V_k < 2\lambda_0$, $\bigcup_{k=1}^{P} V_k \subset V$, and all the neighborhoods V_k $(k = 1, \dots, p)$ are pairwise dsjoint. For each $k = 1, \dots, p$ we denote by t_k a parallel displacement $(t_k : \mathbf{R}^n \to \mathbf{R}^n)$ such that $t_k(O) \in \mathbf{Q}^n$ and $t_k(\operatorname{int} H(r_0)) \supset V_k$ (obviously, such a parallel displacement exists).

A hyperplane $P \subset \mathbf{R}^n$ is called rational if there exist an i $(i = 1, \dots, n)$ and a rational number r' such that P is defined by the equation $x_i = r'$.

We fix k, $k = 1, \dots, p$, and consider the homeomorphism $\Phi_k \overset{\text{def}}{=} \varphi_{r_0} \cdot t_k^{-1}$. According to the definition of φ_{r_0} and t_k, we find that for each rational hyperplane $P \subset \mathbf{R}^n$ intersecting with the cube $G(r_0)$, $\Phi_k^{-1}(P \cap G(r_0))$ is a subset of a sphere belonging to the set Ω. It follows that there exist a number $n(k) \in \mathbf{N}$ and pairwise disjoint (closed) cubes $T_1, \dots, T_{n(k)}$ such that

$$\Phi_k(\mathfrak{B}_k) \subset \bigcup_{j=1}^{n(k)} \operatorname{int} T_j, \qquad \bigcup_{j=1}^{n(k)} T_j \subset \Phi_k(V_k).$$

Consequently,

$$\mathfrak{B} \subset \bigcup_{j=1}^{n(k)} \Phi_k^{-1}(\operatorname{int} T_j), \qquad \bigcup_{j=1}^{n(k)} \Phi_k^{-1}(T_j) \subset V_k,$$

and for each $j = 1, \dots, n(k)$, the set $\Phi_k^{-1}(T_j) \subset V_k$ is homeomorphic to a closed sphere.

It follows that the discontinuum $\mathfrak{B}(v)$ is cell-decomposed in $\mathbf{R}^n$.

Obviously there exists a polygonal line $L(v) \subset \mathcal{W}$ (which is a simple arc) for which $L(v) \cap \mathfrak{B}(v) = \emptyset$, $C_i'(v) \in S^{n-1}(O, R_i)$, $i = 1, 2$, where $C_i'(v)$ are the ends of the polygonal line $L(v)$, and $L(v) \setminus \{C_1'(v), C_2'(v)\} \subset \mathcal{V}$.

Taking into account the simple geometric structure of the discontinua $\mathfrak{B}_k$, $k = 1, \dots, p$ (i.e., the existence of the homeomorphisms Φ_k), the existence of the simple arc $\mathfrak{M}(v)$ with the properties listed in Lemma 6 is proved in the same manner as at the end of the proof of Lemma 4 (it should be noted that the existence of $\mathfrak{M}(v)$ and $L(v)$ can also be proved in a different manner by using Theorem 1 in [**3**]).

Lemma 6 is proved.

PROOF OF THEOREM 2. We denote by $\mathcal{N}_0$ the set of odd natural numbers, and by $\mathcal{N}_j$ ($j \in \mathbf{N}$) certain infinite disjoint subsets of the set of even natural numbers whose union coincides with the set of even natural numbers. Let $\mathfrak{Z} \stackrel{\text{def}}{=} \{Z_1, Z_2 \dots\} \subset S^{n-1}(O,1)$ be a countable set having the following property: for each $i \subset \{0\} \cup \mathbf{N}$, the set $\mathfrak{Z}_i \stackrel{\text{def}}{=} \{Z_k : k \in \mathfrak{N}_i\}$ is dense in $S^{n-1}(O,1)$.

By $\mathcal{H}(Y,k,\varepsilon)$, $Y \in \mathbf{R}^n$, $k \in \mathbf{N}$, $\varepsilon > 0$, we denote the set of k-Lipschitz simple arcs (each has one end coinciding with the point Y) with a base $[Y, Y']$, where Y' is a point such that the vector $\overrightarrow{YY'}$ has the same direction as $\overrightarrow{OZ_k}$, $\rho(Y,Y') = \varepsilon$.

Let $\mathfrak{X} \stackrel{\text{def}}{=} \{X_1, X_2, \dots\}$ be a countable set in $\mathbf{R}^n$ with the following properties: for each $\varepsilon > 0$ there exists $R(\varepsilon) > 0$ such that for each point $X \in \mathbf{R}^n \setminus D^n(O, R(\varepsilon))$ there exists an $i \in \mathbf{N}$ with $\rho(X, X_i) < \varepsilon$; for each $j \in \mathbf{N}$, we have $0 < \rho(O, X_j) < \rho(O, X_{j+1})$.

For each $j \in \mathbf{N}$, let $\mathcal{V}_j \stackrel{\text{def}}{=} V_n(R_1^j, R_2^j)$ (an open layer between concentric spheres (with center O) of radii R_1^j, R_2^j $(0 < R_1^j < R_2^j)$) and $\varepsilon_j > 0$ be such that

$$R_2^j < R_1^{j+1}, \quad X_j \in S^{n-1}\left(O, \frac{1}{2}(R_1^j + R_2^j)\right),$$

$$R_2^j - R_1^j < 2\varepsilon_j < \min\left\{2^{-j}, \frac{1}{2}R_1^j\right\}, \quad (\mathcal{V}_j, \varepsilon_j) \cap (\mathcal{V}_{j+1}, \varepsilon_{j+1}) = \emptyset,$$

and for each plane $P \subset \mathbf{R}^n$ $(1 \le \dim P \le n)$

$$m_P\left(P \cap \left(\bigcup_{k=1}^{\infty} (\mathcal{V}_k, \varepsilon_k)\right)\right) < \frac{1}{4}.$$

Note that this inequality leads to the relation

$$\sum_{k=1}^{\infty} (R_2^k - R_1^k + 2\varepsilon_k) < \frac{1}{8}$$

(for P we should consider a straight line passing through the point O).

The existence of such R_1^j, R_2^j, and ε_j, $j \in \mathbf{N}$, follows directly from the remark on $\eta(r, \delta)$ made at the beginning of Lemma 6.

We fix $j \in \mathbf{N}$ and construct on the segment $[R_1^j, R_2^j]$ a perfect Cantor set $\mathcal{K}_j$ in the same way as the standard Cantor set on the segment $[0,1]$ (we eliminate from $[R_1^j, R_2^j]$ (open) intervals of length $3^{-k}(R_2^j - R_1^j)$ $(k \in \mathbf{N})$.

Let us denote by $\mathcal{A}_k^j$ $(k \in \mathbf{N})$ a set of intervals of rank k (see [**1**], pp. 137–138) that are eliminated from $[R_1^j, R_2^j]$ while constructing $\mathcal{K}_j$. For each $I \in \mathcal{A}_k^j$, we denote by I' the interval obtained from I through homothety having coefficient $1/2$ and center at the center of the interval I. We put $\mathcal{B}_k^j \stackrel{\text{def}}{=} \{I' : I \in \mathcal{A}_k^j\}$. For each interval $\mathcal{J} \in \mathcal{B}_k^j$, we denote by $R_i(\mathcal{J})$ $(i = 1, 2)$ the ends of the interval $\mathcal{J}$ $(R_1(\mathcal{J}) < R_2(\mathcal{J}))$.

For each $k \in \mathbf{N}$ and for each $I \in \mathcal{B}_k^j$, we denote by $Q(j,k,I)$ a finite set of pairwise disjoint oriented blocks having the following properties.

1. The inclusion $k \subset V_n(R_1(I), R_2(I)) \setminus D^n(X_j, \varepsilon_j)$ holds for each point $K \in Q(j,k,I)$.

2. The vector $\overrightarrow{O^1(K)O^2(K)}$ has the same direction as $\overrightarrow{OZ_k}$, $\omega(K) > k$ (recall that $O^i(K)$ is the center of the base K^i, $i = 1, 2$, of the block K).

3. For each point

$$Y \in S^{n-1}\left(O, \frac{1}{2}(R_1(I) + R_2(I))\right) \setminus D^n(X_j, 2\varepsilon_j)$$

and for each simple arc $\Gamma \in \mathcal{H}(Y, k, (4k)^{-1} \operatorname{diam} I)$ there exists an oriented block $K \in Q(j,k,I)$ such that Γ canonically intersects K, where $\Gamma \cap K^2 \neq \emptyset$.

To verify the existence of the set $Q(j,k,I)$, observe that for each $(n-1)$-dimensional cube $T \subset \mathbf{R}^n$ with a hyperplane $A(T)$ orthogonal to the vector $\overrightarrow{OZ_k}$, there exists a number $\nu(a(T)) > 0$ (where $a(T)$ is the edge of the cube T) having the following property: Denote by O'_T the center of the cube T, by $z_k(T)$ a point for which $\overrightarrow{O'_T Z_k(T)} = \overrightarrow{OZ_k}$, and by $\mathcal{P}_T$ the closed half-space whose boundary hyperplane coincides with $A(T)$, $Z_k(T) \in \mathcal{P}_T$. For each point $Y \in D_T \stackrel{\text{def}}{=} D^n(O'_T, \nu(a(T)) \setminus \mathcal{P}_T)$, the arc $\Gamma \in \mathcal{H}(Y, k, a(T))$ intersects canonically the oriented block $K_T \subset \mathcal{P}_T$ whose first base coincides with T and whose height is $(8k)^{-1}a(T)$.

Now it is sufficient to consider the finite set $\mathfrak{T}$ of disjoint $(n-1)$-dimensional cubes whose affine shells are orthogonal to $\overrightarrow{OZ_k}$,

$$\bigcup_{T \in \mathfrak{T}} T \subset V_n(R_1(I), R_2(I)) \setminus D^n(X_j, \varepsilon_j),$$

$$S^{n-1}\left(O, \frac{1}{2}(R_1(I) + R_2(I))\right) \setminus \operatorname{int} D^n(X_j, 2\varepsilon_j) \subset \bigcup_{T \in \mathfrak{T}} \operatorname{int} D_T,$$

$\nu(a(T)) < (8k)^{-1} \operatorname{diam} I$ for each $T \in \mathfrak{T}$, and to take for each $T \subset \mathfrak{T}$ the corresponding block (with a "low" height) contained in K_T.

Let us make the following remarks in preparation to the proof of Theorem 3.

Let $W \subset \mathbf{R}^n$, $n \geq 2$, be an open layer between the spheres $S^{n-1}(O, R_1)$ and $S^{n-1}(O, R_2)$, $0 < R_1 < R_2$. Similarly to the proof of the existence of the sets $Q(j,k,I)$, we can prove that for each $k \in \mathbf{N}$ there exists a finite set $Q(W,k)$ of pairwise disjoint oriented blocks having the following properties.

1. Each block $K \in Q(W,k)$ is contained in W.

2. For each block $K \in Q(W,k)$, the vector $\overrightarrow{O^1(K)O^2(K)}$ has the same direction as $\overrightarrow{OZ_k}$, $\omega(K) > k$.

3. For every point $Y \in S^{n-1}(O, \frac{1}{2}(R_1 + R_2))$ and for every simple arc $\Gamma \in \mathcal{H}(Y, k(4k)^{-1}(R_2 - R_1))$, there exists an oriented block $K \in Q(W,k)$ such that the arc Γ intersects canonically the block K so that $\Gamma \cap K^2 \neq \emptyset$.

Consequently, for every k-Lipschitz, $k \in \mathbf{N}$, simple arc $\Gamma \subset \mathbf{R}^n$ with a base parallel to $\overrightarrow{OZ_k}$, intersecting the layer W canonically, there exists an oriented block $K \in Q(W,k)$ such that $\Gamma \cap K$ is a simple arc intersecting K canonically, $\Gamma \cap K \in \mathcal{H}(K)$. It follows that for every $k \in \mathbf{N}$ there exists a finite set $\mathcal{R}(W,k)$ of pairwise disjoint oriented blocks, with the following properties.

1. Each block $K \in \mathcal{R}(W,k)$ is contained in W.

2. For every block $K \in \mathcal{R}(W,k)$, $\omega(K) > k$.

3. For every k-Lipschitz simple arc $\Gamma \subset \mathbf{R}^n$ intersecting the layer W canonically, there exists an oriented block $K \in \mathcal{R}(W,k)$ such that $\Gamma \cap K$ is a simple arc intersecting K canonically, $\Gamma \cap K \in \mathcal{H}(K)$ (actually, we need only consider in the layer W the disjoint layers $W_{2k+1}, \dots, W_{2k+s}$ $(s \in \mathbf{N})$ between the concentric spheres with center at O such that the set $\{Z_{2k+1}, \dots, Z_{2k+s}\}$ forms a δ-net on the sphere $S^{n-1}(0,1)$ with a sufficiently small δ, and take for $\mathcal{R}(W,k)$ the union $\bigcup_{i=2k+1}^{2k+s} Q(W_i, i)$; then for every k-Lipschitz simple arc $\Gamma \subset \mathbf{R}^n$ intersecting W canonically, there exists a number $i \in \{2k+1, \dots, 2k+s\}$ such that the arc Γ is an i-Lipschitz arc with a base parallel to $\overrightarrow{OZ_i}$.

Let us continue the proof of Theorem 2.

For every $q \in \mathbf{N}$, every interval $I \in \mathcal{B}^j_{2q-1}$, every block $K \in Q(j, 2q-1, I)$, and every real number α, we consider (see Lemma 3) a simple arc $\mathfrak{M}(K) \subset K$ and discontinua $\mathcal{A}(K)$ and $\mathfrak{N}_\alpha(K) \subset \mathcal{A}(K) \subset \mathfrak{M}(K)$. We denote by $\mathfrak{M}(j, 2q-1, I)$ a simple arc such that

(1) $\mathfrak{M}(j, 2q-1, I)$ is the union of the set $\bigcup_{K \in Q(j,2q-1,I)} \mathfrak{M}(K)$ and a finite number of polygonal lines;

(2) the ends $A_i(j, 2q-1, I)$ $(i = 1, 2)$ of the arc $\mathfrak{M}(j, 2q-1, I)$ belong to the boundary of the set

$$V_n\left(R_1(I), R_2(I)\right) \setminus D^n(X_j, \varepsilon_j);$$

(3) we have

$$\mathfrak{M}(j, 2q-1, I) \setminus \{A_1(j, 2q-1, I), A_2(j, 2q-1, I)\} \subset V_n\left(R_1(I), R_2(I)\right) \setminus D^n(X_j, \varepsilon_j)$$

(it can easily be seen that such a simple arc $\mathfrak{M}(j, 2q-1, I)$ exists); for every real number α we put

$$\mathfrak{N}_\alpha(j) \stackrel{\text{def}}{=} \bigcup_{q=1}^{\infty} \bigcup_{I \in \mathcal{B}^j_{2q-1}} \bigcup_{K \in Q(j,2q-1,I)} \mathfrak{N}_\alpha(K).$$

For every $q \in \mathbf{N}$, every interval $I \in \mathcal{B}^j_{2q}$, and every block $K \in Q(j, 2q, I)$ we consider (see Lemma 4) a simple arc $\mathfrak{M}(K, h(K)/2) \subset K$ and a discontinuum $\mathcal{A}(K, h(K)/2) \subset \mathfrak{M}(K, h(K)/2)$. We denote by $\mathfrak{M}(j, 2q, I)$ a simple arc such that

(1′) $\mathfrak{M}(j, 2q, I)$ is a union of the set $\bigcup_{K \in Q(j,2q,I)} \mathfrak{M}(K, h(K)/2)$ and a finite number of polygonal lines;

(2′) the ends $A_i(j, 2q, I)$ $(i = 1, 2)$ of the arc $\mathfrak{M}(j, 2q, I)$ belong to the boundary of the set $V_n(R_1(I), R_2(I)) \setminus D^n(X_j, \varepsilon_j)$;

(3′) $\mathfrak{M}(j, 2q, I) \setminus \{A_1(j, 2q, I), A_2(j, 2q, I)\} \subset V_n(R_1(I), R_2(I)) \setminus D^n(X_j, \varepsilon_j)$.

For every $i \in \mathbf{N}$, we put

$$\mathfrak{G}_i(j) \stackrel{\text{def}}{=} \bigcup_{r \in \mathcal{N}_i} \bigcup_{I \in \mathcal{B}^j_r} \bigcup_{K \in Q(j,r,I)} \mathcal{A}(K, h(K)/2).$$

Finally, for every $\alpha \in \mathbf{R}$, we put

$$\mathfrak{N}_\alpha \stackrel{\text{def}}{=} \bigcup_{j=1}^{\infty} \mathfrak{N}_\alpha(j)$$

and for every $i \in \mathbf{N}$ we put

$$\mathfrak{G}_i \stackrel{\text{def}}{=} \bigcup_{j=1}^{\infty} \mathfrak{G}_i(j).$$

It can easily be seen that each set from $\mathfrak{N}_\alpha$ and σ_i (being the union of a countable set of corresponding discontinua) is homeomorphic to $\mathcal{K} \setminus \{O\}$ (where $\mathcal{K}$ is the Cantor perfect set).

We put

$$\mathcal{A}_0 \stackrel{\text{def}}{=} \bigcup_{j=1}^{\infty} \bigcup_{q=1}^{\infty} \bigcup_{I \in \mathcal{B}^j_{2q-1}} \bigcup_{K \in Q(j,2q-1,I)} \mathcal{A}(K).$$

Now let us assume that $\gamma \subset \mathbf{R}^n$ is an unbounded admissible piecewise-Lipschitz curve. It follows from the unboundedness and admissibility of γ that there exists $j \in \mathbf{N}$ such that $\gamma \cap D^n(X_j, 2\varepsilon_j) = \emptyset$ but $\gamma \cap S^{n-1}(O, R^j_i) \neq \emptyset$ for every $i = 1, 2$. We denote by $\tilde{\mathcal{K}}_j$ the set of all points of the Cantor perfect set $\mathcal{K}_j$ that are not the ends of the intervals removed in the construction of $\mathcal{K}_j$ or the ends of the segment $[R^j_1, R^j_2]$. According to the definition of a piecewise-Lipschitz curve, γ is the union of a countable set of Lipschitz simple arcs. Since $\operatorname{card} \tilde{\mathcal{K}}_j = \mathfrak{c}$, there exist a Lipschitz simple arc $\Gamma \subset \gamma$ and a number $x_0 \in \tilde{\mathcal{K}}_j$ such that $\Gamma \cap S^{n-1}(O, x_0) \neq \emptyset$ and the ends $A_1(\Gamma)$, $A_2(\Gamma)$ of the arc Γ do not lie on the sphere $S^{n-1}(O, x_0)$.

Since $x_0 \in \tilde{\mathcal{K}}_j$, for every $\varepsilon > 0$ there exists $k(\varepsilon) \in \mathbf{N}$ such that for each natural $k > k(\varepsilon)$ there exist intervals $I_1, I_2 \in \mathcal{B}^j_k$ for which $I_1 \subset [x_0 - \varepsilon, x_0]$, $I_2 \subset [x_0, x_0 + \varepsilon]$. Since the set $\mathfrak{Z}_0$ is dense in $S^{n-1}(O, 1)$, there exist a number $k(\Gamma) \in N_0$ and an interval $I(\Gamma) \in \mathcal{B}^j_{k(\Gamma)}$ such that for $i = 1, 2$, $\Gamma \cap S^{n-1}(O, R_i(I(\Gamma))) \neq \emptyset$, and the arc Γ is a $k(\Gamma)$-Lipschitz arc with the base p parallel to $\overrightarrow{OZ_{k(\Gamma)}}$. Consequently, there exist a point $Y(\Gamma) \in \Gamma \cap S^{n-1}(O, \frac{1}{2}(R_1(I(\Gamma)) + R_2(I(\Gamma))))$ and a simple arc $\Gamma' \subset \Gamma$ such that $\Gamma' \in \mathcal{H}(Y(\Gamma), k(\Gamma)(4k(\Gamma))^{-1} \cdot \operatorname{diam} I(\Gamma))$. Considering the relevant block from $Q(j, k(b), I(\Gamma))$, Lemma 3 shows that for every $\alpha \in \mathbf{R}$ we have $\operatorname{card}(\Gamma \cap \mathfrak{N}_\alpha) = \mathfrak{c}$.

Similarly, $m_\Gamma(\Gamma \cap \mathfrak{G}_i) > 0$ for every $i \in \mathbf{N}$ (we must use Lemma 4 and the fact that the set $\mathfrak{Z}_i$ is dense in $S^{n-1}(O, 1)$).

We put $\tilde{\mathcal{W}} \stackrel{\text{def}}{=} \mathbf{R}^n \setminus \bigcup_{k=1}^{\infty} \operatorname{cl}(\mathcal{V}_k, \varepsilon_k)$. Let us suppose that $\tilde{D} \subset \tilde{\mathcal{W}}$ is a closed ball with the center O, such that for every plane $P \subset \mathbf{R}^n$ $(1 \leq \dim P \leq n)$

$$m_P(P \cap \tilde{D}) < 1/4.$$

Obviously, there exists a set $\tilde{S} \subset \tilde{\mathcal{W}} \setminus \tilde{D}$ which is a union of a countable set of concentric spheres with center O such that $\tilde{W} \setminus (\tilde{D} \cup \tilde{S})$ splits into disjoint open "spherical" layers W_j, $j \in \mathbf{N}$, such that $W_j \subset \operatorname{conv} W_{j+1}$ for every $j \in \mathbf{N}$, and the thickness of the layer W_j is smaller than $(2j)^{-1}$ (the thickness of a "spherical" layer is naturally defined as the difference between the radii of the outer and inner spheres bounding the layer).

Recalling the remark concerning $\eta(r, \delta)$ made at the beginning of Lemma 6, we can easily see that for every $j \in \mathbf{N}$ there exist positive numbers δ_j, $R_1(j)$, $R_2(j)$ $(R_1(j) < R_2(j))$ satisfying the following conditions.

1. If we put $\mathcal{U}_j \stackrel{\text{def}}{=} V_n(R_1(j), R_2(j))$, then $W_j = (\mathcal{U}_j, \delta_j)$.

2. For every plane $P \subset \mathbf{R}^n$ $(1 \leq \dim P \leq n)$,

$$m_P\left(P \setminus \bigcup_{k=1}^{\infty} \mathcal{U}_k\right) < \frac{3}{4}; \quad \sum_{k=1}^{\infty} \delta_k < \frac{1}{4}.$$

For each $j \in \mathbf{N}$, we consider (see Lemma 6) a simple arc $\mathfrak{M}(j, \delta_j, \mathcal{U}_j) \subset \mathrm{cl}\,\mathcal{U}_j$, a discontinuum $\mathfrak{B}(j, \delta_j, \mathcal{U}_j) \subset \mathfrak{M}(j, \delta_j, \mathcal{U}_j)$, and a polygonal line $L(j, \delta_j, \mathcal{U}_j) \subset \mathrm{cl}\,\mathcal{U}_j$.

Let $\mathcal{N}^i$, $i \in \mathbf{N}$, be pairwise disjoint infinite subsets of the set $\mathbf{N}$ such that $\bigcup_{i=1}^{\infty} \mathcal{N}^i = \mathbf{N}$, and for every i we have

$$\sum_{q \in \mathcal{N}^i} (R_2(q) - R_1(q)) = \infty$$

(such $\mathcal{N}^i$ obviously exist). For every $i \in \mathbf{N}$, we put

$$\mathfrak{T}_i \stackrel{\text{def}}{=} \bigcup_{q \in \mathcal{N}^i} \mathfrak{B}(q, \delta_q, \mathcal{U}_q).$$

Similalrly to the above, for every $i \in \mathbf{N}$ the set $\mathfrak{T}_i$ is homeomorphic to $\mathcal{K} \setminus \{O\}$. By the properties of the sets $\mathfrak{B}(q, \delta_q, \mathcal{U}_q)$ (see Lemma 6), for every $i \in \mathbf{N}$ and every unbounded curve $\gamma \subset \mathbf{R}^n$ of bounded curvature, we have $m_\gamma(\gamma \cap \mathfrak{T}_i) = \infty$. Put

$$\mathfrak{U} \stackrel{\text{def}}{=} \bigcup_{i=1}^{\infty} \mathfrak{T}_i.$$

Similalrly to the previous case, the set $\mathfrak{U}$ is homeomorphic to $\mathcal{K} \setminus \{O\}$ and for every plane $P \subset \mathbf{R}^n$, $1 \leq \dim P \leq n$, we have $m_P(P \setminus \mathfrak{U}) < 1$.

Let $\mathcal{M} \subset \mathbf{R}^n$ be a nonempty finite set and $v \in \mathbf{R}^n$ a nonzero vector, and $X_0 \in \mathcal{M}$ a point. For every point $X \in \mathcal{M}$ denote by $l(X)$ the straight line passing through X and parallel to v, and by τ_X the parallel displacement $(\tau_X : \mathbf{R}^n \to \mathbf{R}^n)$ by the vector $\overrightarrow{XX_0}$. We put

$$T_0(\mathcal{M}, v) \stackrel{\text{def}}{=} \bigcap_{X \in \mathcal{M}} \tau_X\left(l(X) \cap \mathfrak{U}\right).$$

Since $m_{l(X)}(l(X) \setminus \mathfrak{U}) < 1$ for every point $X \in \mathcal{M}$ and the set $\mathcal{M}$ is finite, we find that $m_{l(X_0)}(T_0(\mathcal{M}, v)) = \infty$.

Let $Y \in l(X_0)$ be a point such that $\overrightarrow{X_0Y} = v$. Let us suppose that $\mu : l(X_0) \to \mathbf{R}$ is an affine transformation such that $\mu(X_0) = 0$, $\mu(Y) = 1$. We put $T(\mathcal{M}, v) \stackrel{\text{def}}{=} \mu(T_0(\mathcal{M}, v))$ and denote by τ^t (where $t \in \mathbf{R}$) a parallel displacement $(\tau^t : \mathbf{R}^n \to \mathbf{R}^n)$ by the vector tv. It is obvious that for every $t \in T(\mathcal{M}, v)$, we have $\tau^t(\mathcal{M}) \subset \mathfrak{U}$.

We put $\mathfrak{R}_0 \stackrel{\text{def}}{=} \{\mathfrak{M}(j, \delta_j, \mathcal{U}_j) : j \in \mathbf{N}\}$. For every $j \in \mathbf{N}$ and $q \in \mathbf{N}$, we put

$$\mathfrak{R}(j, 2q-1) \stackrel{\text{def}}{=} \left\{\mathfrak{M}(j, 2q-1, I) : I \in \mathcal{B}^j_{2q-1}\right\},$$

$$\mathfrak{R}(j, 2q) \stackrel{\text{def}}{=} \left\{\mathfrak{M}(j, 2q, I) : I \in \mathcal{B}^j_{2q}\right\}$$

(for the definitions of $\mathfrak{M}(j, 2q-1, I)$ and $\mathfrak{M}(j, 2q, I)$ see above). We put

$$\mathfrak{R} \stackrel{\text{def}}{=} \left(\bigcup_{j=1}^{\infty}\bigcup_{q=1}^{\infty}\mathfrak{R}(j, 2q-1)\right)\bigcup\left(\bigcup_{j=1}^{\infty}\bigcup_{q=1}^{\infty}\mathfrak{R}(j, 2q)\right)\bigcup\mathfrak{R}_0.$$

Thus, $\mathfrak{R}$ is a countable set of simple arcs. Let us enumerate these arcs by natural numbers: $\mathfrak{R} = \{\mathfrak{M}'_1, \mathfrak{M}'_2, \dots\}$. For every $j \in \mathbf{N}$, we denote by A_1^j, A_2^j the ends of the arc $\mathfrak{M}'_j$. For each $\mathfrak{M}' \in \mathfrak{R}$ we define an open layer $V(\mathfrak{M}')$ between the concentric spheres with center O as follows: if $\mathfrak{M}'$ is a simple arc of the form $\mathfrak{M}(j, 2q-1, I)$ or $\mathfrak{M}(j, 2q, I)$, we put

$$V(\mathfrak{M}') \stackrel{\text{def}}{=} V_n\left(R_1(I) - \frac{1}{3}\operatorname{diam} I, R_2(I) + \frac{1}{3}\operatorname{diam} I\right)$$

(the definitions of $R_1(I)$ and $R_2(I)$ are given above). If $\mathfrak{M}'$ is an arc of type $\mathfrak{M}(j, \delta_j, \mathcal{U}_j)$, we put $V(\mathfrak{M}') \stackrel{\text{def}}{=} (\mathcal{U}_j, \frac{1}{2}\delta_j)$.

We shall refer to an unbounded curve $\mathcal{L} \subset \mathbf{R}^n$ as leading if it has the following properties: (1) $\mathcal{L}$ is homeomorphic to a closed bundle; (2) the end $O(\mathcal{L})$ of $\mathcal{L}$ (belonging to it) satisfies the inequality $\rho(O, O(\mathcal{L})) < \frac{1}{2}\min\{R_1^1, R_1(1)\}$; (3) for every point $X \in \mathcal{L} \setminus \{O(\mathcal{L})\}$, the segment of $\mathcal{L}$ whose ends are the points $O(\mathcal{L})$ and X is a polygonal line, (4) for every $j \in \mathbf{N}$, $\mathcal{L} \cap \operatorname{cl}(\mathcal{V}_j, \varepsilon_j/2)$ is a rectilinear segment lying on a straight line passing through the point O; (5) there exists an open set $\mathcal{W}(\mathcal{Z}) \subset \mathbf{R}^n$ such that $\mathcal{L} \subset \mathcal{W}(\mathcal{L})$ and

$$\mathcal{W}(\mathcal{Z}) \cap \bigcup_{j=1}^{\infty}\mathfrak{M}'_j = \emptyset.$$

Since $\frac{1}{2}(R_2^j - R_1^j) < \varepsilon_j$ for every $j \in \mathbf{N}$, the ball $D^n(X_j, \varepsilon_j)$ "drills a hole in the layer $\mathcal{V}$", i.e., the (rectilinear) segment $l(O, X_j) \cap \operatorname{cl}\mathcal{V}$ (where $l(O, X_j))$ is a bundle with origin O, passing through the point X_j) does not intersect $\operatorname{cl}(\mathcal{V}_j \setminus D^n(X_j, \varepsilon_j))$. Considering this fact as well as the existence of the polygonal lines $L(j, \delta_j, \mathcal{U}_j)$, $j \in \mathbf{N}$, we can easily see that leading curves do exist.

We denote by $\mathcal{L}_0$ a leading curve such that for every $j \in \mathbf{N}$, $X_j \in \mathcal{L}_0$, $L(j, \delta_j, \mathcal{U}_j) \subset \mathcal{L}_0$, $\mathcal{L}_0 \cap ((\operatorname{cl}(\mathcal{U}_j, \frac{1}{2}\delta_j)) \setminus \mathcal{U}_j)$ is a union of two (rectilinear) segments lying on straight lines passing through O, and $\mathcal{L}_0 \cap \operatorname{cl}(\mathcal{V}_j, \frac{3}{4}\varepsilon_j)$ is a rectilinear segment lying on a straight line passing through O.

We shall prove that there exists a curve $\tilde{\mathfrak{M}} \subset \mathbf{R}^n$ which is homeomorphic to a straight line and possesses properties 1–5 (see Theorem 2). We shall subsequently transform $\tilde{\mathfrak{M}}$ into the curve $\mathfrak{M}$ possessing not only properties 1–5, but also property 6. It is more convenient to consider the cases $n \geq 3$ and $n = 2$ separately.

Let $n \geq 3$. It can be easily verified by the standard technique (considering $n \geq 3$) that there exist leading curves $\mathcal{L}_j$ $(j \in \mathbf{N})$ having the following properties:

1. The neighborhoods $\mathcal{W}(\mathcal{L}_j)$ of these curves are pairwise disjoint.
2. $\mathcal{L}_0 \cap (\bigcup_{j=1}^{\infty}\mathcal{W}(\mathcal{L}_j)) = \emptyset$.
3. For each $\varepsilon' > 0$ there exists a $k(\varepsilon') \in \mathbf{N}$ such that for each natural k $(k > k(\varepsilon'))$, $\mathcal{L}_k \subset (\mathcal{L}_0, \varepsilon')$.

4. For each $j \in \mathbf{N}$ there exist polygonal lines l_1^j and l_2^j such that $l_1^j \cap l_2^j = \emptyset$, $l_1^j \cup l_2^j \subset V(\mathfrak{M}'_j)$, $l_1^j \cap (\bigcup_{k=1}^{\infty} \mathcal{W}(\mathcal{L}_k) \cup \mathcal{L}_0) \subset \mathcal{W}(\mathcal{L}_j)$, $l_2^j \cap (\bigcup_{k=1}^{\infty} \mathcal{W}(\mathcal{Z}_k) \cup \mathcal{L}_0) \subset \mathcal{W}(\mathcal{L}_{j+1})$.

5. One end of the polygonal line l_1^j coincides with the end A_1^j of the arc $\mathfrak{M}'_j$, while the other end $\tilde{A}_1^j$ belongs to $\mathcal{L}_j$.

6. $l_1^j \cap \mathcal{L}_j = \{\tilde{A}_1^j\}$.

For l_2^j, the following properties hold.

7. One end of the polygonal line l_2^j coincides with the end A_2^j of the arc $\mathfrak{M}'_j$, while the other end $\tilde{A}_2^j$ belongs to $\mathcal{L}_{j+1}$.

8. $l_2^j \cap \mathcal{L}_{j+1} = \{\tilde{A}_2^j\}$.

9. $l_1^j \cup l_2^j \cup \mathfrak{M}'_j$ is a simple arc. We denote by L_k ($k \in \mathbf{N} \setminus \{1\}$) a polygonal line with ends $\tilde{A}_2^{k-1}$, $\tilde{A}_1^k$ such that $L_k \subset \mathcal{L}_k$. We put $\tilde{L}_k \stackrel{\text{def}}{=} l_2^{k-1} \cup L_k \cup l_1^k$ and denote by L_1 a polygonal line with ends $O(\mathcal{L}_1)$ and $\tilde{A}_1^1$ such that $L_1 \subset \mathcal{L}_1$. We put $\tilde{L}_1 \stackrel{\text{def}}{=} l_1 \cup l_1^1$.

For $n \geq 3$, we put

$$\tilde{\mathfrak{M}} \stackrel{\text{def}}{=} \left(\left(\bigcup_{j=1} \mathfrak{M}'_j \right) \cup \left(\bigcup_{k=1}^{\infty} \tilde{L}_k \right) \right) \setminus \{O(\mathcal{L}_1)\}.$$

It can be easily verified that the curve $\tilde{\mathfrak{M}}$ is homeomorphic to a straight line and has properties 1–5 in Theorem 2.

Let us now suppose that $n = 2$. It should be reiterated that we shall define a polygon as a set $\mathcal{M} \subset \mathbf{R}^2$ such that $\mathcal{M}$ is homeomorphic to $D^2(O,1)$ and $\partial\mathcal{M}$ is a closed polygonal line. It can be easily seen that there exist polygons $\mathcal{M}_j \subset \mathbf{R}^2$ ($j \in \mathbf{N}$) such that for each j we have

$$\rho\left(\mathcal{M}_j \left(\bigcup_{k=1}^{\infty} \mathcal{M}_k \right) \setminus \mathcal{M}_j \right) > 0,$$

$$\mathcal{M}_j \subset V(\mathfrak{M}'_j), \quad \left\{A_1^j, A_2^j\right\} \subset \partial\mathcal{M}_j, \quad \mathfrak{M}'_j \setminus \left\{A_1^j, A_2^j\right\} \subset \operatorname{int} \mathcal{M}_j.$$

For each point $x \in \mathcal{L}_0$ on the segment $l(X)$ of the curve $\mathcal{L}_0$ with ends $O(\mathcal{L}_0)$ and X, we have

$$\rho\left(l(X), \bigcup_{k=1}^{\infty} \mathcal{M}_k \right) > 0.$$

For each $j \in \mathbf{N}$, we denote by λ_j a polygonal line (which is a simple arc) with ends A_2^j, A_1^{j+1} such that $A_1^j \notin \lambda_j$, $A_2^{j+1} \notin \lambda_j$, $\lambda_j \subset V(\mathfrak{M}'_j) \cup V(\mathfrak{M}'_{j+1}) \cup \mathcal{L}_0$. We observe that

$$\rho\left(\lambda_j, \left(\bigcup_{k=1}^{\infty} \mathcal{M}_k \right) \setminus (\mathcal{M}_j \cup \mathcal{M}_{j+1}) \right) > 0.$$

It can easily be seen that there exists a polygonal line $g_1 \subset (\partial\mathcal{M}_1) \cup \lambda_1 \cup \partial\mathcal{M}$, with ends A_2^1, A_1^2, such that $\Gamma_1 \stackrel{\text{def}}{=} \mathfrak{M}'_1 \cup g_1 \cup \mathfrak{M}'_2$ is a simple arc and there exists a polygon $\mathcal{P}_1$ such that $\Gamma_1 \setminus \{A_2^2\} \subset \operatorname{int} \mathcal{P}_1, A_2^2 \in \partial\mathcal{P}_1$, $\rho(\mathcal{P}_1, \bigcup_{k=3}^{\infty} \mathcal{M}_k) > 0$.

We assume that for some $j \in \mathbf{N}$, the polygonal lines $g_1, \dots, g_j$ and the polygons $\mathcal{P}_1, \dots, \mathcal{P}_j$ have been defined with the following properties: for each $k = 1, \dots, j$ the points A_2^k, A_1^{k+1} are the ends of the polygonal lines g_k;

$$\Gamma_k \stackrel{\text{def}}{=} \bigcup_{q=1}^{k+1} \mathfrak{M}'_q \cup \bigcup_{i=1}^{k} g_i$$

is a simple arc with endpoints A_1^1, A_2^{k+1}; $A_2^{k+1} \in \partial\mathcal{P}_k$; $\Gamma_k \setminus \{A_2^{k+1}\} \subset \operatorname{int}\mathcal{P}_k$; $\rho(\mathcal{P}_k, \bigcup_{p=k+2}^{\infty} \mathcal{M}_p) > 0$. If $j > 1$, $\mathcal{P}_t \subset \operatorname{int}\mathcal{P}_{t+1}$, $g_{t+1} \cap \operatorname{int}\mathcal{P}_t = \emptyset$ for each $t = 1, \dots, j-1$.

It can easily be seen that there exists a polygonal line $g'_{j+1} \subset \lambda_{j+1}$ (whose ends are denoted by B^1, B^2) such that $g'_{j+1} \cap \mathcal{P}_j = \{B^1\}$, $g'_{j+1} \cap \mathcal{M}_{j+2} = \{B^2\}$. We denote by g_{j+1} the polygonal line (which is a simple arc) with ends A_2^{j+1}, A_1^{j+2} such that

$$g_{j+1} \subset (\partial\mathcal{P}_j) \cup g'_{j+1} \cup \partial\mathcal{M}_{j+2}, \quad A_2^{j+2} \notin g_{j+1}$$

(obviously, such a polygonal line does exist). In this case, $\Gamma_j \cup g_{j+1} \cup \mathfrak{M}'_{j+2}$ is a simple arc. Since

$$\rho\left(\mathcal{P}_j \cup g_{j+1} \cup \mathcal{M}_{j+2}, \bigcup_{p=j+3}^{\infty} \mathcal{M}_p\right) > 0,$$

there exists a polygon $\mathcal{P}_{j+1}$ such that $\rho(\mathcal{P}_{j+1}, \bigcup_{p=j+3}^{\infty} \mathcal{M}_p) > 0$ and

$$\mathcal{P}_j \cup g_{j+1} \cup \left(\mathcal{M}_{j+2} \setminus \left\{A_2^{j+2}\right\}\right) \subset \operatorname{int}\mathcal{P}_{j+1}, \quad A_2^{j+2} \in \partial\mathcal{P}_{j+1}.$$

Thus, we have defined by induction the polygonal line g_j and the polygon $\mathcal{P}_j$ for each $j \in \mathbf{N}$. We can define Γ_j, $j \in \mathbf{N}$, in the same way as before.

To simplify further analysis, we denote by $\tilde{I}$ a (rectilinear) segment such that $\tilde{I} \subset \operatorname{int}\mathcal{P}_1$, and $\tilde{I} \cup \Gamma_1$ is a simple arc. We denote by $\tilde{O}$ the end of the segment I which does not belong to the arc Γ_1.

We put (for $n = 2$)

$$\tilde{\mathfrak{M}} \stackrel{\text{def}}{=} \bigcup_{j=1}^{\infty} \mathfrak{M}'_j \cup \bigcup_{k=1}^{\infty} g_k \cup \left(\tilde{I} \setminus \{\tilde{O}\}\right).$$

Obviously, the curve $\tilde{\mathfrak{M}}$ is homeomorphic to a straight line (it is sufficient to note that for each point $X \in \tilde{\mathfrak{M}}$ there exists $k \in \mathbf{N}$ such that $\rho(X, \tilde{\mathfrak{M}} \setminus (\Gamma_k \cup (\tilde{I})))$ > 0), the curve $\mathfrak{M}$ possessing properties 1–5 in Theorem 2.

Let us now transform the curve $\tilde{\mathfrak{M}} \subset \mathbf{R}^n$, $n \geq 2$ into the curve $\mathfrak{M}$ having properties 1–5 as well as property 6 in Theorem 2.

A simple arc $L \subset \mathbf{R}^n$, $n \geq 2$, is called quasipolygonal if there exists a homeomorphism $\psi_L : \mathbf{R} \to L \setminus \{A_L^1, A_L^2\}$ (where A_L^1 and A_L^2 are the ends of the simple arc L) such that for each $k \in \mathbf{Z}$, $\psi_L([k, k+1])$ is a rectilinear segment. Such a homeomorphism ψ_L will be called canonical for L.

LEMMA 7. *Let $\Gamma \subset \mathbf{R}^n$, $n \geq 2$, be a simple arc with ends A_1, A_2. Let $\mathcal{V} \subset \mathbf{R}^n$ be an open set such that $\Gamma \setminus \{A_1, A_2\} \subset \mathcal{V}$. Let I_0 be a rectilinear segment with ends B_1 and B_2.*

Then there exist a quasipolygonal line L, a discontinuum $\mathcal{G} \subset I_0$, and a homeomorphism $\tau : I_0 \to L$ having the following properties.

1. *The points A_1 and A_2 are the ends of the quasipolygonal line L.*
2. *$L \setminus \{A_1, A_2\} \subset \mathcal{V}$.*
3. *The linear Lebesgue measure of the discontinuum $\mathcal{G}$ is equal to zero.*
4. *For each point $X \in I_0 \setminus \mathcal{G}$, there exists an (open) interval $I_X \subset I_0 \setminus \mathcal{G}$ such that $X \in I_X$ and the restriction of the mapping τ to I_X is an isometry (in particular, $\tau(I_X)$ is a rectilinear segment).*
5. *$\tau(B_1) = A_1$, $\tau(B_2) = A_2$.*

PROOF. Let $\xi : \mathbf{R} \to \Gamma \setminus \{A_1, A_2\}$ be a homeomorphism. For each $k \in \mathbf{Z}$, put $\tilde{\Gamma}_k \stackrel{\text{def}}{=} \xi([k, k+1])$. Let the open sets $\tilde{\mathcal{V}}_k$ $(k \in \mathbf{Z})$ have the following properties.

1. For each k, $\tilde{\Gamma}_k \subset \tilde{\mathcal{V}}_k \subset \mathcal{V}$, $\operatorname{diam} \tilde{\mathcal{V}}_k < 2 \operatorname{diam} \tilde{\Gamma}_k$.
2. If the integers k_1 and k_2 are such that $|k_1 - k_2| \geq 2$, then $\tilde{\mathcal{V}}_{k_1} \cap \tilde{\mathcal{V}}_{k_2} = \emptyset$

(obviously, such sets $\tilde{\mathcal{V}}_k$ do exist).

By a "polygonal" line, we mean, as usual, a simple (finite-linked) polygonal line which is a simple arc and does not have self-intersections.

It can easily be seen that for each $k \in \mathbf{Z}$, there exists a polygonal line $\lambda_k \subset \mathcal{V}_k$ with ends $\xi(k)$, $\xi(k+1)$. Thus, if the integers k_1 and k_2 are such that $|k_1 - k_2| \geq 2$, then $\lambda_{k_1} \cap \lambda_{k_2} \neq \emptyset$. However, the intersection $\lambda_k \cap \lambda_{k+1}$ of the polygonal lines λ_k and λ_{k+1} for certain $k \in \mathbf{Z}$ may contain points other than their common end. For each $k \in \mathbf{Z}$, we denote by μ_{2k} the smallest (in length) polygonal line contained in λ_{2k} with its ends belonging to the polygonal lines λ_{2k-1} and λ_{2k+1} respectively.

It can now easily be seen that for each $k \in \mathbf{Z}$, there exists a polygonal line $\eta_{2k+1} \subset \lambda_{2k+1}$ such that

$$L_1 \stackrel{\text{def}}{=} \left(\bigcup_{k \in \mathbf{Z}} \mu_{2k} \right) \bigcup \left(\bigcup_{k \in \mathbf{Z}} \eta_{2k+1} \right) \bigcup \{A_1, A_2\}$$

is a quasipolygonal line for which $L_1 \setminus \{A_1, A_2\} \subset \mathcal{V}$ and whose ends are the points A_1 and A_2.

Let $\tau_1 : I_0 \to L_1$ be a homeomorphism for which $\tau_1(B_1) = A_1$ (hence $\tau_1(B_2) = A_2$). Obviously, for each rectilinear segment $I \subset \mathbf{R}^n$ and for any numbers $\varepsilon > 0$ and $\mathcal{N} > 0$, there exists a polygonal line $L(I, \varepsilon, \mathcal{N}) \subset \mathbf{R}^n$ with the following properties.

1. The ends of $L(I, \varepsilon, \mathcal{N})$ coincide with the ends A_I and B_I of I.
2. The length of $L(I, \varepsilon, \mathcal{N})$ is larger than $\mathcal{N}$.
3. For each point $X \in L(I, \varepsilon, \mathcal{N}) \setminus \{A_I, B_I\}$ the angles $\angle X A_I B_I$ and $\angle X B_I A_I$ are smaller than ε.

Hence there exists a homeomorphism $\psi_{L_1} : \mathbf{R} \to L_1 \setminus \{A_1, A_2\}$ that is canonical for L_1 and has the following property: for each $k \in \mathbf{Z}$, there exists a polygonal line $\mathcal{L}_k \subset \mathcal{V}$ with ends $\psi_{L_1}(k)$, $\psi_{L_1}(k+1)$ such that the length of $\mathcal{L}_k$ is larger than the length of the segment

$$t_k \stackrel{\text{def}}{=} \left[\tau_1^{-1}\left(\psi_{L_1}(k)\right), \tau_1^{-1}\left(\psi_{L_1}(k+1)\right) \right],$$

where

$$L \overset{\text{def}}{=} \bigcup_{k \in \mathbf{Z}} \mathcal{L}_k \cup \{A_1, A_2\}$$

is a quasipolygonal line with ends A_1 and A_2.

Consequently, for each $k \in \mathbf{Z}$ there exists a (piecewise linear) homeomorphism $\delta_k : t_k \to \mathcal{L}_k$ such that for each link l of the polygonal line $\mathcal{L}_k$, the length of the segment $\delta_k^{-1}(l)$ is smaller than the length of the link l (if orientations are chosen on t_k and $\mathcal{L}_k$, then there exists a homeomorphism δ_k which induces a chosen orientation).

This leads to the existence of a homeomorphism $\psi_L : \mathbf{R} \to L \setminus \{A_1, A_2\}$ canonical for L and a homeomorphism $\tau_2 : I_0 \to L$ such that for each $k \in \mathbf{Z}$ the length of the segment $q_k \overset{\text{def}}{=} \psi_L([k, k+1])$ is larger than the length of the segment

$$p_k \overset{\text{def}}{=} \left[\tau_2^{-1}(\psi_L(k)), \tau_2^{-1}(\psi_L(k+1))\right], \quad \tau_2(B_1) = A_1.$$

Let us now consider segments p and q such that $|p| < |q|$ (recall that $|s|$ is the length of the segment s), A_p is one of the ends of the segment p and A_q is one of the ends of the segment q.

Using the standard technique, we construct on the segment p a Cantor perfect set $\mathcal{K}_p$, i.e., we discard from p the open middle interval I^1, $|I^1| = \frac{1}{3}|p|$ (rank-1 interval), from the remaining two segments p_1^2 and p_2^2 we discard the (open) middle intervals I_1^2, I_2^2, $|I_1^2| = |I_2^2| = \frac{1}{9}|p|$ (rank-2 intervals), and so on. Naturally, the linear Lebesgue measure of the Cantor set $\mathcal{K}_p$ is equal to zero.

On the segment q we construct the discontinuum $\mathcal{D}_q$ as follows: we discard from q the (open) interval J^1 whose midpoint coincides with the midpoint of q, $|J^1| = |I^1|$. From the remaining two segments q_1^2 and q_2^2, we discard the (open) intervals J_1^2 and J_2^2 whose midpoints coincide with the midpoints of the segments q_1^2 and q_2^2 respectively, $|J_1^2| = |J_2^2| = |I_1^2| = |I_2^2|$, and so on (such a construction is possible since $|q| > |p|$). As a result, we obtain the discontinuum $\mathcal{D}_q \subset q$ whose supplementary intervals are arranged in the same order as the intervals supplementary to $\mathcal{K}_p$ and are respectively equal to these intervals. Naturally, the linear Lebesgue measure of the discontinuum $\mathcal{D}_q$ is equal to $|q| - |p|$.

In accordance with the description given, for example, in [**2**], this leads to the existence of a homeomorphism $\nu : p \to q$ such that $\nu(\mathcal{K}_p) = \mathcal{D}_q$. For each interval $I \subset p$ complementing $\mathcal{K}_p$, the restriction $\nu|_I$ is an isometry, and $\nu(A_p) = A_q$.

For each $k \in \mathbf{Z}$, by appropriately choosing the discontinua $\mathcal{K}_{p_k}$ and $\mathcal{D}_{q_k}$ as well as the homeomorphism $\nu_k : p_k \to q_k$ we can directly verify the existence of the discontinuum $\mathcal{G} \subset I_0$ (which is a union of all discontinua $\mathcal{K}_{p_k}$ ($k \in \mathbf{Z}$) and the ends of the segment I_0) as well as the homeomorphism $\tau : I_0 \to L$ (constructed by using the obvious technique from homeomorphisms of type ν_k ($k \in \mathbf{Z}$), having the properties formulated in Lemma 7.

Lemma 7 is proved.

We put

$$\tilde{\mathcal{A}} \overset{\text{def}}{=} \mathcal{A}_0 \cup \mathfrak{U} \cup \left(\bigcup_{i=1}^{\infty} \mathfrak{G}_i \right)$$

(see above for the definition of the sets on the right-hand side of this equality).

Turning to the construction of the curve $\tilde{\mathfrak{M}}$, we find that $\tilde{\mathfrak{M}} \supset \tilde{\mathcal{A}}$ and there exists a homeomorphism $\sigma : J \to \tilde{\mathfrak{M}}$ (recall that $J \overset{\text{def}}{=}]0,1[$) having the following property: the number $\varepsilon_\sigma \overset{\text{def}}{=} \inf \sigma^{-1}(\tilde{\mathcal{A}})$ belongs to $\sigma^{-1}(\mathcal{A})$, the curve $\sigma(]0, \frac{1}{2}\varepsilon_\sigma[)$ is a rectilinear interval, and $\sup \sigma^{-1}(\mathcal{A}) = 1$.

We denote by $\tilde{C}$ the Cantor perfect set constructed in the standard manner on the segment $[0, \frac{1}{2}\varepsilon_\sigma]$. We put $\tilde{\mathcal{B}} \overset{\text{def}}{=} (\sigma^{-1}(\tilde{\mathcal{A}})) \cup (\tilde{C} \setminus \{O\})$. In this case, $\tilde{\mathcal{B}} \subset J$ is a closed (in J) perfect set which is not dense anywhere, and for each (open) maximal interval $I \subset J \setminus \tilde{\mathcal{B}}$ the ends of I belong to $\tilde{\mathcal{B}}$ (note that $\operatorname{cl} \tilde{\mathcal{B}} \supset \{0,1\}$, where the closure is naturally considered in the topology of $\mathbf{R}$). Obviously, $\tilde{\mathcal{B}} \cup \{0,1\}$ is a discontinuum. It is well known (see, for example, **[2]**) that there exists a homeomorphism $\sigma_1 : J \to J$ such that $\sigma_1(\mathcal{K} \setminus \{0,1\}) = \mathcal{B}$ (recall that $\mathcal{K}$ is the Cantor perfect set). Put $\sigma_2 \overset{\text{def}}{=} \sigma \circ \sigma_1$.

We denote by M the set of (open) maximal intervals contained in $J \setminus \mathcal{K}$ (i.e., M is the set of intervals complementary to $\mathcal{K}$). For each point $X \in J \setminus \mathcal{K}$, we denote by $\tilde{I_X}$ an (open) interval such that $\tilde{I_X} \in M$, $X \in \tilde{I}_X$. We put

$$\varepsilon(X) \overset{\text{def}}{=} \frac{1}{2}\rho\left(\sigma_2(X), \sigma_2(j \setminus \tilde{I_X})\right).$$

For each $I \in M$, we put

$$\mathcal{V}(I) \overset{\text{def}}{=} \bigcup_{X \in I} \operatorname{int} D^n \left(\sigma_2(X), \varepsilon(X)\right).$$

Obviously, the set $\mathcal{V}(I)$ is open for each $I \in M$ and contains $\sigma_2(I)$. If $I_1 \neq I_2$ $(I_1, I_2 \in M)$, then $\mathcal{V}(I_1) \cap \mathcal{V}(I_2) = \emptyset$.

By Lemma 7, for each $I \in M$ there exist a quasipolygonal line $L(I) \subset \mathcal{V}(I) \cup \sigma_2(\partial I)$ (where ∂I is the set of the ends of the interval I), a discontinuum $\mathcal{G}_I \subset \operatorname{cl} I$ and a homeomorphism $\tau_I : \operatorname{cl} I \to L(I)$ such that $\sigma_2(\partial I)$ is the set of the ends of the quasipolygonal line $L(I)$, the linear Lebesgue measure of the discontinuum $\mathcal{G}_I$ is equal to zero, and for each point $X \in I \setminus \mathcal{G}_I$ there exists an open interval $I_X \subset I \setminus \mathcal{G}_I$ such that $X \in I_X$ and the restriction of the mapping τ_I to I_X is an isometry, while for each point $Y \in \partial I$, $\tau_I(Y) = \sigma_2(Y)$. We put

$$\mathfrak{M} \overset{\text{def}}{=} \sigma(\tilde{\mathcal{B}}) \cup \bigcup_{I \in M} L(I),$$

$$\Omega \overset{\text{def}}{=} (\mathcal{K} \setminus \{0,1\}) \cup \bigcup_{I \in M} \mathcal{G}_I.$$

We define $\varphi : J \to \mathcal{M}$ as follows: if $X \in \mathcal{K} \setminus \{0,1\}$, we put $\varphi(X) \overset{\text{def}}{=} \sigma_2(X)$. If there exists an interval $I \in M$ such that $X \in I$, we put $\varphi(X) \overset{\text{def}}{=} \tau_I(X)$. It can be easily seen that φ is a homeomorphism which is an isometry at each point $X \in J \setminus \Omega$, and $m_1(\Omega) = 0$.

It should be observed that the equality $m_n(\mathfrak{M} \setminus \tilde{\mathcal{A}}) = 0$ follows from the construction of the curve $\mathfrak{M}$, since $\mathfrak{M} \setminus \tilde{\mathcal{A}}$ is contained in a union of countable sets of (rectilinear) segments.

Theorem 2 is proved.

The proof of Theorem 3 follows essentially from Lemmas 3, 4, and 6 and the remark about the sets $\mathcal{R}(W,k)$ (see the proof of Theorem 2).

Indeed, let H be a finite set of pairwise disjoint open layers between concentric spheres (with center O) such that the thickness of each layer $\mathcal{U} \in H$ is smaller than $(2M)^{-1}$ and the following equality holds:

$$\bigcup_{\mathcal{U}\in H} \operatorname{cl}\mathcal{U} = \operatorname{cl} V_n(r_1, r_2).$$

In each layer $\mathcal{U} \in H$, we consider a discontinuum $\mathfrak{B}(M, \varepsilon \cdot (\operatorname{card} H)^{-1}, \mathcal{U}) \subset \mathcal{U}$ (see Lemma 6). We put

$$\mathfrak{B}_0 \stackrel{\text{def}}{=} \bigcup_{\mathcal{U}\in H} \mathfrak{B}(M, \varepsilon \cdot (\operatorname{card} H)^{-1}, \mathcal{U}).$$

Let W_j, $j \in G \stackrel{\text{def}}{=} \{1, \dots, 2\mathcal{N}+1\}$, be disjoint open layers between concentric spheres (with center O) such that

$$\bigcup_{j\in G} W_j \subset V_n(r_1, r_2) \setminus D^n\left(O, \max\{\rho(O, X) : X \in \mathfrak{B}_0\}\right).$$

Obviously, for each simple arc $\Gamma \subset \mathbf{R}^n$ which intersects the layer $V_n(r_1, r_2)$ canonically and is a union of $\mathcal{N}$ simple arcs each of which is an M-Lipschitz arc, there exist a natural number $j(\Gamma) \in G$ and an M-Lipschitz simple arc $\Gamma_0 \subset \Gamma$ such that the arc Γ_0 intersects the layer $W_{j(\Gamma)}$ canonically.

For each $j \in G$, let W_j^1 and W_j^2 be disjoint open layers between concentric spheres (with center O) such that $W_j^1 \cup W_j^2 \subset W_j$.

Let $k_0 > M$ be a natural number. By the above remark concerning the sets $\mathcal{R}(W,k)$ of blocks, we consider for each $j \in G$ and for each $i = 1, 2$ the set of blocks $T_j^i \stackrel{\text{def}}{=} \mathcal{R}(W_j^i, k_0)$. We put $T^i \stackrel{\text{def}}{=} \bigcup_{j\in G} T_j^i$, $i = 1, 2$, $\mathfrak{F} \stackrel{\text{def}}{=} \bigcup_{K\in T'} \mathcal{A}(K)$ (see Lemma 3). For each $\alpha \in \mathbf{R}$, we put

$$\mathfrak{C}_\alpha \stackrel{\text{def}}{=} \bigcup_{K\in T^1} \mathfrak{N}_\alpha(K).$$

We define (see Lemma 4)

$$\mathfrak{C} \stackrel{\text{def}}{=} \bigcup_{K\in T^2} \mathcal{A}\left(K, \frac{1}{2}h(K)\right), \qquad \mathfrak{D} \stackrel{\text{def}}{=} \mathfrak{F} \cup \mathfrak{G} \cup \mathfrak{B}_0.$$

It can be easily seen that $\mathfrak{D}$, $\mathfrak{F}$, and $\mathfrak{C}_\alpha$ $(\alpha \in \mathbf{R})$ are discontinua possessing all the properties listed in Theorem 3.

Theorem 3 is proved. In order to prove Corollary 1, it is sufficient to observe that each regular (continuously differentiable) curve (in $\mathbf{R}^n$) is a piecewise Lipschitz curve, and each curve whose projection on at least one hyperplane is not dense everywhere in this hyperplane is admissible.

The proofs of Remark 1 and Corollary 3 are obvious modifications of the proof of Theorem 2.

References

1. P. S. Aleksandrov, *Introduction to the general theory of sets and functions*, GITTL, Moscow, 1948 (Russian); German transl., VEB Deutscher Verlag. Wiss., Berlin, 1956; 2nd ed., 1964.
2. B. Helbaum and J. Olmsted, *Contrexamples in analysis*, Holden-Day, San Francisco, 1964.
3. L. Keldysh, *Embedding of zero-dimensional compact sets in E_n*, Dokl. Akad. Nauk SSSR **147** (1962), 772–775; English transl. in Soviet Math. Dokl. **3** (1962).
4. ———, *Topological embeddings in Euclidean spaces*, Trudy Mat. Inst. Steklov. **81** (1966); English transl., Proc. Steklov Inst. Math. **81** (1968).
5. A. V. Kuz′minykh, Dokl. Akad. Nauk SSSR **305** (1989), 1042–1045; English transl. in Soviet Math. Dokl. **39** (1989).
6. ———, All-Union Conf. Geometry "In the Large", Abstracts of Reports, Novosibirsk, 1987. (Russsian)
7. ———, Ninth All-Union Conf. Geometry, Abstracts of Reports, Kishinev, 1988. (Russian)
8. ———, All-Union Conf. Geometry and Analysis, Abstracts of Reports, Novosibirsk, 1989. (Russian)
9. K. Kuratowski, *Topologie.* Vol. 1 , 4th ed., PWN, Warsaw, 1958.
10. N. N. Luzin, *Theory of functions of a real variable*, 2nd ed., Uchpedgiz, Moscow, 1948. (Russian)

NOVOSIBIRSK-54, 630054, RUSSIA

Amer. Math. Soc. Transl.
(2) Vol. **176**, 1996

Structure of the Neighborhood of an Isolated Zero of the Lipshits–Killing Curvature on a Surface Which Is Not 0-tight

V. A. Kuznetsov

Introduction

The notion of a canonical neighborhood of a point on a surface in E^3 was introduced by Efimov [**1**]. He proved that an isolated zero of the Gaussian curvature has a canonical neighborhood, and the intersection of a canonical neighborhood of a point on the surface with the tangent plane consists of an even number of simple arcs glued at this point.

In the present article, we shall define the canonical neighborhood of a point on an m-dimensional surface in an $(m+q)$-dimensional Euclidean space, which is defined by a certain normal at the given point (Definition 1.2). The statements generalizing Efimov's theorems (Theorems 1.3 and 1.4) will be proved.

1. Definition and formulation of the main results

1.1. Projection function. Let a complete m-dimensional surface F in the Euclidean space E^n, $n = m+q$, be defined by the C^r-imbedding

$$f : W \to E^n$$

of the m-dimensional C^∞-manifold W in E^n, where $r = m-1$ for $m > 2$ and $r = 2$ for $m = 2$.

We denote the intrinsic metric of the surface F by ρ. Let $x_0 \in W$, and let U be a neighborhood of x_0. Without any loss of generality, we assume that U is a sphere in the tangent plane at x_0 and the surface F in U is defined by the equations

$$x^i = f^i(x^1, \dots, x^m), \tag{1}$$

where the functions $f^i \in C^r$ and

$$f^i(0, \dots, 0) = \frac{\partial f^i}{\partial x^j}(0, \dots, 0) = 0$$

1991 *Mathematics Subject Classification.* Primary 53A07.

for all $i = m+1, \dots, n$ and $j = 1, \dots, m$. In other words, the vector-valued function η defining F in U has the form

$$\eta = (x^1, \dots, x^m, f^{m+1}, \dots, f^n)$$

(see [**2**]).

Let us take an arbitrary straight line l from the bundle of straight lines p^{n-1} with the center at the origin. We assume that l is the number axis. Let ν be a directional vector of the axis l. We consider the orthogonal projection π_ν of the space E^n on the l-axis. The function $f^\nu = \pi_\nu \circ f$ is called the projection function. The point $x \in U$ is a critical point of the function f^ν if and only if ν is orthogonal to the tangent space T_x at the point x. We denote by A_ν the set of the critical points of the function f^ν. The level sets $U_\nu(h) = \{x \in U | f^\nu(x) = h\}$ in the vicinity of each point $x \notin A_\nu$ are $(m-1)$-dimensional submanifolds in U. The trajectories orthogonal to them in the metric ρ are called gradient lines. Only one maximal gradient line can be drawn through each point $x \notin A_\nu$ (see [**3**]).

Let ν_0 be a unit normal at the point $x_0 \in F$ and let the straight line l_0 be parallel to ν_0. Then x_0 is a critical point of the function f^{ν_0}. Let us suppose that the origin coincides with x_0. Then $f^{\nu_0} = \langle \eta, \nu_0 \rangle$. The functions f^i are obviously projection functions on the coordinate axes: $f^i = \langle \eta, e_i \rangle$, where e_i are the vectors of the orthonormal basis of the normal space N_{x_0}, which are defined by the system of coordinates.

The function f^{ν_0} defines at each point $\overline{x} = f(x)$ the vector field of unit normals. Indeed, let us suppose that $\nu_0 = \sum_{m+1}^n \lambda^i e_i$. Then $f^{\nu_0} = \sum_{m+1}^n \lambda^i f^i$. The vector field $\nu(x) = (y^1(x), \dots, y^n(x))$, where

$$y^i(x) = \frac{-\frac{\partial f^{\nu_0}}{\partial x^i}(x)}{\sqrt{1 + \operatorname{grad}^2 f^{\nu_0}(x)}} \tag{2}$$

for $i = 1, \dots, m$, and

$$y^i(x) = \frac{\lambda^i}{\sqrt{1 + \operatorname{grad}^2 f^{\nu_0}(x)}}$$

for $i = m+1, \dots, n$, is orthogonal to the tangent plane at every point $x \in U$. According to the construction of the field, it does not depend on the bases of the spaces N_{x_0} and T_{x_0}.

This means that at each point $x \in U$, we have defined the Lipshits–Killing curvature $K(x, \nu) = k_1(x, \nu) \dots k_m(x, \nu)$, where $k_i(x, \nu)$ are the principal curvatures of F at the point x relative to the normal ν. The curvature $K(x, \nu)$ coincides to within a nonzero factor (see [**4**]) with

$$\det \left(\frac{\partial^2 f^{\nu_0}}{\partial x^i \partial x^j} \right).$$

DEFINITION 1.2. The neighborhood U of a point x_0 said to be canonical if there exists a vector $\nu_0 \in N_{x_0}$ such that $\nu_0 \neq \nu(x)$ for all $x \in U$ and $x \neq x_0$.

The straight line $l_0 \in p^{n-1}$ with the direction ν_0 will be called a canonical direction. If U is a canonical neighborhood of the point x_0 and l_0 is a canonical direction, the function f^{ν_0} has no other canonical points in U than the point x_0.

THEOREM 1.3. *Let an m-dimensional surface F be defined by equattions* (1), *f^{ν_0} the projection function on the direction $\nu_0 \in N_{x_0}$, and ν the field* (2). *Assume that the Lipshits–Killing curvature satisfies the condition $K(x,\nu) \neq 0$ for $x \neq x_0$. Then the point x_0 has a canonical neighborhood.*

The proof of Theorem 1.3 is literally the same as the proof of the similar theorem formulated for two-dimensional surfaces by Efimov (see [**1**]).

THEOREM 1.4. *Suppose that the conditions of Theorem* 1.3 *are satisfied. Assume that the level set $U_{\nu_0}(0) \setminus \{x\}$ contains more than one component. Then there exists a neighborhood V of the point x_0 at which the set $V_{\nu_0}(0) = V \cap U_{\nu_0}(0)$ is homeomorphic to a cone over $B = \partial V \cap U_{\nu_0}(0)$, and each component of the set B is diffeomorphic to the sphere S^{m-2}.*

The surfaces that cannot be divided into more than two components by a hyperplane are known as 0-tights (see [**5**]). In Theorem 1.4, surfaces which do not belong to the class of 0-tights are considered.

2. Gaussian mapping of a canonical neighborhood

Let U be a canonical neighborhood of a point x_0 relative to the normal $\nu_0 \in N_{x_0}$. We consider the space $N(U)$ of normal foliation and assume that $N_1 = \{\,(x,\nu) \in N \mid |\nu| = 1\,\}$. The Gaussian mapping $\mu : N_1 \to S^{n-1}$ is defined by the equality $\mu(x,\nu) = \nu$. We denote by μ_0 the restriction of the Gaussian mapping on $U \times \{\nu(x)\}$, where $\nu(x)$ is the continuation (2) of the vector ν_0 to the neighborhood U. Then μ_0 is the map of U in S^{n-1}. Obviously, we can write $\mu_0(x_0) = \nu_0$. The map μ_0 is called a canonical Gaussian map. The standard metric of the sphere S^{n-1} is denoted by ρ^*.

LEMMA 2.1. *Let V be a neighborhood of the point x_0, $\overline{V} \subset U$. There exists a neighborhood $V_1 \subset V$ of x_0 such that $\mu_0(V_1) \cap \mu_0(U \setminus V) = \emptyset$.*

PROOF. The neighborhood V is also canonical. Consequently

$$\min_{x \in U \setminus V} \rho^*(\nu_0, \nu(x)) = \alpha_0 > 0.$$

Let us consider the numerical sequence $\varepsilon_k \to 0$, $\varepsilon_k > 0$, and suppose that D_k are balls with center at x_0 of radius ε_k in the domain U. Then

$$\beta(\varepsilon_k) = \max_{x \in D_k} \rho^*(\nu_0, \nu(x)) \to 0$$

for $\varepsilon_k \to 0$. The function $\beta(\varepsilon_k)$ does not increase, i.e., $\beta(\varepsilon_k) \leq \beta(\varepsilon_p)$ for $\varepsilon_k > \varepsilon_p$. There exists a number k_0 such that the inequality $2\beta(\varepsilon_k) < \alpha_0$ holds for $k > k_0$. The set $\mu_0(U \setminus V)$ lies outside the geodesic ball of radius α_0 with center at ν_0 on the sphere S^{n-1}. The set $\mu_0(D_k)$ is contained in the geodesic ball with center at ν_0 and radius $\beta_0 = \beta(\varepsilon_{k_0}) < \alpha_0/2$. We put $V_1 = D_{k_0}$. The lemma is proved.

COROLLARY 2.2. *Suppose that the vector $\nu \in \mu_0(V_1)$. Then all critical points of the function f^ν lie in V.*

2.3. Construction of a regular canonical neighborhood. Let U be a canonical neighborhood of the point x_0, l_0 the canonical direction with the directional vector ν_0, and μ_0 a canonical Gaussian map.

We introduce the following notation:

$$\begin{aligned} U_\nu(h_1, h_2) &= \{x \in U | h_1 \leq f^\nu(x) \leq h_2\}, \\ U_\nu(h) &= \{x \in U | f^\nu(x) = h\}, \\ g(x) &= \sqrt{1 + \operatorname{grad}^2 f^{\nu_0}(x)}, \end{aligned}$$

A_ν being the set of critical points of the function f^ν, and $G_\nu(M)$ the set of points belonging to the gradient lines passing through points of the set M or having limit points in M:

$$A_\nu(h) = G_\nu(A_\nu) \cap U_\nu(h).$$

If h is a noncritical value, the set $A_\nu(h)$ consists of points whose gradient lines have limit points in A_ν. Obviously, the set $A_\nu(h)$ is closed.

According to the general theorem on t-regularity (see [**6**]), we can choose the ball D so that $\overline{D} \subset U$, and the sphere ∂D intersects transversally the manifold $U_{\nu_0}(0) \setminus \{x_0\}$ and all manifolds $U_{\nu_0}(h)$ for $|h| < h_1$, where h_1 is a sufficiently small number. The set $B = U_{\nu_0}(0) \cap \partial D$ consists of a finite number of $(m-2)$-dimensional submanifolds of the sphere $S^{m-1} = \partial D$.

Let γ be a gradient line passing through the point $b \in B$, and $x = x(s)$ be its natural parametrization. The function f^{ν_0} is strictly monotonic in s. Indeed, if $\partial f^{\nu_0}/\partial s = 0$ for a certain s_0, the vector tangent to γ at a point $x(s_0)$ is orthogonal to ν_0, whence $x(s_0)$ is a new critical point of the function f^{ν_0}. This contradicts the choice of the neighborhood U. Then $h_2 = \min_{x \in M} |f^{\nu_0}(x)| > 0$, where $M = G_{\nu_0}(B) \cap \partial U$.

Let us suppose that $h_0 < \min(h_1, h_2)$. We choose the ball $D_1 \subset U_{\nu_0}(-h_0, h_0)$ so that the sphere ∂D_1 intersects the set $U_{\nu_0}(0) \setminus \{x_0\}$ transversely. Every component of the set $D_1 \cap U_{\nu_0}(0)$ has points in common with ∂D_1. Indeed, if there exists in D_1 a component M of the level $U_{\nu_0}(0)$ which does not intersect ∂D_1, then M bounds a compact set in D_1. In this case, there exists in D_1 one more critical point of the function f^{ν_0}, which contradicts the definition of a canonical neighborhood.

Since the number of components of the set B is finite, the number of components of the set $(U_{\nu_0}(0) \cap D_1) \setminus \{x_0\}$ is also finite. From the set $U_{\nu_0}(0) \cap D_1$, we eliminate components that do not contain x_0. The remaining set is denoted by $\Omega_{\nu_0}(0)$.

Let the intersection $\Omega_{\nu_0}(0) \cap \partial D_1$ consist of the manifolds $B_1, \dots, B_p$. Then the gradient lines passing through the points of the manifold B_i intersect the levels $U_{\nu_0}(h_0)$ and $U_{\nu_0}(-h_0)$ within U along the manifolds $B_i(h_0)$ and $B_i(-h_0)$ respectively, where $i = 1, \dots, p$.

The manifolds $B_i(h_0)$ bound on $U_{\nu_0}(h_0)$ a certain manifold $\Omega_{\nu_0}(h_0)$ with boundary each of whose components contains points of the set $A_{\nu_0}(h_0)$. Similarly, we obtain the manifold $\Omega_{\nu_0}(-h_0)$ on the set $U_{\nu_0}(-h_0)$ (Figure 1).

Let us consider a neighborhood of the point x_0 whose boundary consists of the manifolds $\Omega_{\nu_0}(h_0)$ and $\Omega_{\nu_0}(-h_0)$ glued by the cylinders $G_{\nu_0}(B_i)$ from gradient lines. We denote this neighborhood by Ω_{ν_0} and define it as a regular canonical neighborhood of x_0 for the canonical direction l_0.

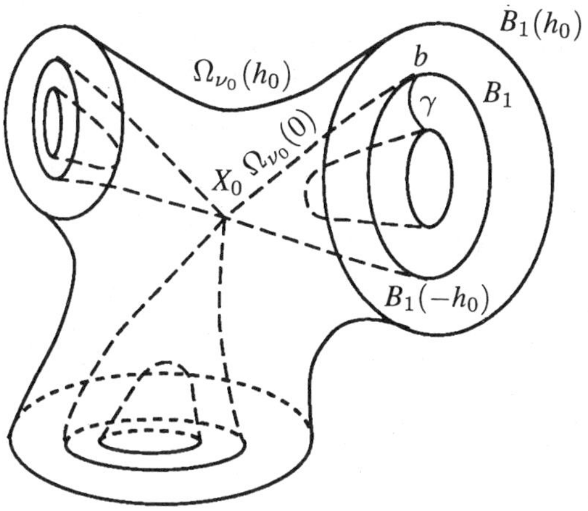

FIGURE 1

According to Morse theory, all sets $\Omega_{\nu_0}(h)\setminus A_{\nu_0}(h)$ in a regular canonical neighborhood are homeomorphic. The homeomorphism is determined by gradient lines.

LEMMA 2.4. *Let Ω_{ν_0} be a regular canonical neighborhood of the point x_0. Then for any neighborhood W of the set $A_{\nu_0}(h_0)$ at the level $\Omega_{\nu_0}(h_0)$ and for any neighborhood $V \subset G_{\nu_0}(W) \cup G_{\nu_0}(x_0)$ of the point x_0, there exists a neighborhood V^* of the point ν_0 in $\mu_0(U)$ such that $G_\nu(V) \cap \Omega_{\nu_0}(h_0) \subset W$ for all $\nu \in V^*$.*

PROOF. For the neighborhood V, we construct the neighborhood V_1 in accordance with Lemma 2.1 so that $\mu_0(V_1) \cap \mu_0(\Omega_{\nu_0} \setminus V) = \emptyset$. All critical points of the function f^ν for $\nu \in \mu_0(V_1)$ lie in V.

Let us consider a system of differential equations that defines the gradient lines of the function f^ν in the canonical neighborhood U with initial values on ∂V. This system has the form

$$\frac{dx^i}{dh} = \varphi(x, \nu), \quad x = (x^1, \dots, x^m).$$

All singular points of the system lie in V. Therefore, all the conditions of the theorem on existence and uniqueness are satisfied in the domain $\overline{U \setminus V}$. According to the theorem on the dependence of solutions of the system on parameters, there exists a neighborhood X_1^* of the point ν_0 for which the solutions depend uniformly and continuously on ν in a closed domain $(U \setminus V) \times V_1^*$.

We denote by W_1 the union of balls of radius ε with centers at points $x \times A_{\nu_0}(h_0)$. The value of ε is chosen so that $W_1 \cap \Omega_{\nu_0}(h_0) \subset W$. Then there exists a neighborhood V_2^* of the point ν_0 such that $G_\nu(y) \cap W_1 \neq \emptyset$ for all $y \in \partial V$ and $\nu \in V_2^*$. The neighborhood V_2^* can be chosen independently of y. Then $V^* = \mu_0(V_1) \cap V_2^*$ satisfies the required condition, and the lemma is proved.

LEMMA 2.5. *Let V^* be the neighborhood constructed in Lemma 2.4, $\nu \in V^*$, and Φ an open neighborhood of the set $G_\nu(A_\nu) \cap \Omega_{\nu_0}(h_0)$ on $\Omega_{\nu_0}(h_0)$. Assume that all points of the set A_ν are nondegenerate and have equal Morse indices. Then $G_\nu(\Phi)$ contains the point x_0.*

PROOF. We assume the converse is true. Let us suppose that $x_0 \notin G_\nu(\Phi)$. We consider the neighborhood Φ' of the set Φ on $\Omega_{\nu_0}(h_0)$ such that $\overline{\Phi} \subset \Phi'$, $\Phi' \setminus \Phi = \partial\Phi \times (0,1)$ and $x_0 \notin G_\nu(\Phi')$.

At each point x of the set $G_\nu(\Phi' \setminus \Phi)$, the vectors

$$\tau_0(x) = (\partial f^{\nu_0}/\partial x^1, \dots, \partial f^{\nu_0}/\partial x^m)$$

and

$$\tau(x) = (\partial f^{\nu}/\partial x^1, \dots, \partial f^{\nu}/\partial x^m),$$

which are tangent to the corresponding gradient lines, are not collinear. Indeed, let us suppose that $x_1 \in \mu_0^{-1}(\nu)$. Then

$$f^\nu(x) = \langle \eta(x)\nu(x)\rangle = \frac{1}{g(x)_1}\left(-x^1\frac{\partial f^{\nu_0}}{\partial x^1}(x_1) - \dots - x^m\frac{\partial f^{\nu_0}}{\partial x^m}(x_1) + f^{\nu_0}(x)\right).$$

Therefore,

$$\tau(x) = \frac{1}{g(x_1)}\left(-\frac{\partial f^{\nu_0}}{\partial x^1}(x_1) + \frac{\partial f^{\nu_0}}{\partial x^1}(x) - \dots - \frac{\partial f^{\nu_0}}{\partial x^m}(x_1) + \frac{\partial f^{\nu_0}}{\partial x^m}(x)\right).$$

It follows from the last equality that $\tau(x) = \frac{1}{g(x_1)}(\tau_0(x) - \tau_0(x_1))$. If $\tau(x) = \lambda\tau_0(x)$ at some point x, then $\tau_0(x_1) = (1 - \lambda g(x_1))\tau_0(x)$, which can occur only for $x \in \mu_0^{-1}(\nu)$, i.e., outside the domain $G_\nu(\Phi' \setminus \Phi)$.

Let us define the vector field τ_1 in Ω_{ν_0}, which coincides with τ_0 in $\Omega_{\nu_0} \setminus G_\nu(\Phi')$, with τ in $G_\nu(\Phi)$, and with $t\tau_0 + (1-t)\tau$ at points $(x,t) \in G_\nu(\partial\Phi \times [0,1])$. Obviously, τ_1 is a smooth field having zeros at the points of the set $A_\nu \cup \{x_0\}$.

The sum of the indices of zeros of the field τ_1 is given by

$$\chi(\Omega_{\nu_0}) - \chi(\Omega_{\nu_0}(-h_0)),$$

where χ is the Euler–Poincaré characteristic (see [**8**]).

On the other hand, the index of the point x_0 can be calculated by this formula. Then the sum of the indices of nondegenerate points is equal to zero, which contradicts the hypothesis. The lemma is proved.

LEMMA 2.6. *Let the conditions of Lemma 2.5 be satisfied. Then for any point $x_1 \in A_{\nu_0}(h_0)$ and any sequence $\{\nu_n\} \to \nu_0$, $\nu_n \in \mu_0(U)$, there exists a sequence of points $y_n \in G_{\nu_n}(A_{\nu_n}) \cap \Omega_{\nu_0}(h_0)$ converging to x_1.*

PROOF. We assume that the statement of the lemma does not hold, and none of the sequences y_n from the set $\bigcup_n (G_{\nu_n}(A_{\nu_n}) \cap \Omega_{\nu_0}(h_0))$ converges to x_1.

Let W be an arbitrary neighborhood of the set $A_{\nu_0}(h_0)$ on $\Omega_{\nu_0}(h_0)$, and V^* a neighborhood constructed according to Lemma 2.4. We consider the neighborhood $W_1 \subset W$ of x_1 that does not contain points from $\bigcup_n (G_{\nu_n}(A_{\nu_n}) \cap \Omega_{\nu_0}(h_0))$.

Let us choose an open neighborhood Φ_n of the set $G_{\nu_n}(A_{\nu_n}) \cap \Omega_{\nu_0}(h_0)$ so that $W_1 \cap \Phi_n \neq \emptyset$ and $\Phi_n \subset W$. In this case, $G_{\nu_n}(W/\overline{W}_1) \supset G_{\nu_n}(\Phi_n)$ and $G_{\nu_n}(\Phi_n) \cap G_{\nu_n}(W_1) = \emptyset$. According to Lemma 2.5, $G_{\nu_n}(\Phi_n) \ni x_0$. Consequently, we can choose the neighborhood V, $\mu_0(V) \subset V^*$, of the point x_0 so that $V \subset G_{\nu_n}(W \setminus \overline{W}_1)$. Then $G_{\nu_n}(V) \subset G_{\nu_n}(W \setminus W_1)$ for every ν_n.

On the other hand, there exists ν_k such that $G_{\nu_k}(W_1) \cap V \neq \emptyset$. This contradiction proves the lemma.

3. Proof of Theorem 1.4

Let U be a neighborhood of the point x_0 for the direction l_0 in which $K(x, \nu) \neq 0$; $\Omega_{\nu_0} \subset U$ a regular canonical neighborhood of the point x_0, ν the field (2), and V^* the neighborhood constructed according to Lemma 2.4.

LEMMA 3.1. *Let $x_1, \dots, x_p$ be critical points of the function f^ν, $\nu \in V^* \setminus \{\nu_0\}$. Then the points $x_1, \dots, x_p$ are nondegenerate, and the Morse indices, i.e., the indices of the Hessian $\|\partial^2 f^\nu / \partial x^i \partial x^j\|$, are equal at these points.*

PROOF. At the points $x_1, \dots, x_p$, we have

$$\nu(x_i) = \frac{1}{g(x_i)} \left(-\frac{\partial f^{\nu_0}}{\partial x^1}(x_i), \dots, -\frac{\partial f^{\nu_0}}{\partial x^m}(x_i), \lambda^{m+1}, \dots, \lambda^n \right),$$

where $\nu_0 = \sum_{m+1}^n \lambda^i e_i$, $e_i \in N_{x_0}$ and $\nu(x_1) = \dots = \nu(x_p)$.

Then

$$f^\nu(x) = \frac{1}{g(x_1)} \left(-\sum_{k=1}^m x^k \frac{\partial f^{\nu_0}}{\partial x^k}(x_1) + f^{\nu_0}(x) \right).$$

Consequently

$$\frac{\partial^2 f^\nu}{\partial x^i \partial x^j} = \frac{1}{g(x_1)} \frac{\partial^2 f^{\nu_0}}{\partial x^i \partial x^j}.$$

We connect the points x_1 and x_2 by a smooth curve that does not pass through x_0 and assume that $x = x(s)$ is its smooth parametrization. Then the coefficients of the matrix

$$\left\| \frac{\partial^2 f^{\nu(s)}}{\partial x^i \partial x^j}(x(s)) \right\|$$

at points $x(s)$ depend smoothly on s; hence the principal minors of the matrix also depend smoothly on s. Since the principal minors do not vanish, the Jacobi formula for the reduction of a quadratic form to the canonical form implies that the principal curvatures $k_i(x(s), \nu(s))$ depend continuously on s. If some curvature k_i changes sign, there exists a point s_0 at which $k_i = 0$. The latter assumption contradicts the nondegeneracy of the quadratic form. The lemma is proved.

LEMMA 3.2. *Let the set $A_{\nu_0}(-h_0)$ (see 2.3) consist of isolated points $x_1(-h_0)$, $\dots, x_p(-h_0)$, and let $D_i(-h_0)$ be neighborhoods of them diffeomorphic to a ball of the neighborhood. Then the sets $S_i(h_0) = G_{\nu_0}(\partial D_i(-h_0)) \cap \Omega_{\nu_0}(h_0)$, which are manifolds diffeomorphic to a sphere of dimension $m-2$, bound on $\Omega_{\nu_0}(h_0)$ a connected manifold B.*

PROOF. We assume that B is not connected and contains more than two components. According to the construction, each component contains points from $A_{\nu_0}(h_0)$. We choose one of the components and glue to it the relevant balls $D_i(-h_0)$ by the cylinders $G_{\nu_0}(\partial D_i)$ formed by gradient lines. We obtain a closed manifold bounding a neighborhood Q of x_0. Then the gradient lines passing through the points of $A_{\nu_0}(h_0)$ on other components of B must intersect ∂Q. We arrive at a contradiction which proves the lemma.

LEMMA 3.3. *Let the function f^ν have a nondegenerate critical point x_1. Construct for the point x_1 a regular canonical neighborhood $\Omega_\nu(h_1, h_2)$ and assume that $\Omega_\nu(h_1)$ consists of two components, while $\Omega_\nu(h_2)$ consists of only one component. Then the Morse index of the point x_1 is equal to unity, and the set $A_\nu(h_1)$ consists of two points.*

PROOF. During the transition through a nondegenerate critical point of index λ from the level $\Omega_\nu(h_1)$ to the level $\Omega_\nu(h_2)$, a handle $H^m_\lambda = D^\lambda \times D^{m-\lambda}$, where D^q is a ball of dimension q, is glued to a neighborhood of a sphere $s^{\lambda-1}$ of dimension $\lambda - 1$ at the level $\Omega_\nu(h_1)$ (see [**7**]).

According to the conditions of the lemma, the two components of the level $\Omega_\nu(h_1)$ must be glued by the handle H^m, since it is only S^0 which is an unconnected sphere. The lemma is proved.

For the set $B(-h_0)$ at the level $\Omega_{\nu_0}(-h_0)$, we denote by $B(h_0)$ the intersection of the cylinder $G_{\nu_0}(B(-h_0))$ with $\Omega_{\nu_0}(h_0)$.

LEMMA 3.4. *Let Q_1 and Q_2 be two components of the set $\Omega_{\nu_0}(-h_0)$. There exist components $B_1(-h_0) \subset \partial Q_1$ and $B_2(-h_0) \subset \partial Q_2$ such that the corresponding sets $B_1(h_0)$ and $B_2(h_0)$ lie on the same component of $\Omega_{\nu_0}(h_0)$.*

PROOF. We suppose that the converse is true. Let the sets $B_1(-h_0)$, $B_3(-h_0)$, $\dots$,, $B_k(-h_0)$ form the boundary of Q_1.

The manifolds $B_1(h_0), B_3(h_0), \dots, B_k(h_0)$ are contained in the boundary of certain manifolds $Q'_1, \dots Q'_t$ on $\Omega_{\nu_0}(h_0)$. By the hypothesis, the manifolds $Q'_1, \dots, Q'_t$ do not contain other components of the boundary. We glue them by the cylinders from the gradient lines $G_{\nu_0}(B_i(-h_0))$ to the manifold Q_1.

This gives a manifold Φ which bounds a neighborhood of x_0. Consequently, $\Omega_{\nu_0}(-h_0)$ does not contain components differing from Q_1. This contradicts the conditions of the lemma.

The lemma is proved.

3.5. Proof of Theorem 1.4. According to Lemma 2.1, all critical points of the function f^ν for $\nu \in \mu_0(V_1)$ lie on the closure of the neighborhood $\overline{V}_1$. All these points are nondegenerate and have the same Morse index (see Lemma 3.1). Since $\overline{V}_1$ is compact, the number of such points is finite.

We choose $h_1 < h_0$ so that $\overline{V} \subset \Omega_{\nu_0}(-h_1, h_1)$, where h_0 is chosen in the construction of a regular canonical neighborhood. Let the number $\varepsilon > 0$ satisfy the following conditions: $\overline{V} \subset \Omega_{\nu_0}(-h_1 + \varepsilon, h_1 - \varepsilon)$ and $h_1 + \varepsilon < h_0$. Then for the vector $\nu \in \mu_0(V)$ satisfying the condition

$$\rho^*(\nu_0, \nu) < \arcsin(\varepsilon/d),$$

the level $\Omega_\nu(h')$ is diffeomorphic to $\Omega_0(h_1)$, where

$$h' = h_1 \cos \rho^*(\nu_0, \nu),$$

$d = \max_y |y_0 y|$, y_0 is a point on the axis $l_0 \| \nu_0$ at a distance h_1 from the origin x_0, and y is a point on the boundary of Ω_{ν_0} in the layer $\Omega_{\nu_0}(h_1 - \varepsilon, h_1 + \varepsilon)$. The diffeomorphism is established from the gradient lines of the function f^{ν_0} in the layer $\Omega_{\nu_0}(h_1 - \varepsilon, h_1 + \varepsilon)$.

According to the conditions of the theorem, one of the sets $\Omega_{\nu_0}(-h_0)$ or $\Omega_{\nu_0}(h_0)$ contains more than one component. Then one of the sets $\Omega_\nu(-h')$ or $\Omega_\nu(h')$ respectively also contains more than one component. For the sake of definiteness, we shall assume that $\Omega_{\nu_0}(-h_0)$ contains more than one component.

According to Lemmas 3.1, 3.3, and 3.4, all critical points of the function f^ν are nondegenerate and have Morse index equal to unity, while the set $A = G_\nu(A_\nu) \cap \Omega_{\nu_0}(-h')$ consists of a finite number of points.

According to Lemma 2.6, points of the set $G_\nu(A_\nu) \cap \Omega_{\nu_0}(-h_1)$ converge to the set $A_{\nu_0}(-h_1)$ so that a certain sequence of points from $G_\nu(A_\nu) \cap \Omega_{\nu_0}(-h_1)$ converges at each point of the set. Consequently, $A_{\nu_0}(-h_1)$ consists of a finite number of points.

Let us assume that $A_1, \dots, A_p$ are points of the set $A_{\nu_0}(-h_1)$ and $D_1, \dots, D_p$ are their spherical neighborhoods on $\Omega_{\nu_0}(-h_1)$. We construct a regular canonical neighborhood V_{ν_0} of x_0 proceeding from the level $D = D_1 \cup \dots \cup D_p$. According to Lemma 3.2, the sets $G_{\nu_0}(\partial D_i)$ bound at some level $\Omega_{\nu_0}(h_1)$ a compact manifold with a boundary which consists of only one component. It should be noted that each component of the closure of $V_{\nu_0}(0) \setminus \{x_0\}$ is homeomorphic to the ball D_i. The homeomorphism is established by the gradient lines. The theorem is proved.

References

1. N. V. Efimov, *Qualitative problems of deformation of surfaces "in the small"*, Trudy Mat. Inst. Steklov. **30** (1949). (Russian)
2. M. M. Postnikov, *Introduction to Morse theory*, "Nauka", Moscow, 1971. (Russian)
3. J. Milnor, *Lectures on the h-cobordism theorem*, Princeton Univ. Press, Princeton, NJ, 1965.
4. Yu. D. Burago and V. A. Zalgaller, *Geometrical inequalities*, "Nauka", Leningrad, 1980; English transl., Springer-Verlag, Berlin, 1988.
5. T. F. Banchoff, *High codimesional 0-tight maps on spheres*, Proc. Amer. Math. Soc. **29** (1971), 133–137.
6. B. A. Dubrovin, S. P. Novikov, and A. T. Fomenko, *Modern geometry. Methods and applications*, "Nauka", Moscow, 1979; English transl., Parts I, II, Springer-Verlag, Berlin, 1984, 1985.
7. ———, *Modern geometry. Methods of homology theory*, "Nauka", Moscow, 1984; English transl., Springer-Verlag, Berlin, 1990.
8. V. A. Kuznetsov, *The Kronecker–Morse formula for nonclosed complete m-dimensional surfaces in a Euclidean space*, Investigations in the Theory of Riemannian Manifolds and Their Imbeddings (A. I. Verner, editor), Leningrad. Gos. Ped. Inst., Leningrad, 1985, pp. 32–41. (Russian)

46 Pionerskaja, Apt. 68, Khabarovsk 680023, Russia

Amer. Math. Soc. Transl.
(2) Vol. **176**, 1996

Space-like Convex Surfaces in Pseudo-Euclidean Spaces

A. D. Milka

The results of the theory of convex surfaces obtained by Aleksandrov and Pogorelov [**1,2,6,21**] have been generalized to pseudo-Euclidean space in a number of papers [**10,16,24**]. The present paper is devoted to similar analysis of polyhedral space-like caps. We prove theorems on the realizability of the convex metric, on the unique definiteness, and on the continuous flexibility and stability of space-like caps, which are generalizations of classical theorems. The realizability and stability are investigated similarly to the results obtained by Volkov [**8,9**] for Euclidean surfaces. These results are present in a new and more comprehensive form. This approach to realizability is adapted for space-like caps, while the implementation of Aleksandrov's classical method of realizability involves considerable difficulties. For comparison, we recall that another method of realizability of convex polyhedral metrics based on their regular approximation has been proposed by Lusternik (see [**11**]) and that the realizability of arbitrary polyhedral metrics through not necessarily convex polyhedra has been proved by Burago and Zalgaller [**7**]. Here the stability of a cap is considered with respect to the variation of a certain integral characteristic introduced in [**9**]. Its continuous flexibility is studied in the class of space-like convex polyhedra isometric to the cap. In this respect, the method used here is similar to that proposed by the author in [**17**]. The uniqueness theorems are considered separately for isometric polyhedra and for combinatorially equivalent polyhedra formed by equal faces. The former are equivalent to the uniqueness theorems for general isometric convex surfaces. Their proofs are based on the generalizations of the principles of maxima for isometric surfaces with a Pogorelov's border. For combinatorially equivalent polyhedra, other uniqueness theorems are actually realized by generalizing the classical results of Cauchy [**27**] and Minkowski [**28**]. It is quite natural that their proofs are essentially based on lemmas on the deformation of closed convex polygons (on the Cauchy and Aleksandrov lemmas). Here we give new proofs for these lemmas on spheres in Euclidean and pseudo-Euclidean spaces. On the whole, the results obtained for such a natural class of space-like polyhedra as convex caps are analogs of basic classical theorems. They emphasize the real possibility of constructing a general synthetic theory of convex

1991 *Mathematics Subject Classification*. Primary 53A35.

surfaces in a pseudo-Euclidean space which turns out to be as meaningful as the classical theory. The necessity of such a construction is especially appealing since the Euclidean and pseudo-Euclidean spaces are geometrically at least as similar as classical spaces of constant curvature. Therefore, a comparison of the corresponding results is useful both for the subsequent development of these results and for getting a deeper insight into their meaning. The basic results are obviously dual to each other, their specific distinctions are more diverse and expressive, and the space-like convex surfaces in question offer a convenient realization of dual metrics (i.e., with positive and negative intrinsic curvature). As in Euclidean space, this construction in pseudo-Euclidean space can be based on the method developed by Aleksandrov, which consists in approximating a general convex surface by convex polyhedra. Further, we consider surfaces in coordinate spaces in which the distances are determined by the Euclidean form $x^2 + y^2 + z^2$ and the pseudo-Euclidean form $x^2 + y^2 - z^2$. It is assumed in particular, that the surfaces under investigation are situated in the half-spaces $z \geq 0$, the border of a specific cap being situated in the xy-plane.

1. Existence

In the Euclidean theory of convex polyhedra, the problems associated with the realizability of metrics with nonnegative curvature are traditionally considered much later than the corresponding uniqueness theorems. If the analysis is confined to polyhedral caps, then Volkov's method allows us to reverse this order. This approach is adopted in the present work for the following reasons. First, it is not necessary here to follow the method of continuous extension in a parameter (see [**1,2**]). Therefore, the proof of the existence theorems is only slightly more complicated than that of the uniqueness theorems. Second, certain ideas used in the proof of the realizability are applied in a broader context, in the proofs of the existence theorems as well. To elucidate the starting points of these proofs, an earlier presentation of the theorem on realizability is preferable. Third, the issues of continuous flexibility of convex polyhedra are also associated with the problem of uniqueness. It is, therefore, natural not to separate the study of these issues. This requires the theorem on realizability to be prepared in advance.

1.1. Prismatic polyhedra. A simply connected convex polyhedron whose border belongs to a plane and whose projection onto this plane does not leave the boundary is called a *cap*. Since we consider polyhedra in pseudo-Euclidean space, below we assume that all faces of these polyhedra are space-like. It has already been noted that the proof of the theorem on realizability for a space-like cap is based on the technique developed by Volkov in [**8**]. Without violating the basic idea of this technique, we essentially change the elements of the proof at the principal stages because specific features of the pseudo-Euclidean space metric call for such a modification. On the other hand, new schemes introduced here are fully applicable to the Euclidean case as well (of course, with appropriate corrections in the statements and proofs associated with one of the most important concepts, namely, the curvature). This allows us to make comparisons in the course of the presentation and note the changes that are introduced in the Euclidean case by the proposed approach. The definitions of basic concepts are borrowed entirely from [**8**], the only difference being that the terms "prismatic" and "normal" are used instead of "cylindrical" and "special" for polyhedra belonging to the class in question. In addition to this, the so-called double and multiple prisms and prismatic pyramids are introduced for convenience. The corresponding definitions will be given in this section.

A space-like cap, as well as its development, i.e., the set of plane Euclidean polygons from which the cap is glued, will both be denoted by the symbol S. Let S be a convex cap and $\overline{S}$ its projection onto the plane containing the boundary Γ of the cap. Further, referring to projections of objects, we shall mean the projections onto the plane containing the boundary. It is this plane that will be taken as the coordinate plane xy. It is obvious that Γ is a convex polygonal line. It can easily be proved that the rotation of the polygonal line Γ on the cap itself is positive at every vertex. A solid convex polyhedron M bounded by the surfaces S and $\overline{S}$ can be considered as glued from the polyhedra M_i (they will be called prisms) that are bounded by the faces T_i of the cap from above, by the projections $\overline{T}_i$ of these faces from below and by time-like cylindrical polyhedral surfaces connecting the boundaries T_i and $\overline{T}_i$ from the lateral sides. It is quite natural that the developed surface S is referred to as the upper base of the polyhedron M and the set of polygons $\overline{T}_i$ with the gluing rule induced in it by the developed surface S as its lower base. This lower base is considered as the developed surface $\overline{S}$. It can easily be proved that the developed surface S considered as a metric has a nonpositive curvature and in this sense the curvature of $\overline{S}$ is equal to zero. The set $\{M_i\}$ of polyhedra will be naturally called a solid polyhedron realizing the developed surface S. In the correspondence $S \to M$, the set $\{h\}$ of the altitudes of the vertices of the developed surface over the plane xy corresponds to the developed surface S. It should be noted that in the set $\{M_i\}$, dihedral angles of the polyhedron M at the inner edges of the developed surface S do not exceed the straight solid dihedral angle, since we take into account the convexity of a cap and notice that for each inner vertical edge of M, the sum of dihedral angles which belong to the prisms M_i of the polyhedron and which meet at this edge is equal to 2π.

Now we consider a certain developed surface S, i.e., a finite set of Euclidean convex polygons with a prescribed gluing rule which converts it into a two-dimensional metric. Suppose that a set of prisms $\{M_i\}$, where a space-like polygon T_i from S situated in the half-space $z \geq 0$ serves as the upper base of each specific prism, is put in correspondence with the developed surface S. In the set $\{M_i\}$, the gluing rule for the vertical faces of the prisms M_i for which the gluing vertical edges are assumed to have equal length, is naturally induced by the developed surface S. This set together with the gluing rule is called a prismatic polyhedron. A prismatic polyhedron M is called convex if at each inner edge of its upper base S, the union of dihedral solid angles of two prisms meeting at this edge forms a solid dihedral angle which does not exceed the straight solid angle. A prismatic polyhedron M is said to have nonpositive curvature if, for each vertical edge at an inner vertex of S, the sum of the dihedral angles of the prisms M_i belonging to the set of prisms constituting M and meeting at this edge is not smaller than 2π. It is obvious that this sum coincides with the sum of plane angles of the polygons $\overline{T}_i$ of the developed surface $\overline{S}$ of the lower base of M, which meet at the projection of the corresponding vertex of S. Later we consider such convex prismatic polyhedra M with nonpositive curvature for which all altitudes of the boundary vertices of the upper base of S

are equal to zero. Such polyhedra are called normal if they satisfy the following additional requirements.

(i) The developed surface S is homeomorphic to a circle, and each of its boundary vertices rotates in the positive direction.

(ii) At each inner vertex, S has a negative curvature (in this case, all inner vertices of S are said to be real vertices).

(iii) All faces of the developed surface S are Euclidean triangles.

A prismatic polyhedron formed by gluing two prisms whose upper bases are triangles is called a double prism. If CAB and DAB are the triangles of the upper base, then the double prism would be denoted by $CADB$, which should not cause confusion. More than two prisms with triangular upper bases can also be glued consecutively at the vertex A. Such a prismatic polyhedron will be called a multiple prism. Finally, a prismatic polyhedron whose upper base is composed of triangles glued together consecutively and cyclically at the single inner vertex of the base in the same way as the faces of an ordinary pyramid are glued with each other is called a prismatic pyramid. Let us draw a section through each prism constituting the Euclidean pyramid with the help of a sphere of unit radius with its center at the common vertex of the triangles of the upper base. While carrying out this procedure, we can extend certain faces of the prisms if necessary. The union of all spherical triangles formed is the developed surface with the gluing rule induced by the pyramid. This developed surface will be called the spherical section of the pyramid at its vertex.

1.2. Development with the help of a cap. Let S be a developed surface homeomorphic to a circle, having nonpositive curvature and real inner vertices, and rotating along the boundary in the positive direction. The realization of the developed surface by a cap is reduced to the proof of the existence of a space-like cap isometric to this developed surface. We can always assume that the developed surface S is formed by Euclidean triangles and that all inner vertices of S are real. In general, such a representation of S is not unique, and a transition from one triangulation to another in S is called retriangulation. When applied to the pseudo-Euclidean case, Volkov's method consists in choosing the polyhedron with the largest sum of altitudes of the vertices of S among all normal polyhedra whose upper base is isometric to S. It turns out that such an extremal polyhedron exists in the class of normal polyhedra, and it is just this polyhedron that realizes the developed surface S, i.e., its upper base serving as the required space-like cap. It should be emphasized that an extension of Volkov's result to pseudo-Euclidean space is far from trivial. In this section, we formulate and prove the basic theorem on the realization of the cap, reproducing completely only the line of reasoning in [**8**]. An account of certain auxiliary results which are used in the proof and which differ significantly from Volkov's approach is given in the subsequent sections. In 1.3, we prove the existence of the external polyhedron and formulate the lemma which simplifies investigations in the Euclidean case. This lemma allows us to avoid complex constructions of [**8**] associated with the degenerate and nondegenerate parts of the limited polyhedron. In 1.4, we consider the problems arising in connection with the retriangulation of the developed surface S as we go over to the limited polyhedron. In particular, we give the complete proof (missing in [**8**]) of the existence of such a retriangulation. Finally, in 1.5, we study the behavior of the curvature under the finite deformation of a prismatic pyramid. This approach is more constructive and consistent than the corresponding approach used in [**8**], which is based on the principles of differential geometry.

THEOREM. *Every developed surface S of negative curvature homeomorphic to a disk and having a positive rotating boundary is realized by a space-like cap.*

PROOF. We consider all possible developed surfaces $\{S_k\}$ isometric to S which are formed by plane triangles only, and whose vertices are real (i.e., with negative curvature) inner vertices of S and its boundary vertices. It is easy to prove that the number of such developed surfaces is finite. Further, let E_k be the class of normal polyhedra such that the developed surface S serves as the upper base for each of them. Let us denote by M_k the polyhedron of this class with the maximum sum $\overline{h}_k = \sum h$ of the altitudes of the vertices. The class E_k is nonempty, since it contains, for one, the normal polyhedron whose altitudes h are all equal to zero. The fact that the quantity $\overline{h}_k$ is bounded, as well as the fact that the polyhedron M_k exists in this class, need special substantiation. This proof is given in 1.3. It should be emphasized here that the proof for pseudo-Euclidean space at this stage differs significantly from that in Euclidean space. Further, there exists a polyhedron M among the polyhedra $\{M_k\}$ such that the sum $\overline{h}$ of its altitudes has the maximum value. The polyhedron M is a normal polyhedron with zero curvature at the vertical inner edges. This statement follows from the lemma stated below. It now becomes obvious that the upper base of M is just the space-like cap with the developed surface S (for this purpose it is sufficient to carry out a proper gluing of vertical edges of the prisms constituting M).

The theorem is proved.

LEMMA. *If M is a normal polyhedron of negative curvature with upper base S, then there exists a normal polyhedron $\widetilde{M}$ such that its upper base is isometric to S and the sum of the altitudes at the inner vertices of its upper base is larger than the corresponding value for M.*

PROOF. Let us isolate those inner vertices of the upper base of the polyhedron M for which the curvature at the corresponding vertical edge is negative. The curvatures at the remaining vertical inner edges (if such edges exist) equal zero. Then, using the former upper base S as a metric, we shall construct a new normal polyhedron M with the lengths of vertical edges at the isolated vertices increased by the same small quantity ε and with the former lengths of the vertical edges at the remaining inner vertices. For an appropriate choice of ε, the curvatures at the extended edges of the new polyhedron remain negative. The polyhedron $\widetilde{M}$ turns out to be normal if the following conditions are satisfied. First, the dihedral angles at the non-boundary edges of the upper base of the polyhedron must not exceed the straight solid angle. For brevity, we say that they must not exceed π. This ensures the convexity of the new polyhedron. The violation of this condition can take place only at the edges at which the dihedral angles of M are equal to π. In this case, lifting of the isolated vertices must be preceded by a special retriangulation of the upper base. This retriangulation is naturally accompanied by an appropriate cutting-gluing operation for the composing prisms. The problem associated with the existence of such retriangulation is elucidated in 1.4. Second, the curvatures at the other inner vertical edges whose lengths remain unchanged must be nonpositive. The curvature of the polyhedron at these edges is equal to zero. The fact that the deformation introduced by us really possesses this property is proved in 1.5. This follows from the conservation of the convexity of the polyhedron under deformation.

The lemma is proved.

1.3. Elimination of singular faces. First let us consider the case of a Euclidean polyhedron. There can exist, in principle, vertical faces belonging to the upper base of the limit polyhedron M. Although it is intuitively clear that such faces must adjoin only the boundary of the upper base, the study of this fact and other facts associated with it carried out in [**8**] is rather complicated. On the other hand, peculiar features of the arrangement of the vertical faces become quite clear if we establish the extrinsic geometric shape of the lines in the developed surface S for the normal polyhedron. In this developed surface, the lines in question are quasigeodesic polygonal lines. It turns out that the convexity property similar to the convexity property of the geodesic on the convex surface formed by projecting a line by a cylinder is valid for any polyhedron belonging to the class E_k for such quasigeodesics. This property of quasigeodesics for normal polyhedra will be called Liberman's principle in order to distinguish it from the property presented by the well-known lemma for geodesics on convex surfaces [**14**]. Quasigeodesics on convex surfaces have been studied in this respect by Pogorelov [**22**].

LEMMA 1. *Let V be a convex prismatic pyramid with vertex A and with nonnegative curvature of the upper base, and let $\tilde{p}\tilde{q}$ be a two-linked polygonal line on this base with the inner vertex A such that it is a quasigeodesic line in the metric of the base. Denote the angles between the links $\tilde{p}$ and $\tilde{q}$ of this polygonal line and the inner vertical edge of the pyramid at the vertex A by p and q respectively. Then $p+q \leq \pi$.*

PROOF. Suppose the converse is true, i.e., $p+q>\pi$. We consider the section of the pyramid V by a sphere of unit radius with its center at the vertex A. This spherical section is a certain developed surface G composed of the spherical triangles meeting cyclically at the common vertex O situated at the point of intersection of the sphere and the inner vertical edge of the pyramid. The boundary of the spherical developed surface rotates nonnegatively at each of its vertices since the pyramid V is convex. On the boundary of G, the lines $\tilde{p}$ and $\tilde{q}$ mark the points P and Q respectively, thus dividing this boundary into two polygonal lines a_1 and a_2. On the developed surface G, the segments OP and OQ belong to certain triangles and have the lengths p and q respectively. Let us assume that on the developed surface G all gluings are carried out between the sides of spherical triangles arranged cyclically around O and that a cut is drawn along the segment OP with banks p_1 and p_2.

Then we can define a new spherical developed surface $\widetilde{G}$ which has no vertices inside and which is bounded by the polygonal line $p_1a_1a_2p_2$. For simplicity, the lengths of the segments of this polygonal line will be denoted henceforth by the same symbols p_1, a_1, a_2, and p_2, and, accordingly, the segment OQ will be denoted by the symbol q. Let us assume that the developed polygon G lies on a unit sphere. The following notation is introduced for the angles on the developed surface $\widetilde{G}$: α_1 is the angle between p_1 and a_1, α_2 the angle between p_2 and a_2, β_1 the angle between q and a_2 and, finally, γ_1 and γ_2 are the angles between q and p_1 and q and p_2 respectively. Since the boundary of G is convex, $\alpha_1 + \alpha_2 \leq \pi$ and $\beta_1 + \beta_2 \leq \pi$. In accordance with our assumption, we get the following relations for the lengths of polygonal lines and their parts: $p_1 = p_2 = p$ and $p + q > \pi$. In addition to this, $a_1 \leq \pi$ and $a_2 \leq \pi$ since the polygonal line $\tilde{p}\tilde{q}$ is quasigeodesic in the developed surface, i.e., in the upper base of V, with nonnegative curvature. Considering the parts of G and the polygons a_1qp_1 and a_2qp_2 and comparing the lengths of the polygonal lines a_1 and qp_1 with a_2 and qp_2, we find that $\gamma_1 < \pi$ and $\gamma_2 < \pi$ since, otherwise, we would get $p_1 + q \leq \pi$ or $p_2 + q \leq \pi$. Then it follows from the convexity of these polygons and from the inequality $p + q > \pi$ that $a_1 < \pi$ and $a_2 < \pi$.

Let us now deform G, deforming the polygons a_1qp_1 and a_2qp_2 appropriately without violating their convexity in such a way that the angles γ_1 and γ_2 increase and the segments a_1 and a_2 are transformed into rectilinear segments. We denote the new developed surface again by G, preserving the previous notation for its elements, in particular, for the angles α and β. It follows from the well-known Cauchy lemma (see Lemma 2 in 2.3) on deformations of convex spherical polygons without violation of convexity that the inequalities $\alpha_1 + \alpha_1 \leq \pi$ and $\beta_1 + \beta_2 \leq \pi$ are valid for the new developed surface $\widetilde{G}$ as well as for G, since the corresponding angles α and β can only decrease under the deformation in question.

Now we prove the inequality $p + q \leq \pi$, which will lead to a contradiction with our assumption. If the inequalities $\alpha_1 + \alpha_2 < \pi$ and $\beta_1 + \beta_2 < \pi$ are simultaneously satisfied in the quadrangle $a_1p_1p_2a_2$, then we can continuously deform this quadrangle by increasing its diagonal q, because the inequality $p + q > \pi$ is not violated. Under such a deformation, a situation arises in which the strict equality is satisfied in at least one of the inequalities $\alpha_1 + \alpha_2 \leq \pi$ and $\beta_1 + \beta_2 \leq \pi$. We shall assume that, say, $\beta_1 + \beta_2 = \pi$. In the opposite case, we can go over to a new quadrangle drawing a cut in $\widetilde{G}$ along the segment q and gluing together p_1 and p_2.

The equality $\alpha_1 = \alpha_2$ is valid in the isosceles spherical triangle $p_1a_1a_2p_2$, and hence $\alpha_1 \leq \pi/2$ and $\alpha_2 \leq \pi/2$, since $\alpha_1 + \alpha_2 \leq \pi$. Suppose that $\gamma_1 + \gamma_2 \leq \pi$ at the vertex O. Let us reflect this triangle on a sphere with respect to the side a_1a_2. In the convex quadrangle formed in this way, the line q and its image cut out a smaller convex quadrangle, whose perimeter satisfies the inequality $2q + 2p \leq \pi$, i.e., $p + q < \pi$. This contradicts the initial assumption. Thus, we must assume that $\gamma_1 + \gamma_2 > \pi$. In this case, we shall extend the lines p_1 and p_2 beyond the point O inside the triangle up to their intersection with the side a_1a_2, denoting the extended segments by b_1 and b_2 respectively. It is obvious that $b_1 = b_2$ for the lengths of these segments. It is important that $p_1 + b_1 = p_2 + b_2 = \pi$. In the isosceles convex triangle formed by the segments b_1 and b_2 and a part of the side a_1a_2, the angles on this side are equal to one another and satisfy the inequality $\alpha_1 = \alpha_2 \leq \pi/2$. Let us reflect this triangle on a sphere with respect to the side a_1a_2. Note that the line q belongs to the triangle, since $\gamma_1 < \pi$ and $\gamma_2 < \pi$. In the resulting convex quadrangle the line q and its reflection cut out a smaller convex-concave quadrangle. It follows from a comparison of the lengths of the parts of this quadrangle that $2q \leq 2b_1 = 2b_2$, i.e., $q \leq b_1$, and hence $q + p_1 \leq b_1 + p_1 = \pi$. Thus, in either case, we arrive at a contradiction with the assumption that $p + q > \pi$. Hence, this assumption is incorrect.

The lemma is proved.

It was mentioned above that Liberman's principle expressed by this lemma allows us to simplify considerably the part of the study of normal polyhedra from class E_k which deals with the structure of the corresponding limit polyhedron. This lemma is also interesting in itself in the study of such generalized objects as Euclidean normal polyhedra. We shall now go over to the problem of the existence of the extremal polyhedron M_k among space-like polyhedra belonging to the class E_k.

LEMMA 2. *There exists an extremal polyhedron M_k in the class E_k of space-like normal polyhedra.*

PROOF. Recall that we must prove the existence of the polyhedron with the maximum value of the sum $\overline{h}_k = \sum h$ of the altitudes for the inner vertices of the upper base S_k.

Let $\{M^j\}$ be a sequence of polyhedra from E_k for which the values of the sums $\{h^j\}$ of the altitudes coverage to a maximum value $\overline{h}_k$ which, in principle, may be equal to infinity. Suppose that the inclinations of all upper faces of all polyhedra M^j to the coordinate plane xy are uniformly bounded from above. In other words, we shall assume that all external normals to these faces of the polyhedra are directed to one and the same circular cone in the half-space $z \geq 0$ which is situated strictly inside the light cone. In this case, let M_k be the limit polyhedron for the converging subsequence of polyhedra from $\{M^j\}$. The polyhedron M_k is obviously the normal polyhedron, and the sum $\overline{h}_k$ of its altitudes has the maximum value.

Thus, if the extremal polyhedron does not exist in the class E_k, we must assume that the inclinations of at least certain faces of the upper base of the polyhedra from the set $\{M^j\}$ to the plane xy are unbounded, i.e., the planes of certain faces tend to the planes tangent to the light cone. This, in turn, is equivalent to the statement that the altitudes of certain vertices (having the same notation on the developed surface S_k) of the upper base of the sequence $\{M^j\}$ of the polyhedra increase infinitely. Let us show that in this case altitudes of all internal vertices of the upper base S_k of polyhedra $\{M^j\}$ do the same.

Suppose that the developed surface S_k contains inner vertices whose altitudes tend to infinity in the sequence $\{M^j\}$, and vertices whose altitudes are uniformly bounded. Then we can find neighboring vertices A and B in S_k which are distinguished from one another by the behavior of their altitudes as indicated above and which are connected by the edge AB of the developed surface. For definiteness, let the altitudes of A be bounded while those of B increase. Since the length of AB is finite and remains constant in the sequence $\{M^j\}$ and the difference of the altitudes tends to infinity, the ray AB tends to the light-like direction. Consider the specific polyhedron M^j. Let us choose a certain direction for going about A and then glue consecutively the prisms that constitute this polyhedron and that are adjacent to the vertex A, along the corresponding vertical faces. For the initial prism, we choose the second prism encountered in this direction and containing the edge AB. Since the curvature of the polyhedron at the vertical edge passing through A is nonpositive, the glued prisms cover an angle not smaller than 2π on the plane xy (this angle is measured from the projection of the ray AB). Then the vertical plane through ray AB intersects the upper base of the glued prisms along a certain segment CA whose inclination to the plane xy is not smaller than the slope of AB. This follows from the convexity of M^j. Hence, the ray A also tends to the light-like direction. This contradicts the uniform boundedness of the altitudes of the points C and A over the plane xy. Thus, we must assume that the altitudes of all inner vertices of the upper base S_k of the polyhedron M^j increase infinitely.

Let T be a boundary vertex of the developed surface S_k lying in the plane xy. Let TR and TS be the links of boundary of S_k which lie in the same plane. We consider the multiple prism formed by the prisms of the polyhedron M^j adjacent to T. The inner edges of the upper base of the prism which diverge from T are directed toward the inner vertices of the developed surface S_k. It follows that these inner edges tend to the same light-like direction e as the number j increases. In any case, this can be attained by choosing an appropriate subsequence from the set of the corresponding multiple prisms. It follows from the convexity of the space-like polyhedron M^j that the projection $\bar{e}$ of the direction e onto the plane xy (this projection is measured from the vertex T) is situated strictly inside the angle RTS. Further, since the faces of M^j adjacent to the segments TR and TS are space-like, contain segments tending to the light-like direction e, and are projected uniquely inside the angle RTS, the angles between $\bar{e}$ and the rays RT and TS tend to $\pi/2$, while the angle between these rays tends to π.

Now let us consider the developed surface $\overline{S}_k^j$ which is the lower base of the polyhedron M^j. The curvature of this developed surface is nonpositive. The angle RTS introduced above is the angle at an arbitrary vertex situated on the boundary of this developed surface. Since the number of such vertices is finite, the complete rotation of the boundary of $\overline{S}_k^j$ tends to zero as j increases. According to the Gauss–Bonnet formula this rotation is not smaller than 2π since the curvature of the $\overline{S}_k^j$ is nonpositive. Thus, we arrive at a contradiction. Hence, our assumption that there are no extremal polyhedra in the class E_k is false.

The lemma is proved.

It should be noted that this lemma is also true in an extended class of prismatic polyhedra, i.e., such polyhedra for which all the requirements of normality are satisfied, except the restriction that the curvature have the same sign at the inner vertical edges. The proof given above is fully applicable to the general case. For this purpose the gluing of the prisms for obtaining the segment AC on completing one rotation about the vertex A is continued, if necessary, by a subsequent motion in the same direction. This refinement of the lemma is useful for the study of the integral curvature associated with problems of stability in an extended class of polyhedra.

1.4. Retriangulation of the developed surface. The proof of the theorem in 1.2 on the realization of a space-like cap is based on the fact that the extremal normal polyhedron M has zero curvature. The fact that its curvature is zero follows from the lemma according to which otherwise the polyhedron M is transformed to the normal polyhedron $\widetilde{M}$ for which the sum of the corresponding altitudes of vertices is larger than that for M. As we go over from M to $\widetilde{M}$, the altitudes of the vertices for which the curvatures at the inner vertical edges are equal to zero remain unchanged, while inner vertical edges with nonzero curvature increase in length by one and the same small number ε. The polyhedron M belongs naturally to a certain class E_k. If under such a transformation we impose the requirement that $\widetilde{M}$ be preserved in the same class, i.e., we deform M with the same developed surface S_k, the convexity of the polyhedron is necessarily violated since othewise M_k is not an extremal polyhedron. To avoid this situation in the proof of the existence of the necessary transformation $M \to \widetilde{M}$ we must carry out an appropriate retriangulation of the developed surface S_k, i.e., transfer the polyhedron M to another class from $\{E_k\}$ before deforming this polyhedron. As applied to the proof of the theorem in 1.2, such a transfer of M to another class is also impossible since for M the sum of the altitudes is already maximum within the set of sums $\{\overline{h}_k\}$ of the altitudes of the union of polyhedra $\{M_k\}$. In this connection, the lemma in 1.2 together with the constructive elements of its proof acquires independent significance for studying problems associated with deformations of normal polyhedra.

Suppose that the convexity of the polyhedron M with the developed surface S_k is violated at the edge AB of the developed surface under the deformation $M \to \widetilde{M}$ in the class E_k. Among the prisms belonging to the polyhedron M, we consider the double prism whose upper base has the edge AB as its inner edge and is formed by the triangles CAB and DAB of the developed surface S_k. It is clear that these triangles lie in the same space-like plane. It is also obvious that one of the vertices A or B (let it be A) is not lifted under the transformation $M \to \widetilde{M}$, since the convexity is lost and the lifting vertices of the triangles are shifted by one and the same amount ε. Then the angle A in the quadrangle $ACBD$ turns out to be smaller than π. This follows from the fact that A is either the boundary vertex of S_k (the positive rotation of S_k), or the inner vertex of S_k (and the curvature of the inner vertical edge of M at this vertex is equal to zero). In the latter case the prisms belonging to M and meeting at the vertex A are glued along the corresponding vertical faces in such a way that their upper bases represent an ordinary strictly convex polyhedral angle with the vertex A. The angle CAD is smaller than π, since it belongs to some face of this polyhedral angle. In Lemma 2, which is formulated and proved below, we deem it necessary to extend this important property of the angle CAD to more general prismatic pyramids similar to that under consideration with the vertex A. It should be noted that the limiting case is also possible in the Euclidean version when the vertex A is situated in the space over the plane boundary of the developed surface S. In this case, the polyhedron M is an ordinary convex polyhedral angle with the vertex A in the vicinity of the segment $A\overline{A}$ which now belongs to S. Here $\overline{A}$ is the projection of A onto the plane xy. This statement is derived from Liberman's principle. Let us prove the following lemma on the angle at the vertex B of the quadrangle $ACBD$ in question.

LEMMA 1. *Under the deformation of M leading to the loss of convexity at the edge AB, but preserving the class from $\{E_k\}$, only three cases are possible, in each of which the angle CBD is necessarily smaller than π:* (i) *the vertex B is not lifted and at least one of the vertices C and D is lifted,* (ii) *three vertices B, C, and D are lifted, and* (iii) *the vertex B and only one of the vertices C and D are lifted. Let it be the vertex D in the latter case. Then in the quadrangle $ACBD$, the vertex C is situated farther from the straight line BD than the vertex A.*

PROOF. In the first case, the angle at B is smaller than π for the same reason as the angle at A is.

Consider the second case. If $\angle B = \pi$, the quadrangle $ACBD$ is transformed again into a plane quadrangle upon deformation, and convexity is not lost at the edge AB. Suppose that $\angle B > \pi$. Let us draw the segment CD, continue AB up to the intersection with CD at the point E, and lay off segments of length ε from the points C, E, and D along the vertical in the direction $z > 0$. By E' we denote the upper end of the midpoint of this segment. Draw the segment AE' as well and denote by B' its point of intersection with the vertical drawn from B. The resulting deformation of the quadrangle $ACBD$ can be carried out in two stages. First, we displace the quadrangle as a whole without changing the altitude of A and lifting the vertices C and D by ε. Under this deformation, the altitude of the vertex B increases by the quantity $BB' < EE' = \varepsilon$. At the second stage, we lift only the displaced vertex B, making its total displacement equal to ε. The altitudes of the other vertices of the quadrangle remain unchanged at this stage. The convexity of M at the edge AB of the quadrangle is not violated either. Thus, it is necessary to assume in the case under consideration that $\angle B < \pi$. Then the convexity is really lost at the edge AB.

Let us consider the third case. We lay off segments of length ε from the points B and D along the vertical in the direction $z > 0$ and pass a plane through the ends of these segments and the point A. We shall denote by l the line of intersection of this plane with the plane of the quadrangle $ACBD$. As in the second case, the resulting deformation of the quadrangle can be carried out in two stages with the difference that at the second stage, we displace only the image of the vertex C, returning this vertex along its altitude to its former position. Here the convexity at the edge AB is violated only if the segment BD and the vertex C lie in the plane of the quadrangle $ACBD$ on different sides of the straight line l. Notice that at the first stage, the field of the vertical components of displacements of the points of the quadrangle is represented by the auxiliary plane which we have drawn. For such a location of C, the distance from it to the straight line BD is really larger than the distance from A to BD.

The lemma is proved.

Thus to carry out necessary deformation $M \to \widetilde{M}$ with an increase in the sum of the altitudes of the polyhedron, i.e., the deformation mentioned in the lemma in 1.2, it is necessary to retriangulate. At the initial stage, the retriangulation is reduced to a single act of replacing a certain edge of the upper base of the polyhedron M by another edge. It transfers the polyhedron in $\{E_k\}$ from one class to another. In the cases considered in Lemma 1, the edge AB of the convex quadrangle $ABCD$ is replaced by the edge CD and the triangles ACB and ADB of the developed surface M are replaced by the triangles ACD and BCD. It is obvious that now the convexity at the edge CD is not violated under the deformation of M. After that, an attempt is made to deform M in the new class as prescribed by the lemma in 1.2. It may turn out that in this case, too, the convexity of M is violated at a certain edge of the developed surface S other than CD. Then we retriangulate again and replace the inappropriate edge as described above. Finally, after a finite sequence of local retriangulations, we transfer the polyhedron M to a class from $\{E_k\}$ within which the deformation $M \to \widetilde{M}$ transfers M to a normal polyhedron, i.e., the convexity of M is not violated under such a deformation.

However, in retriangulations as a sequence of M-transitions from one class to another in $\{E_k\}$ the sequence may turn out to be cyclic. This means that after a finite number of steps, M returns to a class in which it has already been. In this case, the required deformation $M \to \widetilde{M}$ is not realized. It will be shown below that such a situation cannot be ruled out.

The developed surface S_k is formed by triangles. In the potential deformation $M \to \widetilde{M}$ in the class E_k, this developed surface contains triangles in which only one or only two or all three vertices are lifted. We shall denote respectively by σ_1, σ_2, and σ_3 the sums of the surface areas of such triangles, and associate the number $\sum_k = \sigma_1 + 2\sigma_2 + 3\sigma_3$ with the class E_k. We assert that the sum $\sum$ increases as a result of local retriangulation and the respective transition of M to another class denoted by E_{k+1}. We prove this by using Lemma 1 and studying separately each case of the loss of convexity at edge AB listed in the formulation of this lemma. Let us draw the diagonal CD in the convex quadrangle $ACBD$ and denote by σ_A, σ_C, σ_B, and σ_D the surface areas of the four triangles into which $ACBD$ is divided by the diagonals AB and CD. In this notation, σ_A corresponds to the triangle with side AC, σ_C to the triangle with side CB, σ_B to the triangle with side BD, and σ_D, to the triangle with side DA.

First case. Let only the vertex C be lifted. Then the contribution of the quadrangle to the sum $\sum_k$ is given by the term $\sigma_A + \sigma_B$, whereas the corresponding contribution to the sum $\sum_{k+1}$ is described by the larger term $(\sigma_A + \sigma_D) + (\sigma_C + \sigma_B)$. Let both vertices C and D be lifted. Then we get the term $(\sigma_A + \sigma_C) + (\sigma_B + \sigma_D)$ in $\sum_k$ and the larger term $2(\sigma_A + \sigma_D) + 2(\sigma_C + \sigma_B)$ in $\sum_{k+1}$.

Second case. We have the term $2(\sigma_A + \sigma_C) + 2(\sigma_D + \sigma_B)$ in the sum $\sum_k$ and the larger term $2(\sigma_A + \sigma_D) + 3(\sigma_C + \sigma_B)$ in $\sum_{k+1}$.

Third case. We have the term $(\sigma_A + \sigma_C) + 2(\sigma_D + \sigma_B)$ in the sum $\sum_k$ and the term $(\sigma_A + \sigma_D) + 2(\sigma_C + \sigma_B)$ in $\sum_{k+1}$. The second term is larger than the first one since according to Lemma 1 the surface area of the triangle CBD is larger than that of ABD, and hence $\sigma_C > \sigma_D$.

Let us now turn to the proof of Lemma 2. It is similar to Lemmas 3 and 6 from [**9**] and corresponds to Lemma 10 from [**8**], in which Lemmas 9–12 are used to study the structure of the Euclidean polyhedron which is limiting for the sequence of normal polyhedra.

LEMMA 2. *If V is a convex prismatic pyramid with negative curvature of its upper base and nonnegative curvature of its lower base, the pyramid does not contain a multiple prism with the vertex at the vertex A of the pyramid with a plane upper base and with the total angle at A of this base no smaller than π.*

PROOF. Suppose the lemma is not true. We can assume that the curvature of V at the inner vertex of the lower base is positive. It has already been shown just before the proof of Lemma 1 that, otherwise, the upper base of V is an ordinary strictly convex polyhedral angle one of whose plane angles is at least π. This is obviously impossible. Let us cut the pyramid V into two multiple prisms V_1 and V_2 with common vertex A, such that the total angle at the vertex A of the upper base for the first of them and the total angle at the vertex of the lower base which is the projection of A are both equal to π. The existence of the prism V_1 follows from our assumption and from the conditions of the lemma. It follows from the same conditions that V_1 does not exhaust the whole prism V, so V_2 also exists. In V_2, the total angle at A of the lower base which is the projection of A is smaller than π, since the curvature of the lower base of V is positive.

Let us demonstrate that the total angle at the vertex A of the upper base of the prism V is smaller than π. For simplicity, we assume that V_2 is a double prism. We can obtain the same result in the general case by applying the methods used for the double prism a finite number of times.

Let $ACBD$ be the upper base of V_2. We shall extend the segment DA along the straight line beyond the point A in such a way that its length increases by the length of the segment AC, and denote by C' the end of the extended segment. We also pass a time-like plane τ through the vertical AA'. This plane is the symmetry plane for the time-like half-planes $CA\overline{A}$ and $C'A\overline{A}$ having the common straight line $A\overline{A}$. The angles $CA\overline{A}$ and $C'A\overline{A}$ are equal since the vertical faces of the prisms V_1 and V_2 are glued at the point A. Hence, the points C and C' are symmetric with respect to τ. Since the angle at the vertex $\overline{A}$ of the lower base of V_2 is smaller than π, the points B and C' are situated on different sides of the plane τ. It follows that $C'B^2 > CB^2$ since the plane τ is time-like; hence $C'B > CB$ because the segment CB is space-like. Comparing the triangles ACB and $AC'B$, we find that $\angle CAB < \angle C'AB$ and $\angle CAB + \angle DAB < \angle C'AB + \angle DAB$.

Thus, the total angle at the vertex A of the upper base of the prism V is smaller than 2π. This contradicts the conditions of the lemma. Hence, the assumption that V contains a multiple prism whose plane angle at the vertex A of its upper base is not smaller than π is not true.

The following method of proving the statement for the prism V_2 can also be generalized to multiple prisms. Let us denote by $\widetilde{C}$ the point of intersection of the vertical $C\overline{C}$ with the plane of the triangle ABD. The inclination of the ray $A\widetilde{C}$ to the plane xy is larger than that of the ray AC', whereas the inclination of AC' is smaller than that of AC. Hence, there exists a ray AE between the rays AC and $A\widetilde{C}$ in the plane ABD such that its inclination with respect to the plane xy is equal to that of the ray AC. The points C and E are symmetric with respect to the bisecting plane for the half-planes $CA\overline{A}$ and $EA\overline{A}$. Since the segment CE is space-like and the angle at the vertex A of the lower base of V is smaller than π, we find that $BC < BE$, $\angle BAC < \angle BAE$ and $\angle DAB + \angle BAC < \angle DAE < \pi$. This method is also applicable for proving the statement for a prism V in Euclidean space.

The lemma is proved.

1.5. Behavior of prismatic curvature. Now we must complete the proof of the lemma and hence of the theorem in 1.2. Namely, it is necessary to establish a decrease in the curvature at the inner vertex of the lower base of a prismatic pyramid if this pyramid is deformed while preserving the convexity in such a way that the altitude of the inner vetex of the upper base remains unchanged, whereas the altitudes of the boundary vertices of the upper base can only increase. By assumption, the curvature of the lower base is equal to zero, but we consider the more general case of nonnegative curvature. The curvature at the inner vertex of the lower base of a prismatic polyhedron is called the prismatic curvature. The lemmas below generalize Lemma 8 from [**9**] and Lemma 4 from [**8**].

Let $ACBD$ be a convex double prism for which the total angle at the vertex A of the upper base is equal to π and the vertical time-like faces $DA\overline{A}$ and $CA\overline{A}$ are the complementary angles of the time-like half-plane. In the Euclidean case, the latter condition corresponds to the condition that the sum of the plane angles formed by the vertical faces meeting at A is equal to π. Such a double prism is called a degenerate prism. It turns out that the structure of a degenerate double prism is quite simple. It always represents the dihedral angle which is in the pseudo-Euclidean case formed by two faces: the vertical time-like face and the inclined space-like face. To confirm this statement it is sufficient to eliminate the case when the vertical faces of the prism do not belong to the same plane and the dihedral angle formed by them is smaller than π. If this angle is larger than π, we must go over to the supplementary double prism, extending the edge AB beyond the point A. In this case, as was mentioned in the proof of Lemma 2 in 1.4, the angle at the vertex A of the upper base of the prism is smaller than π. This result does not agree with the degeneracy of the prism. This property of degenerate prisms is of paramount importance for studying problems associated with the deformation of prismatic pyramids.

LEMMA 1. *Let V_1 and V_2 be nondegenerate convex double prisms whose upper space-like bases $A_1C_1B_1D_1$ and $A_2C_2B_2D_2$ are identically composed from corresponding equal faces, and let $A_1\overline{A_1} = A_2\overline{A_2}$, $C_1\overline{C_1} = C_2\overline{C_2}$, $D_1\overline{D_1} = D_2\overline{D_2}$, and $B_1\overline{B_1} < B_2\overline{B_2}$. Then the angles φ_1 and φ_2 at the vertices $\overline{A_1}$ and $\overline{A_2}$ of the lower bases of the prisms satisfy the inequality $\varphi_1 < \varphi_2$.*

It follows from the conditions of the lemma that the dihedral angles at the edges A_1C_1 and A_1D_1 of the prism V_1 are smaller than the corresponding dihedral angles at the edges A_2C_2 and A_2D_2 of V_2. Since we are dealing here with dihedral angles formed by time-like and space-like planes, it is necessary to dwell on the above relation between the angles at greater length and elucidate its meaning. For instance, for the angles at the edges A_1D_1 and A_2D_2, it implies that upon displacement of the prisms bringing the faces $D_1A_1\overline{A_1}$ and $D_2A_2\overline{A_2}$ into coincidence, the face $D_2A_2B_2$ of V_2 lies strictly over the face $D_1A_1B_1$ of V_1, if, of course, we exclude the points of their edges A_1D_1 and A_2D_2 which are brought into coincidence. It is obvious that this relation between the dihedral angles and the inequality $B_1\overline{B_1} < B_2\overline{B_2}$ are equivalent, but it turns out that the inequality $\varphi_2 > \varphi_1$ is also equivalent to them. The latter equivalence becomes obvious if we do the following.

(i) We establish that the congruence $V_1 \equiv V_2$ and, in particular, the equality of the corresponding dihedral angles of the prisms is obtained from the equality $\varphi_2 = \varphi_1$; this follows from the nondegeneracy of the prisms V_1 and V_2 and from an elementary geometric statement of the type "a convex trihedral angle is uniquely defined by its angles"; we shall not consider this statement in detail because of lack of space, but simply observe that it is well known for Euclidean space, whereas in pseudo-Euclidean space it is established by simple methods of analytical geometry.

(ii) We prove the converse of the lemma, i.e., the statement that the corresponding inequalities for the dihedral angles of V_1 and V_2 follow from the inequality $\varphi_1 < \varphi_2$. This will be proved below. The proof is essentially the same in the Euclidean and pseudo-Euclidean cases and is carried out without resorting to the theorems of spherical trigonometry used in the analytical works [**8,9**]. Notice that there is a difference in the formulation of the converse lemma between the Euclidean and pseudo-Euclidean cases, i.e., as the angle increases, the dihedral angles increase in the pseudo-Euclidean case and decrease in the Euclidean case.

We need the following auxiliary statement about double prisms.

LEMMA 2. *Let $ABCD$ be a convex double prism such that the union of its vertical faces $DA\overline{A}$ and $CA\overline{A}$ is equal to or exceeds the straight space-like angle and the angle φ at the lower base at the vertex $\overline{A}$ is not equal to π. Denote by ψ the angle at the vertex A on the developed surface of the upper base of the prism. If $\varphi < \pi$, then $\psi < \pi$, and if $\varphi > \pi$, then $\psi > \pi$.*

The proof of this statement can easily be reduced to the case where the union of the vertical faces is equal to a straight angle. For this purpose, assuming that the upper base of the prism is not plane, we extend the plane of the face $DA\overline{A}$ beyond the edge $A\overline{A}$ and obtain the intersection AC' of the plane of the face ABC with the extended half-plane $DA\overline{A}$. The sum of the angles $C'A\overline{A}$ and $DA\overline{A}$ is smaller than a straight angle. If we assume that the sum of the angles $C''A\overline{A}$ and $DA\overline{A}$ is larger than a straight angle, then there exists a ray AC'' between the rays AC and AC' in the plane BAC such that the sum of the angles $C''A\overline{A}$ and $DA\overline{A}$ is equal to a straight angle. The prism $AC''BD$ satisfies the conditions of the formulated statement, so that for the corresponding angles of this prism we find that $\psi < \pi$ if $\varphi < \pi$ (here $\angle C''AB > \angle CAB$) and $\psi > \pi$ if $\varphi > \pi$ (here $\angle C''AB < \angle CAB$). For the particular case of the statement, the proof is carried out by the same method (for both $\varphi < \pi$ and $\varphi > \pi$) described in the proof of Lemma 2 of the previous section. It is worth noting that the present statement, together with the outlined proof, can be extended completely to the Euclidean case.

PROOF OF LEMMA 1. We prove the converse of the lemma. It is sufficient to consider two cases: (i) $\varphi_2 > \pi$, $\varphi_1 \leq \pi$ and (ii) $\varphi_1 < \varphi_2 < \pi$. Further, as we carry out the proof of the lemma for space-like double prisms, we shall supplement it with elucidating remarks concerning the Euclidean version of the lemma.

Case 1. Suppose that $\varphi_1 < \pi$. Let us extend the faces $DA\overline{A}$ of the prisms beyond the straight line $A\overline{A}$ and plot the angle $C'A\overline{A}$ from the ray $A\overline{A}$ in the extended half-planes in such a way that it is equal to the angle $CA\overline{A}$. Needless to say, all constructions are carried out separately for each of the prisms V_1 and V_2, and the notation used for the new points C carries the corresponding indices. We assert that the angle DAC' measured toward the point $\overline{A}$ is smaller than a straight angle. Indeed, it follows from the above Lemma 2 that otherwise the total angle ψ_1 at the vertex A_1 of the upper base of V_1 is smaller than π, and the similar total angle ψ_2 for the prism V_2 is larger than π. This is a contradiction, since it follows from the isometry of the upper bases of the prisms that $\psi_1 = \psi_2$.

Suppose that the dihedral angle at the edge A_2D_2 of the prism V_2 is smaller than the dihedral angle at the edge A_1D_1 of the prism V_1, or equal to it. Let us displace the trihedral angle $A_2C_2'B_2D_2$ until the plane angles $D_2A_2C_2'$ and $D_1A_1C_1'$ coincide at the corresponding points in such a way that the face $D_2A_2B_2'$ is not higher than the face $D_1A_1B_1$ in the same half-space relative to the plane $D_1A_1C_1'$ containing the prism V_1. The new position of the point B_2 is denoted by B_2'. Let us write down the sequence of inequalities for the squared lengths of the segments of the type BC. These inequalities are based on the peculiarities of the mutual arrangement of the points B and C, which will be fixed. For the prism V_2, the pair of points B_2 and C_2' and the point C_2 belong to different half-spaces with respect to the time-like plane passing through the straight line $A_2\overline{A_2}$ in such a way that it serves as the bisecting plane for the half-planes $C_2A_2\overline{A_2}$ and $C_2'A_2\overline{A_2}$ bounded by their common straight line $A_2\overline{A_2}$. For the prism V_1, the pair of points B_2' and C_1', and the point B_1 belong to different half-spaces with respect to the time-like plane passing through the straight line A_1D_1 in such a way that it serves as the bisecting plane for the half-planes $B_2'A_1D_1$ and $B_1A_1D_1$ bounded by their common straight line A_1D_1 (notice that the coincidence $B \equiv B_2'$ is possible here). For the prism V_1 the pair of points B_1 and C_1 and the point C_1' belong to different half-spaces with respect to the timelike plane passing through the straight line $A_1\overline{A_1}$ in such a way that it serves as the bisecting plane for the half-planes $C_1A_1\overline{A_1}$ and $C_1'A_1\overline{A_1}$. We get

$$C_2B_2^2 > C_2'B_2^2 = C_1'B_2'^2 \geq C_1'B_1^2 > C_1B_1^2.$$

In the nonstrict inequality, it is taken into account that the segment $B_2'B_1$ is the time-like normal to the corresponding bisecting plane. Similarly, the strict inequalities take into account the fact that the segments C_2C_2' and C_1C_1' are space-like. This contradicts the initial equality $C_2B_2 = C_1B_1$. Hence, in this case our assumption about the relations between dihedral angles at the edges AD and AC of the prisms V_1 and V_2 is not false. In the Euclidean version, the proof is carried out in the same way with the exception that (i) the opposite relation for the dihedral angles is established for this case, and (ii) the face $D_2A_2B_2$ after its displacement turns out to be not lower than the face $D_1A_1B_1$ in the corresponding geometric construction when we give the proof by contradiction.

Now let us put $\varphi_1 = \pi$. Since V_1 is convex, the plane angle $C_1A_1D_1$ measured toward $\overline{A_1}$ is smaller than, or equal to, a straight angle. The angle cannot be straight since the prisms V_1 and V_2 are nondegenerate. Then it is sufficient to repeat the arguments used in the proof under the assumption that $\varphi_1 < \pi$. Naturally, we shall not do so here.

Case 2. $\varphi_1 < \varphi_2 < \pi$. Suppose that the dihedral angle at the edge A_2D_2 of the prism V_2 is smaller than, or equal to, the dihedral angle at the edge A_1D_1 of the prism V_1. We shall not describe the corresponding procedure in detail. Let us displace the angle of V_2 appropriately until the faces $D_1A_1\overline{A_1}$ and $D_2A_2\overline{A_2}$ coincide. The pair of points C_1 and B_1 and the point C_2' lie on different sides of the corresponding time-like bisecting plane passing through the straight line $A_1\overline{A_1}$. The pair of points C_2' and B_2' and the point B_1 also belong to opposite sides of the corresponding bisecting plane passing through the straight line A_1D_1 (here the coincidence $B_1 \equiv B_2'$ is of course possible). Such an arrangement of the points C_2', B_2', and B_1 follows from the convexity of V_2, i.e., from the fact that the point C_2' belongs to the dihedral convex angle formed by the half-planes $D_1A_1\overline{A_1}$ and $D_1A_1B_2'$ intersecting along the edge A_1D_1. It is this fact that allows us to ignore the value of the angle $CA\overline{A}$ in the present case. This gives the following system of inequalities, which also leads to a contradiction:

$$C_1B_1^2 < C'_2B_1^2 \leq C'_2B'^2_2 = C_2B_2^2.$$

In the Euclidean version, the scheme of the proof remains practically unchanged with the exception that, in the proof by contradiction, the point B_2' must not be lower than B_1. If we consider here the case $\varphi_1 < \varphi_2 < \pi$, there are no sufficient grounds to establish the arrangement of the points B_1, B_2', and C_2' necessary for proving the nonstrict inequality, since now the point C_2' must be inside the convex dihedral angle formed by the faces $D_1A_1B_1$ and $D_1A_1\overline{A_1}$. Therefore, we must pass from the prisms V_1 and V_2 to the supplementary double prisms, by extending the edges A_1B_1 and A_2B_2 beyond the points A_1 and A_2, and consider the case $\varphi_2 > \varphi_1 > \pi$, where the necessary arrangement of the points under consideration is already satisfied. In this case the arrangement of B_1, C_1, and C_2' necessary for proving the strict inequality may not take place if we make the edges A_1D_1 and A_2D_2 coincide. Then it is sufficient to turn the displaced prism V_2 around the edge $A_1\overline{A_1}$ and make the faces $C_2'A_1\overline{A_1}$ and $C_1A_1\overline{A_1}$ coincide. Thus, for V_1 and V_2, we must compare the dihedral angles at the edges AC. Here we must take into account the fact that the dihedral angles at the edges AD and AC are simultaneously larger or smaller in V_2 than in V_1.

The lemma is proved.

LEMMA 3. *Let V_1 and V_2 be convex prismatic pyramids such that* (i) *their upper bases are identically composed of equal faces and have the same curvature, and* (ii) *the inner vertices of their bases are situated at equal distances from the plane xy. Suppose that the curvature of the lower base of V_1 is nonnegative, and the altitudes at the corresponding vertices over the plane xy along the boundaries of the upper bases of the pyramids are not smaller in V_2 than in V_1 but may be larger at certain vertices. Then the curvature of the lower base in V_1 is larger than that in V_2, i.e., $\omega_1 > \omega_2$.*

PROOF. We can assume that the upper bases of the pyramids are isometric and put the points of these bases in isometric correspondence. Further, we shall consider separately two cases: (i) the altitudes of the corresponding points along the boundaries are larger in V_2 than in V_1, and (ii) the boundaries of the pyramids contain points of equal altitude that correspond to one another according to the isometry.

Case 1. Let us move V_1 in the direction $z > 0$ parallel to itself in such a way that the relation between the altitudes of the pyramids along the boundary remains unchanged and the altitude of the inner vertex A_1 of the upper base of V_1 becomes larger than the altitude of the corresponding vertex A_2 of V_2. Then the polygonal lines $\Gamma_1 \subset V_1$ and $\Gamma_2 \subset V_2$ with the following properties appear on the upper bases: they correspond to one another according to the isometry, they are situated at equal distances from the plane xy, their vertices can lie only on the edges of the pyramids, they encircle the vertices A_1 and A_2, and they bound simply connected domains on the bases. Inside these domains, the points of V_1 lie farther from the plane xy than the corresponding (according to the isometry) points of V_2. Further, we assume that these domains exhaust the upper bases. Then, the boundaries of the lower bases of the pyramids corresponding to the projections of the polygonal lines Γ_1 and Γ_2 are denoted by $\overline{\Gamma}_1$ and $\overline{\Gamma}_2$. The isometry $\Gamma_1 \leftrightarrow \Gamma_2$ naturally induces the isometry $\overline{\Gamma}_1 \leftrightarrow \overline{\Gamma}_2$ as the correspondence between the projections $\overline{C} \in \overline{\Gamma}$ of points $C \in \Gamma$ corresponding to one another according to the isometry. Let $C_1 \in \Gamma_1$ and $C_2 \in \Gamma_2$ be vertices of the polygonal lines Γ and let $\overline{C_1} \in \overline{\Gamma}_1$ and $\overline{C_2} \in \overline{\Gamma}_2$ be the corresponding vertices of $\overline{\Gamma}$. We denote by τ_C^1 and τ_C^2 the rotations of the boundaries on the lower bases of the pyramids from the side of these bases at the points $\overline{C}$. Let us compare the neighborhoods of the pyramids V_1 and V_2 along the segments $C_1\overline{C_1}$ and $C_2\overline{C_2}$. These neighborhoods can be considered as double prisms. If these prisms are degenerate, then $\tau_C^1 = \tau_C^2 = 0$, i.e., in the vicinity of the vertices $\overline{C}$, the lines $\overline{\Gamma}_1$ and $\overline{\Gamma}_2$ are straight, and the same is true for the lines Γ_1 and Γ_2 in the neighborhood of the vertices C. Such vertices C and $\overline{C}$ will be called singular vertices. The pyramid V_1 satisfies the conditions of Lemma 2 of 1.4. Then it follows from Lemma 2 of the present section that the singular vertices do not exhaust all vertices of $\overline{\Gamma}_1$. According to Lemma 1, the inequality $\tau_C^1 < \tau_C^2$ is valid in a nonsingular vertex, and hence the rotations τ_1 and τ_2 of the polygonal lines $\overline{\Gamma}_1$ and $\overline{\Gamma}_2$ from the side of the lower bases of the pyramids V_1 and V_2 satisfy the same inequality $\tau_1 < \tau_2$. Applying the Gauss–Bonnet formula to the lower bases, we find that $\omega_1 > \omega_2$. This is the desired result.

Case 2. Among the prisms constituting V_1 and V_2, we can find multiple prisms W_1 and W_2 in which the upper bases with the vertices A_1 and A_2 are identically composed of equal triangular faces, the corresponding (according to the isometry) boundary sides A_1C_1, A_1D_1 and A_2C_2, A_2D_2 of these bases are situated at the same distance from the plane xy, and the remaining points corresponding to one another according to the isometry are arranged in such a way that the points of V_2 lie higher than the points of V_1 over the plane xy. To prove the lemma, it is sufficient in this case to establish that the angles φ at the vertices $\overline{A_1}$ and $\overline{A_2}$ of the lower bases of W_1 and W_2 satisfy the inequality $\varphi_1 < \varphi_2$.

Let us displace W_1 along the axis $z > 0$ itself. Then, the polygonal lines Γ_1 and Γ_2 having the following properties appear on the upper bases of the prisms W_1 and W_2: they correspond to one another according to the isometry, they are situated at equal distances from the plane xy, and their inner vertices lie on the inner edges of the upper bases. The altitude of the displacement is taken sufficiently small in order to ensure the intersection of the polygonal lines Γ_1 and Γ_2 with all inner edges of the upper bases. The outer links of these polygonal lines obviously belong to the faces of the upper bases of W_1 and W_2 adjoining the edges A_1C_1, A_1D_1 and A_2C_2, A_2D_2 respectively. Moreover, these links are respectively parallel to these sides of the bases. Let us go over to the projections $\overline{\Gamma}_1$ and $\overline{\Gamma}_2$ of Γ_1 and Γ_2. Studying the rotations τ_i of the polygonal lines $\overline{\Gamma_i}$ in the lower bases of the prisms W_1 and W_2 from the side $\overline{A}$, we find, as in the previous case, that $\tau_1 < \tau_2$. It follows from the same considerations that there exist nonsingular vertices on the polygonal lines Γ_i and $\overline{\Gamma_i}$. Further, using the Gauss–Bonnet formulas we conclude that the inequality $\omega_1 > \omega_2$ holds since the outer links of the polygonal lines $\overline{\Gamma_i}$ are parallel to the segments $\overline{AC}$ and $\overline{AD}$, respectively. It is sufficient to observe that the pyramids V_1 and V_2 are formed by prisms of the type W and perhaps by congruent multiple prisms as well.

The lemma is proved.

In conclusion, we present a statement which follows immediately from Lemma 3 and is well known in the Euclidean case [**21**].

Let V_1 and V_2 be ordinary convex (space-like in the pseudo-Euclidean case) cones formed identically by equal dihedral angles with their convexities pointing in the direction $z > 0$. Let the edges of the cone V_2 and the direction $z < 0$ form angles which are not smaller than the angles formed by the corresponding edges of V_1 and the direction $z < 0$. Then the cones V_1 and V_2 are congruent and the angles formed by their edges with the direction $z < 0$ are respectively equal to one another.

2. Uniqueness

The classical uniqueness theorems proved by Cauchy [**27**] and Minkowski [**28**] for convex polyhedra with given local geometric properties form the basis of the theory of general convex surfaces in spaces of constant curvature. The theory was developed by Aleksandrov [**1,2**] and Pogorelov [**21**]. Its results also include the uniqueness theorems for caps and surfaces isometric to caps. In the present section, the results will be extended to the case of convex space-like polyhedra. Two approaches to the problem of the uniqueness of polyhedra can be singled out in view of the specific features of the concept of isometry for polyhedra, an analysis and comparison of the original proof of the Cauchy theorem with the proof of the Minkowski theorem, and an analysis of possible generalizations of these theorems to a spherical space considered by the author in [**15**]. These approaches are usually considered as equivalent and lead to uniqueness theorems which differ from one another. The refinement of the formulation of these well-known theorems is the second problem considered in this section. In solving this problem, we have found new proofs of the well-known lemmas of Cauchy and Aleksandrov on deformations of convex polygonal lines. Incorporation of space-like surfaces in the geometric study along with Euclidean surfaces makes it possible to get a comprehensive view of the existence theorems.

Ordinary convex surfaces are considered throughout this chapter. In 2.1, we establish theorems on the equality of isometric space-like polyhedra brought into a one-to-one correspondence. Pogorelov's well-known principles of maxima for isometric convex polyhedra and the corresponding corollaries [**21**] are extended to the case of pseudo-Euclidean polyhedra. The theorem on the continuous flexibility of convex space-like polyhedra isometric to caps is proved. Here we use the generally accepted concept of isometry introduced by Cohn-Vossen [**13**]; surfaces mapped onto one another are isometric if all curves transformed to one another under this map have equal lengths. Such a concept of isometry makes it possible to incorporate convex polyhedra into the metric class of arbitrary general convex surfaces. The first theorems conforming to this concept were Aleksandrov's theorem on the equality of closed convex polyhedra with equal developed surfaces [**1**] and Olovyanishnikov's theorem [**20**] on the equality of a closed general convex surface and a closed convex polyhedron that are isometric to each other. In 2.1, by isometric polyhedra we shall mean polyhedra with equal developed surfaces. This eliminates the topological requirement that the structure of the set of natural edges and faces of the compared polyhedra be the same, i.e., the requirement in the Cauchy theorem. We shall also assume that equally developed surfaces of isometric polyhedra are brought into an appropriate one-to-one correspondence.

The closed convex polyhedra considered in the Cauchy theorem are isometric in the sense indicated above. However, the isometry is associated with their natural developed surfaces. This seems to be a strong additional requirement, which entails especially far-reaching consequences in the formal extension of the Cauchy theorem to spherical and hyperbolic spaces [**1**]. For these cases, the theorem is overdetermined, and it is established in [**15**] that closed combinatorially equivalent polyhedra are equal. Notice that the initial conditions under which this statement is valid are essentially common for the Cauchy and Minkowski theorems. As a result, the use of natural combinatorial structure for comparing polyhedra supplements the requirement about their isometry with additional details. The theorems of Cauchy and Minkowski, combinatorially equivalent polyhedra in the pseudo-Euclidean space and on its spheres, and the corresponding lemmas on the deformations of convex polygonal lines are considered in 2.2 and 2.3.

2.1. Isometric surfaces. It has been mentioned that here we shall consider ordinary surfaces. However, for convenience, the surfaces are interpreted as prismatic polyhedra throughout this subsection except for Theorem 5. This allows us to use the results of the previous section directly.

THEOREM 1. *Let P_1 and P_2 be isometric prismatic convex polyhedra. Consider points X which belong to the upper bases of P_1 and P_2 and are isometric to one another, and denote by $z_1(x)$ and $z_2(x)$ the altitudes of these points over the plane xy. Then the function $\delta z(x) = z_1(x) - z_2(x)$ is a constant, or attains its maximum positive or minimum negative value only on the boundary of the polyhedron.*

PROOF. Suppose that the converse is true. Let δz be not equal to a constant and suppose, for example, that the value $h = \max \delta z > 0$ is reached at certain internal points of the upper bases of the polyhedra P_1 and P_2. Then there exists a real inner vertex, say A, at which $z_1(A) - z_2(A) = h$. Hence, the inequality $z_1(X) - z_2(X) \leq h$ is valid in a sufficiently small vicinity of this vertex. Moreover, there exist points at which $z_1(X) - z_2(X) < h$. Consider the prismatic pyramids V_1 and V_2 formed by the constituent polyhedra of prisms meeting at the vertex A. Let us increase the altitudes of all points of the upper base of V by one and the same value h, i.e., lift all triangles of the upper base by h. Then the pyramid V_1 and the transformed pyramid V_2 satisfy the conditions of Lemma 3 of 1.5. According to that lemma, the prismatic curvature of the pyramid V_2 along the vertical edge at the vertex A must be negative. However, this curvature is equal to zero, and we arrive at a contradiction. Hence, our assumption is not true.

The theorem is proved.

So far, we have considered convex polyhedra "whose concave sides face the plane xy." Obviously, this concerns the arrangement of the upper bases only. Naturally, we can also define prismatic polyhedra "whose concave sides face away from the plane xy". The statement that identical sides of two convex polyhedra face a certain point or a certain plane does not require additional explanation.

THEOREM 2. *Let P_1 and P_2 be isometric convex prismatic polyhedra whose identical sides face the plane xy and which are uniquely projected with the help of the time-like rays passing through the origin of coordinates O. Consider points X that belong to the upper bases of P_1 and P_2 and correspond to one another according to the isometry. Denote by $r_1^2(X)$ and $r_2^2(X)$ the squared radius vectors of these points. Assume that $z_1(X) + z_2(X) > 0$ everywhere on the bases. Then the function*

$$\delta r(X) = \frac{r_1^2(X) - r_2^2(X)}{z_1(X) + z_2(X)}$$

is either constant or reaches its maximum positive or minimum negative value only on the boundaries of the polyhedra.

PROOF. Suppose that the converse is true. Let δr be not equal to a constant and suppose, for example, that the value $2h = \max \delta r > 0$ is attained at certain internal points of the upper bases of the polyhedra P_1 and P_2. Then there exists an inner vertex, say A, at which $\delta r(A) = 2h$, and in an arbitrarily small neighborhood of A there exist points X at which $\delta r(X) < 2h$. Let us increase the altitudes of all points of the upper base of P_1 and decrease the altitudes of all points of the upper base of P_2 by one and the same value h. Then for the transformed polyhedra P_1 and P_2, we get

$$\delta \tilde{r} = \frac{\tilde{r}_1^2 - r_2^2}{\tilde{z}_1 + \tilde{z}_2} = \frac{r_1^2 - r_2^2}{z_1 + z_2} - 2h = \delta r - 2h.$$

Hence at the point A, we have the local imbedding of the polyhedra, i.e., $\tilde{r}_1^2(A) = r_2^2(A)$, and in a sufficiently small neighborhood of this point, $\tilde{r}_1^2(X) - \tilde{r}_2^2(X) < 0$. Moreover, there exist points X at which $\tilde{r}_1^2(X) - r_2^2(X) < 0$. Since the upper base of the polyhedron P_1 is lifted in the positive direction of the axis z by the quantity $h > 0$, $\tilde{r}_1^2(A) < 0$; hence the vector $\tilde{r}_1(A)$ is time-like. The vector $r_2(A)$ is time-like as well, since $\tilde{r}_2(A) = r_1(A)$. Through the point O we pass space-like planes τ_1 and τ_2 orthogonal to the vectors $\tilde{r}_1(A)$ and $\tilde{r}_2(A)$ respectively. It is obvious that in the vicinity of the point A in the new arrangement, identical sides of the polyhedra P_1 and P_2, i.e., the sides which faced the plane xy before the displacement, now face the point O and the planes τ_1 and τ_2 respectively. Therefore, the neighborhoods of A on the polyhedra P_1 and P_2 can be considered as convex prismatic pyramids over the planes τ_1 and τ_2 with isometric upper bases. It follows from the relation between the radius-vectors $\tilde{r}_1(X)$ and $\tilde{r}_2(X)$ of the upper bases of the prismatic pyramids that these pyramids satisfy the corollary of Lemma 3 in 1.5. Hence the pyramids are equal, and we arrive at a contradiction. Thus, our assumption is not true. The theorem is proved.

Let us present the direct corollaries of the principles of maxima.

THEOREM 3. *Let the conditions of Theorem 1 be satisfied and let $\delta z(X)$ be constant along the boundary of the polyhedra. Then the polyhedra are equal and $\delta z(X) \equiv \text{const}$ everywhere on their upper bases. In particular, convex isometric polyhedral caps are equal.*

THEOREM 4. *Let the conditions of Theorem 2 be satisfied and let $\delta r(X)$ be constant along the boundaries of the polyhedra. Then the polyhedra are equal and $\delta r(X) \equiv \text{const}$ everywhere on their upper bases. In particular, convex isometric polyhedra fixed relative to the point O are equal, if $r_1^2(X) \equiv r_2^2(X)$ at the corresponding boundary points.*

In the Euclidean version, Theorem 4 for surfaces fixed relative to a point has been established by Aleksandrov and Senkin in [**3**]. They proposed a proof based on the continuous displacement of one of the surfaces up to their local imbedding. The method has been refined in [**4**], where small displacements of one of the surfaces were used. Here the proof is based on Zalgaller's well-known lemma on the deformation of a closed polygonal line in a plane of constant curvature [**12**]. The method of local imbedding was used in [**18**] for the proof of a theorem that states that general closed surfaces are uniquely defined by the metric in the hyperbolic space. In the de Sitter space, the same theorem for space-like surfaces has been established under certain restrictions [**19**]. As in [**18**], its proof is based on Pogorelov's well-known mapping of a pair of isometric objects [**21**]. Both theorems can be considered as uniqueness theorems on spheres in a pseudo-Euclidean space. The new method of simultaneous displacement of isometric surfaces introduced in the proof of Theorem 2 can apparently be used in the Euclidean case as well.

Now that we have proved theorems on the realizability and the unique definiteness of space-like caps, we can naturally study the problem of continuous flexibility of surfaces isometric to them. In Euclidean space, the problem of flexibility of one convex surface to another surface isometric to it was posed by Pogorelov (see [**21**]) for surfaces which are homeomorphic to a circle, whose orientation is the same, and for which the sign of the rotation on their borders is the same. The solution of this problem has been obtained by Shor [**26**] and the author [**17**] by using the method of deformations and trivial and nontrivial gluing (see also [**1**]). In the following theorem, we consider this problem for surfaces isometric to space-like caps.

THEOREM 5. *Isometric space-like convex polyhedra with a positive rotation along the border, which are homeomorphic to a circle and have the same orientation, can be transformed to one another by means of continuous bending in the same class of surfaces.*

PROOF. It is sufficient to establish that each of the polyhedra in question (they are assumed to have real inner vertices) is transformed to a cap by means of continuous bending. Let P be one such polyhedron. We shall construct the appropriate bending of P in two stages. The first stage is trivial. The polyhedron is transferred with the help of a continuous motion into the position when the plane of support (in the strict sense) parallel to the plane xy can touch P at a certain inner vertex. Further, we shall assume that the polyhedron P belongs to the half-space $z > 0$. It is such a polyhedron that will be transformed to a cap. For this purpose, we shall use the technique of deformation of nontrivial gluing similar to that proposed in [**17**] for the Euclidean case. Let us denote by Γ the polygonal line forming the boundary of P.

Let M be a convex solid polyhedron equal to the intersection of the half-spaces bounded by the planes of the faces of P, $\overline{\Gamma}$ a closed convex polygonal line forming the boundary of the intersection of M and the plane xy, $\overline{M} \subset M$ the convex hull of the union of P and $\overline{\Gamma}$, and $\overline{P}$ the convex space-like cap with the boundary $\overline{\Gamma}$, which includes P, and which is a part of the boundary of the polyhedron $\overline{M}$. Finally, let Φ be the zone on the cap $\overline{P}$ between the polygonal lines Γ and $\overline{\Gamma}$. The zone Φ can be developed and does not contain any of the real vertices of the cap. The polygonal line $\overline{\Gamma}$ rotates in the positive direction with respect to Φ at any vertex since $\overline{P}$ is a cap, and Γ rotates in the nonpositive direction since P is isometric to the cap and the intrinsic curvature of $\overline{P}$ is non-positive at any point. It follows that the zone Φ is divided into curvilinear trapezoids and triangles whose lateral sides are the shortest lines on $\overline{P}$ that connect points on Γ and $\overline{\Gamma}$ in Φ. In this case the following conditions are satisfied. For each trapezoid, the lower base is one of the links of Γ, and the lateral sides are orthogonal to it on $\overline{P}$. The lateral sides themselves cut out on $\overline{\Gamma}$ a polygonal line (in the general case) which forms the upper base of the trapezoid. It is clear that in the metric of the trapezoid itself, the upper base is a convex polygonal line which forms angles smaller than π with the lateral sides. For each triangle, one of the vertices is a vertex of the polygonal line Γ. The lateral sides of the triangle meeting at this vertex are also the lateral sides of the neighboring trapezoids. These sides cut out on $\overline{\Gamma}$ a polygonal line (in the general case) which forms the curvilinear base of the triangle. The third side of the triangle forms angles smaller than π with the lateral sides.

Let us now construct a continuous family of metrics $\{R_t, 0 \leq t \leq 1\}$ for which the imbedding $R_t < R_{t'}$ is valid if $t' < t$. Here R_0 coincides with the metric of the cap $\overline{P}$ and R_1 with the metric of the polyhedron P. These metrics are realized by the family of caps $\{P_t, 0 \leq t \leq 1\}$, where $P_0 = \overline{P}$ and P_1 is isometric to P, which transforms P to a polyhedral cap in the class of convex space-like polyhedra through continuous bending. The realizability of the metrics follows from the existence theorem for caps proved in 1.2. The continuity of bending follows from Theorem 3 about the uniqueness of the caps. Then the theorem which we are trying to prove also follows from the same theorem and from the possibility of constructing the caps P_t. It remains for us to present the appropriate family of metrics or the family of the developed surfaces of the caps $\{P_t\}$.

Let us fix the number t in the interval $0 < t < 1$ and subject each triangle and each trapezoid lying in the zone Φ to a special transformation under which the line $\overline{\Gamma}$ goes over to a certain closed polygonal line $\overline{\Gamma}_t \subset \Phi$. A trapezoid is transformed by affine contraction toward the lower base with coefficient $(1-t)$ to a new trapezoid with a curvilinear upper base. A triangle is transformed to a new triangle by a similar contraction (directed toward the vertex lying on Γ) with coefficient $(1-t)$. The polygonal line $\overline{\Gamma}_t$ is formed by the curvilinear bases of new trapezoids and new triangles. Denote by $\Phi_t \subset \Phi$ the zone between the lines $\overline{\Gamma}$ and $\overline{\Gamma}_t$, and consider the line Γ_t which lies in this zone and which has the minimum length among the lines enveloping $\overline{\Gamma}_t$. It is obvious that Γ_t is a polygonal line that does not have common points with Γ, but has at least one common point with $\overline{\Gamma}_t$. The latter follows, e.g., from the fact that the intrinsic curvature of the polyhedron P is negative. In each vertex, the polygonal line Γ_t rotates in the positive direction on the cap $\overline{P}$ with respect to the domain P, since Γ_t is the minimal polygonal line. The polygonal line Γ_t tends to Γ as $t \to 1$ because Γ_t is minimal and Γ turns in the positive direction with respect to the zone Φ. Let us put $\Gamma_1 \equiv \Gamma$. By construction, $\Gamma_t \to \overline{\Gamma}$ as $t \to 0$. Let us put $\Gamma_0 \equiv \overline{\Gamma}$. For each t, the number of the links of the polygonal line Γ_t do not exceed the number of the links of $\overline{\Gamma}$ and Γ. By R_t we denote the part of the developed surface of $\overline{P}$ which is bounded by Γ_t and contains P. The family of developed surfaces $\{R_t, 0 \leq t \leq 1\}$ is continuous in the parameter, and it is the required developed surface.

In conclusion, it should be noted that the method of continuous bending of convex surfaces developed in [**17**] cannot be extended to space-like caps. The theorem is proved.

The problem of constructing a homeomorph of a cap without self-intersections in such a way that it admits non-trivial bending through the sliding of theborder along a certain plane without violation of the combinatorial structure is of interest in connection with well-known studies of closed flexible polyhedra carried out by Brickar, Connelly, and Stephen [**5,29**]. The uniqueness theorems for convex caps give sufficient reasons to hope that such flexible polyhedra exist in both the Euclidean and the pseudo-Euclidean case.

2.2. The Cauchy and Minkowski theorems. In this section, we consider classical uniqueness theorems for Euclidean polyhedra.

The polyhedra P and P' are called combinatorially equivalent, or identically composed, or isomorphic, if their elements, i.e., vertices, edges and faces, can be brought into a one-to-one correspondence preserving their inclusion relations.

THEOREM 1. *The closed convex polyhedra P and P' identically composed of equal faces are equal to one another, i.e., can be made to coincide with the help of a motion.*

This is the Cauchy theorem, which historically was the first result of paramount importance established for convex polyhedra in the theory of convex surfaces. The second such result is the Minkowski theorem. Its formulation will be preceded by the following definition.

The convex polyhedra P and P' are considered to be in correspondence according to the parallelism of external normals if their faces can be brought into a one-to-one correspondence in such a way that external normals to the corresponding faces are equal to one another.

THEOREM 2. *Let P and P' be closed convex polyhedra brought in correspondence with one another according to parallelism of external normals. Suppose that the surface areas of the corresponding faces are equal to one another. Then the polyhedra P and P' are equal and can be made to coincide via parallel translation.*

Minkowski did not associate his result, obtained in the study of the theory of volumes, with the Cauchy result. However, his theorem is, in a certain sense, a natural generalization of the Cauchy theorem. This can be perceived in well-known proofs, namely, in the original proof by Cauchy himself and in Aleksandrov's version of the proof of the Minkowski theorem which is close in spirit to Cauchy's proof. At a certain stage of his analysis Cauchy approaches the Minkowski theorem closely.

Consider the scheme of proof proposed by Cauchy. By hypothesis, two requirements are actually imposed on the polyhedra P and P', namely that the corresponding plane angles are equal and the corresponding edges are equal. If the polyhedra P and P' are not equal to one another, then there can be edges in P at which the dihedral angles are larger than those at the corresponding edges of P' and edges at which the dihedral angles are smaller. The edges of the former and the latter type are allocated the signs "plus" and "minus" respectively. According to the Cauchy lemma on the flexibility of a polyhedral angle, at least four changes of sign occur as we go around each vertex of P from which the edges labeled with a sign diverge. According to the Cauchy combinatorial lemma, it is impossible to ascribe signs to the edges of a closed convex polyhedron in such a manner. Hence, all corresponding dihedral angles of the polyhedra P and P' are equal to one another. Notice that this conclusion has been derived only from the first requirement imposed on the polyhedra. Let us now find the external geometric characteristics of the polyhedra P and P' without insisting on the fulfilment of the second requirement at this stage. It is obvious that P and P' admit a mutual arrangement such that these polyhedra can be brought into correspondence according to the parallelism of the external normal, and this correspondence is consistent with combinatorial equivalence. We shall assume that this arrangement actually takes place. Let k be the number of the edges of the polyhedron and f the number of its faces. Then the failure of the two such polyhedra to be equal to one another is characterized by f parameters, namely, supporting numbers of faces of the polyhedron P'. Thus, if instead of the second requirement we impose on the polyhedra other conditions and present them as f intrinsic geometrical conditions, then the equality of P and P' must follow from these conditions, too. The equality of the surface areas for f pairs of the corresponding faces of the polyhedra is the most natural such condition. Thus, we have obtained a nonstandard version of the Cauchy theorem, which includes the Minkowski theorem for the particular case of combinatorially equivalent polyhedra P and P' as well. The initial second requirement imposed on P and P' is more stringent, since though independent of the mutual arrangement of the polyhedra, it involves the fulfilment of $2k \geq 3f \sim k \gg f$. The Cauchy theorem is reformulated as well, and its reformulation generates a condition for the equality of combinatorial structures of isometric polyhedra.

The particular version of the Minkowski theorem can be proved by Aleksandrov's method, which follows the same scheme of Cauchy. In this case, Cauchy's scheme is applied to the network of lines on the polyhedron P, dual to the network of edges. The Cauchy lemma on the deformation of a convex polyhedral angle, or the equivalent lemma on the deformation of a convex spherical polygon with the preservation of the lengths of sides, is replaced here by Aleksandrov's dual lemma on the deformation of a plane convex polygon with the preservation of the angles if the surface area is preserved. The particular version proves to be sufficient for deriving the general statement of the theorem as its simple corollary. Indeed, consider the convex polyhedra $P_t = tP + (1-t)P'$ and $P_{1-t} = (1-t)P + tP'$ in the conditions of Lemma 2. These polyhedra are obtained by the vector summation of P and P' according to Minkowski for $0 < t < 1$. It can easily be verified that the polyhedra P_t and P_{1-t} are combinatorially equivalent. At the same time, they satisfy the condition of Theorem 2. Hence the particular version of the theorem is fulfilled for them, i.e., P_t and P_{1-t} can be made to coincide by parallel translation. Then the polyhedra P and P' can also be made to coincide by parallel translation as the limits $P' = \lim P_t$ and $P = \lim P_{1-t}$ as $t \to 0$. The method of reducing the theorem to the case of combinatorially equivalent polyhedra is also applicable to the version of the Minkowski theorem formulated by Aleksandrov. In this version, the perimeters of the faces corresponding to one another according to parallelism of external normals are equal to one another. The condition of the equality of perimeters is preserved as we go over to the polyhedra P_t and P_{1-t}. In the course of the proof carried out according to Aleksandrov's scheme, the corresponding lemma on four sign changes caused by the change in the lengths of the sides of the faces is established with the help of Lemma 5 of 2.3.

Let us formulate a new theorem, which can be considered as intermediate between the Cauchy and Minkowski theorems.

Let P and P' be closed convex polyhedra combinatorially equivalent to one another, and let the corresponding plane angles and the surface areas of the corresponding faces of these polyhedra be equal to one another. Then P and P' are equal to one another and can be brought into coincidence by a motion.

Instead of the Cauchy theorem, we can now generalize this theorem to spherical, and not only to spherical, space: a more convenient formulation is in elliptic space. In the next section, it will be extended to spheres in pseudo-Euclidean space, i.e., Lobachevskiĭ and de Sitter spaces. In the conditions of the theorem generalized in such a way, the requirement that the surface areas of the corresponding faces be equal to one another becomes superfluous since it is satisfied automatically. The theorem is a generalization of both the Minkowski and Cauchy theorems.

THEOREM 3. *Let P and P' be closed convex polyhedra in elliptic space that are combinatorially equivalent to one another and let the corresponding plane angles of these polyhedra be equal to one another. Then P and P' are equal to one another and can be made to coincide by a motion.*

PROOF. A convex polyhedron in elliptic space is interpreted as the intersection of a strictly convex polyhedral cone in four-dimensional Euclidean space with a unit hypersphere, the vertex of the cone being situated at the center of the sphere. The convex polyhedral cone with center at the same sphere constructed so that it is dual to the cone defining the initial polyhedron in turn defines the new polyhedron on the hypersphere. This new polyhedron is called polar to the initial polyhedron. Mutually polar polyhedra have dual combinatorial structures in the spherical space. A plane angle of one of the polyhedra corresponds to the supplementary plane angle of the other polyhedron. It follows from the duality property that the plane angles of one polyhedron that converge at the common vertex correspond to plane angles of the entire face of the other polyhedron. The edges of one polyhedron supplement the values of the corresponding dihedral angles of the other polyhedron up to the angle π.

Let us consider P and P' as polyhedra in the spherical space. According to the Cauchy result, the equality of the corresponding plane angles of P and P' entails the equality of the corresponding dihedral angles. The polyhedra polar to P and P' also satisfy the conditions of the theorem. Hence the corresponding dihedral angles are equal to one another in these polyhedra, too. This means that the corresponding edges of P and P' are equal to one another. Thus, P and P' are equal.

The use of polar polyhedra makes it possible to prove equality of the dihedral angles and edges of P and P' only with the help of Cauchy's auxiliary results. One can also do without polar polyhedra. In this case, we must use Aleksandrov's auxiliary lemma on the deformations of polyhedra and his scheme of proof of the Minkowski theorem to prove the equality of the edges. This is, of course, equivalent to the use of polarity. However, it is necessary to resort both to Cauchy's and Aleksandrov's auxiliary results if we want to generalize the theorems to spheres in pseudo-Euclidean space. This is because polar polyhedra, like polar spherical polygons, belong to metrically different spheres.

2.3. Combinatorially equivalent caps. Theorem 3 of 2.2 together with its proof is extended to the case of closed convex combinatorially equivalent polyhedra on unit spheres of pseudo-Euclidean space, i.e., Lobachevskiĭ and de Sitter spaces. We consider space-like de Sitter polyhedra. Here we can confine ourselves only to the formulation of the corresponding theorem. In view of duality, it is sufficient to prove this theorem only for one of the spaces. Only auxiliary results on the deformations of the closed convex polyhedral lines in the spherical, Lobachevskiĭ, and de Sitter planes are required for actually carrying out this proof. These auxiliary results, which are naturally referred to as the Cauchy and Aleksandrov lemmas, will be described in this section along with some other similar results. In view of duality, polygonal lines are not considered in the de Sitter plane since it is sufficient to obtain the corresponding results in the Lobachevskiĭ plane. We do not present the formulation of the lemmas (of Cauchy type) on deformations of the convex polyhedral cones which preserve the values of planes or dihedral angles. The lemmas on deformations of polygonal lines represent a simple reformulation of these lemmas. For instance, the analog of the Cauchy lemma on the change of signs under an isometric deformation of a space-like convex polyhedral cone with preservation of the values of plane angles is equivalent to Aleksandrov's lemma on the deformation of a closed convex polygon in a hyperbolic plane with the preservation of its angles. This section also contains some new results, including uniqueness theorems for space-like caps and the corresponding auxiliary results on deformations of non-closed polygonal lines. For completeness, we also present two simple uniqueness theorems for combinatorially equivalent polyhedra with nonintrinsic geometrical conditions.

Below, by the angle between the faces of a convex dihedral space-like angle whose convex side is oriented in the direction $z > 0$ we mean the angle between the external normals to the faces. By the angle of inclination of the boundary face of a space-like cap to the plane xy we mean the angle between the external normal to the face and the direction $z > 0$, taken with a minus sign. For space-like polyhedra in de Sitter space, the concept of the angle of inclination of a face and that of a dihedral angle correspond to these definitions.

THEOREM 1. *Let P and P' be closed convex combinatorially equivalent polyhedra on unit spheres in the pseudo-Euclidean case, i.e., in Lobachevskiĭ or de Sitter space. Also let the corresponding plane angles of these polyhedra be equal to one another. Then the polyhedra P and P' are equal to one another and can be made to coincide by a motion.*

Let us now formulate the Cauchy lemmas.

LEMMA 1. *If all angles of a convex polygonal line do not decrease under an isometric deformation on a sphere or in a spherical space and at least one of its angles increases, then the distance between the ends of the polygonal line increases.*

PROOF. For simplicity we shall consider four-link polygonal lines, since our arguments are applicable to polygonal lines with an arbitrary number of links. Let $L = ABCDK$ and $L' = A'B'C'D'K'$ be polygonal lines on a sphere in which all corresponding sides are equal to one another; all corresponding angles satisfy inequalities of the same form, e.g., $\angle C \leq \angle C'$; among the corresponding angles there are unequal angles, say $\angle C < \angle C'$. Let us show that if the polygonal line L is convex, the inequality $AK < A'K'$ is valid for the chords closing the polygonal lines L and L'.

Let ABB_1 and AKB_1, $A'B'B'_1$ and $A'K'B'_1$ be great semicircles on the sphere, BCC_1 and CDD_1 the links of the polygonal line L continued up to the intersection with the arc KB_1, and $B'C'C'_1$ and $C'D'D'_1$ the corresponding links of L' continued by segments of the same lengths.

Comparing the triangle KDD_1 with the triangle $K'D'D'_1$, the triangle D_1CC_1 with the triangle $D'_1C'C'_1$, and the triangle C_1BB_1 with the triangle $C'_1B'B'_1$, we find that $KD_1 \geq K'D_1'$, $D_1C_1 > D_1'C_1'$ and $C_1B_1 \geq C_1'B_1'$. Summing these inequalities, we obtain $KB_1 > K'B_1'$. In view of the equality $AK + KB_1 = A'K' + K'B_1'$, we find that $AK < A'K'$. The lemma is proved.

This result is more general than the Cauchy lemma. It is interesting to note that neither Cauchy nor other mathematicians who repeatedly discussed the original proof given by Cauchy himself [**5**] could find the optimal proof presented above. Cohn-Vossen [**13**] remarks that Cauchy's proof must be considered as a proof by induction with respect to the number of sides of the polygons.

Cauchy deduced the following lemma from Lemma 1. We also propose a new proof independent of Lemma 1. This proof is equally applicable in any (spherical, Euclidean or hyperbolic) plane. It is based on the method of the "hinged" quadrangle, which is often used in solving problems associated with isoperimetry.

Let $L = AB \dots C \dots D$ and $L' = A'B' \dots C' \dots D'$ be nontrivially isometric convex polygons. Let us use the following convention to ascribe the plus and minus signs to the corresponding angles of L and L' that are not equal to one another. For corresponding vertices C and C', the vertex C is ascribed the plus sign if the angle C is larger than the angle C', and the minus sign in the opposite case. Hence C' has a sign opposite to that of C. It can easily be seen that at least four vertices in each of the polygons L and L' are ascribed signs. Otherwise these polygons are equal to one another.

LEMMA 2. *There are not fewer than four sign changes as we go around each of the polygons L and L' along a closed contour.*

PROOF. Suppose that the number of sign changes is less than four for the polygons L and L'. Let n be the number of labeled vertices in each of these polygons. We shall demonstrate that we can go from the pair L and L' to another pair of non-trivially isometric convex polygons with fewer labeled vertices but with a number of changes of the sign that is also less than four in each of the polygons. Such a transformation is carried out with the help of the hinge deformation of L and L'. It is important that under the proposed transformation the number of sign changes does not increase in each of the polygons, some of the labeled vertices inevitably survive and the number of the labeled vertices decreases. As a result of a finite number of such transformations, we can obtain isometric convex polygons in which the number of labeled vertices is not larger than three in each of the polygons. Such polygons must be equal to one another and they cannot contain vertices labeled with signs. Thus we arrive at a contradiction. Hence, our assumption is false. So now we must construct appropriate deformations of L and L'.

Let $n = 4$. Since the number of changes of sign is less than four, in one of the polygons, say in L, four labeled vertices H, R, T, and E serve as vertices of a convex quadrangle and three of the vertices, say H, T and E, are ascribed a minus sign. Let us subject the polygon L to a continuous isometric deformation defined by a hinge deformation of the quadrangle $HRTE$ with a decrease in the diagonal HE accompanied by a decrease in the values of the angles R and T. All other angles of L, except E and H, do not undergo any change, and the polygon L remains convex. Under such a deformation, each of the four polygonal lines in which L is divided by labeled vertices is subjected to a motion. The vertices of the polygon subjected to such a transformation preserve the sign they had before the start of the deformation. The deformation is completed when at least one of the angles, say E, or H, or R, which had a plus sign before the deformation, becomes equal to the corresponding angle of L'. At this moment, the vertex of such an angle and the corresponding vertex of L become unlabeled. The new convex polygon $\overline{L}$ obtained from L is isometric, but not equal, to L', since the angle at the vertex T decreases as a result of the deformation. As we go over from L, L' to $\overline{L}, L'$, the number of sign changes does not increase in either polygon, and remains less than four. The number of labeled vertices decreases in each of the polygons and becomes less than $n = 4$.

Let $n > 4$. Since the number of sign changes is less than four, among five labeled vertices arbitrarily chosen in one of the polygons or among five corresponding vertices of the other polygon, there exist four vertices with the same arrangement of the signs as in the case $n = 4$, i.e., three vertices labeled with the minus sign. Repeating the procedure of deforming the pair L, L' outlined above, we obtain a new pair of isometric convex polygons with the required arrangement of signs. In particular, the number of labelled vertices is less than n in each of the polygons belonging to the new pair. The lemma is proved.

The following lemma is dual to Lemma 1. Similarly to Aleksandrov's lemma, it is proved for polygons lying in an arbitrary (spherical, Euclidean or hyperbolic) plane.

Let $A \dots BC \dots D$ and $A' \dots B'C' \dots D'$ be convex polygonal lines which lie inside the angles AOD and $A'O'D'$ respectively, face the vertices of these angles both by their convex sides or both by their concave sides, and form the polygons $L = OA \dots BC \dots D$ and $L' = O'A' \dots B'C' \dots D'$ together with the intercepts of the sides of the corresponding angles. We shall assume that, in the segments $A \dots BC \dots D$ and $A' \dots B'C' \dots D'$ of the polygonal lines L and L', all corresponding angles are equal to one another ($\angle A = \angle A', \dots, \angle D = \angle D'$), all corresponding sides satisfy the same inequalities ($\dots, BC \geq B'C', \dots$) and there exist strictly unequal sides among the corresponding sides, e.g., $BC > B'C'$. In the angles AOD and $A'O'D'$ the polygons L and L' bound close domains whose surface areas are denoted by $\sigma(L)$ and $\sigma(L')$ respectively.

LEMMA 3. $\sigma(L) > \sigma(L')$.

PROOF. For simplicity we consider polygons with three link sections, since our method of proof is applicable to polygons with an arbitrary number of links. Connecting the vertex O with all vertices of the polygonal line $ABCD$ by rectilinear segments, we divide L into triangles. On each link of the polygonal line $A'B'C'D'$, we construct a triangle whose angles at the vertices of the links coincide with the angles of the corresponding triangle constructed for L. It is obvious that the set Σ of triangles constructed for the polygon L' covers the interior of this polygon.

By construction, the surface area of each of the triangles realizing the partitioning of L is not smaller than the surface area of the corresponding triangle from the set Σ. The surface area of one of the triangles, namely of the one with the side BC, is larger than the surface area of the corresponding triangle (the one with the side $B'C'$). It follows hence that $\sigma(L) > \sigma(L')$. The lemma is proved.

Let $L = A \ldots BC \ldots D$ and $L' = A' \ldots B'C' \ldots D'$ be convex polygons in which all corresponding angles are equal to one another, i.e., $\angle A = \angle A', \ldots \angle D = \angle D'$, etc., and there are unequal sides among the corresponding sides of these polygons, e.g. $BC \neq B'C'$. Suppose, in addition, that the surface areas of these polygons are equal to one another if we deal with polygons in Euclidean space. We ascribe the plus and minus signs to unequal corresponding sides of the polygons according to the following rule. The plus sign is ascribed to the side BC, and the minus sign to the side $B'C'$ if $BC > B'C'$, and vice versa if $BC < B'C'$. At least four sides of the polygons L and L' are ascribed signs. Otherwise these polygons turn out to be equal.

LEMMA 4. *There are at least four changes of sign as one goes around each of the polygons along a closed path.*

PROOF. Suppose that the number of sign changes is less than four for each of the polygons. This means it can be equal to two or zero.

If there are no sign changes for the polygons L and L', the application of Lemma 3 to them leads to a contradiction. Indeed, the equality of the surface areas of L and L' in the Euclidean case is ensured by imposing an additional condition. For polygons in the spherical or hyperbolic planes, this equality follows from the equality of the corresponding angles of these polygons. According to Lemma 3, the surface area of one of the polygons is larger as the polygons are not equal to one another. Thus we arrive at a contradiction.

Suppose that the number of sign changes is equal to two. In this case, the polygon L is divided into two polygonal lines $A \ldots BC$ and $AD \ldots C$. All sides ascribed the plus sign belong to one of these polygonal lines and those with the minus sign belong to the other line. Let A and C be the vertices common to the polygonal lines and let OA and OC be straight lines supporting L at the points A and C. The surface areas of the polygons L and $OAD \ldots C$ are denoted by s and σ respectively. Let $O'A'$ and $O'C'$ be the supporting straight lines for L', which are drawn in such a way that at the vertices A and C, these lines form angles with the sides of L' which are equal to the angles formed by the corresponding supporting straight lines with the sides of L. Let σ' be the surface area of the polygon $O'A'D' \ldots C'$. Suppose that the convex side of the polygon $A'D' \ldots C'$ faces the point O' and the convex side of the polygon $AD \ldots C$ faces the point O. Let us ascribe the minus sign to the sides of the polygon $AD \ldots C$ and the plus sign to those of the polygon $A'D' \ldots C'$. Clearly, certain sides may not be labeled at all. Comparing the polygons $OA \ldots BC$ and $O'A' \ldots B'C'$, we find from Lemma 3 that $s + \sigma > s + \sigma'$. According to the same lemma applied to the polygons $OAD \ldots C$ and $O'A'D' \ldots C'$, we have $\sigma < \sigma'$. Thus we come to a contradiction.

In the case considered above, we have not studied the important issues associated with mutual arrangement of the straight lines that serve as supporting lines to a polygon (L or L'). These issues do not arise if we consider the polygons in a spherical plane. However, for a hyperbolic or Euclidean plane, there can exist supporting straight lines that even do not intersect one another. The following additional comments can be made in this connection.

Let α and γ be nonintersecting straight lines that are supporting lines for the polygon L at the vertices A and C. Let us draw the straight lines α' and γ' through

A and C in such a way that they serve as supporting lines for L' and with its sides form angles equal to the angles formed by α and γ with the sides of L. Suppose that α' and γ' also do not intersect. We assume that the sides of the polygonal line $AD\ldots C$ are labeled with the minus sign and those of $A\ldots BC$ with the plus sign. Comparing the polygonal line $AD\ldots C$ situated between the straight lines α and γ with the convex polygonal line $A'D'\ldots C'$ situated between the straight lines α' and γ', we conclude from Lemma 5 (presented below) that the distance between α' and γ' is greater than that between α and γ. Applying the same lemma to compare the convex polygonal lines $A\ldots. BC$ and $A'\ldots B'C'$, we find that these distances satisfy the opposite inequality. Thus, we come to a contradiction. Let us now suppose that the straight lines α and γ intersect. These straight lines are denoted by OA and OC respectively. We assume that the convex side of the polygon $AD\ldots C$ faces the point O and that the edges of this polygon are ascribed the plus sign. The straight lines α' and γ' also intersect (we denote the point of intersection by O'). Moreover, it turns out that the convex side of the polygon $A'D'\ldots C'$ faces the point O. To prove these statements it is sufficient to construct the system of triangles projecting the polygonal line $AD\ldots C$ from the point O and to construct, as in the proof of Lemma 3, the set Σ of corresponding triangles on the polygonal line $A'D'\ldots C'$. A simple geometrical analysis of Σ shows that α' and γ' do indeed intersect, that the point O of intersection belongs to the region of the plane covered by the set Σ, and that the convex side of the polygonal line $A'D'\ldots C'$ faces the point O.

Thus, in any case, the assumption that there are fewer than four sign changes on the polygons L and L' leads to a contradiction. Hence, this assumption is not true. The lemma is proved.

Let $L = AB\ldots CD\ldots E$ and $L' = A'B'\ldots C'D'\ldots E'$ be convex polygonal lines in a Euclidean or hyperbolic plane. Suppose that L is situated between disjoint straight lines α_A and α_E passing through the points A and E, and L' lies between the straight lines $\alpha_{A'}$ and $\alpha_{E'}$ passing through A' and E'. We shall assume that for these polygonal lines, all corresponding angles are equal to one another, i.e., $\angle B = \angle B', \ldots, \angle C = \angle C'$, etc., all corresponding sides satisfy the same inequalities, e.g., $CD \geq C'D'$ etc., and there exist strictly unequal corresponding sides, e.g., $CD > C'D'$. Suppose that at the end points of these polygonal lines from the convex side, they form equal angles with the straight lines α_A and α_E, $\alpha_{A'}$ and $\alpha_{E'}$.

LEMMA 5. *The distance h between the straight lines α_A and α_E is greater than the distance h' between the straight lines $\alpha_{A'}$ and $\alpha_{E'}$* $(h > h')$.

PROOF. Let us draw mutually non-intersecting straight lines $\alpha_{B'}, \ldots, \alpha_{C'}, \alpha_{D'}, \ldots$ passing through the inner vertices $B', \ldots, C', D', \ldots$ of the polygonal line L' in such a way that they do not intersect the straight lines $\alpha_{A'}$ and $\alpha_{E'}$ and are orthogonal to the common perpendicular to the straight lines $\alpha_{A'}$ and $\alpha_{E'}$. The distances between the straight lines $\alpha_{A'}$ and $\alpha_{B'}$, $\alpha_{B'}$ and $\ldots, \ldots, \alpha_{C'}$ and $\alpha_{D'}, \ldots$ are successively denoted by $h_{A'}, h_{B'}, \ldots, h_{C'}$, etc. Let us also draw the straight lines $\alpha_{B'}, \ldots, \alpha_{C'}, \alpha_{D'}, \ldots$ through the vertices $B, \ldots, C, D, \ldots$ of the polygonal line L in such a way that the angles formed at these vertices by these lines and the links of L are equal to the angles formed by the straight lines

$\alpha_{B'}, \ldots, \alpha_{C'}, \alpha_{D'}, \ldots$ and the corresponding links of L'. The distances between the straight lines $\alpha_A, \alpha_B, \ldots, \alpha_C, \ldots$ are denoted consecutively by $h_A, h_B, \ldots, h_C, \ldots$ as for the straight lines $\alpha_{A'}$, $\alpha_{B'}$, etc. It follows from the conditions of the lemma and from the construction of the straight lines that $\alpha_A, \alpha_B, \ldots, \alpha_C, \ldots$ are mutually disjoint. The inequalities $h_A \geq h_{A'}, h_B \geq h_{B'}, \ldots, h_C > h_{C'}, h_D \geq h_{D'}, \ldots$ are valid on the same grounds. Then $h > h'$ since $h \geq h_A + h_B + \cdots + h_C + \ldots$ and $h' = h_{A'} + h_{B'} + \cdots + h_C + \ldots$. The lemma is proved.

Needless to say, this lemma is trivial for the Euclidean plane. However, if we carry out this simple proof step by step, it is naturally generalized to the proof given for the case of the hyperbolic plane. Lemma 5 is an analog of Lemma 3 if the latter is reformulated for the spherical and hyperbolic plane with the help of the Gauss–Bonnet theorem as the statement about the comparison of the angles O and O' of the polygons L and L'. Together with the proof given above, Lemma 3 is dual to Lemma 1. This indicates a direct connection of Cauchy's lemma with Lemmas 3 and 5.

Naturally, the uniqueness theorems for combinatorially equivalent caps are proved in Euclidean, spherical and hyperbolic spaces with the help of the corresponding uniqueness theorems for closed polyhedra. For this purpose, each cap is joined with the cap symmetric to it with respect to the base plane. In pseudo-Euclidean and de Sitter spaces, such a union of a space-like cap with the cap symmetric to it leads to a conventional convex surface for which the polyhedral cones are not space-like along the gluing line. The sum of the internal angles at the vertices of these cones is less than 2π. In spite of this fact, we do not have sufficient grounds for using the Cauchy lemma on the deformations of polyhedral cones. Here we need independent results on deformations of conventional convex cones in which we must consider and compare dihedral angles introduced at the beginning of this section. In this case, we can use the technique developed by Cauchy. We present the uniqueness theorem for combinatorially equivalent caps in pseudo-Euclidean and de Sitter spaces. The proofs are omitted since they are similar to that of Theorem 3 from 2.2. In addition to Lemmas 2 and 4 we also require statements characterizing deformations of nonclosed cones along the border of the cap. Dual equivalents of these statements, i.e., Lemmas 6 or 7 on deformations of plane polygonal lines, are formulated below.

THEOREM 2. *Let P and P' be combinatorially equivalent space-like caps in the pseudo-Euclidean and de Sitter space. Let the corresponding plane angles of these caps be equal to one another and let the surface areas of the corresponding faces be equal to one another for the case of pseudo-Euclidean polyhedra. Then the caps P and P' are equal to one another and can be made coinciding by a motion.*

To prove this theorem by the Cauchy–Aleksandrov method we must compare dihedral angles at the inner and boundary edges of boundary faces belonging to nonclosed polyhedral cones of the caps in question. Two cases are possible for a specific cone of this type. In the first case, all inner edges of the cap which diverge from the vertex of the cone are ascribed the same sign. Then, according to Lemma 6, the boundary edges must be ascribed opposite signs. In the second case, some inner edges diverging from the cone vertex are ascribed one sign and others are ascribed the opposite sign. The arrangement of the signs is such that there is only one sign change as we go around the vertex along the inner edges. Then at least one of the boundary edges is ascribed the sign opposite to that ascribed to the inner edges of the cone which are situated on the side of this boundary edge. The latter statement follows from Lemma 7. The possibility of realizing such a configuration of the signs allows us to transfer the signs to the corresponding edges of symmetric caps without any change and thereby obtain the arrangement of signs on closed conventional polyhedra. In this case, there are no fewer than four sign changes in a conventional polyhedron, as go around the vertex situated on the gluing line.

Let us formulate the lemmas necessary for the proof of Theorem 2.

LEMMA 6. *Under the conditions of Lemma 3, let $L = OA \dots BCD$ and $L' = O'A' \dots B'C'D'$ be convex-concave polygons situated in a Euclidean or hyperbolic space. If the convex sides of the polygonal lines $A \dots BCD$ and $A' \dots B'C'D'$ face the points O and O' respectively, then $OA > O'A'$ and $OD > O'D'$.*

PROOF. For simplicity, we consider polygons with three link parts $ABCD$ and $A'B'C'D'$, since our method is applicable to polygons with an arbitrary number of links. Let us plot the segment BA on the side BA in such a way that it is equal to $B'A'$ and draw the segment $\overline{A}O_1$ ($O_1 \in OD$), where $\angle O_1\overline{A}B = \angle OAB$. Let us continue the link CB of the polygonal line $DCBA$ of the polygon L until it intersects the segment $\overline{A}O_1$, and denote the segment obtained by CBB'. We also continue the link $C'B'$ of the polygonal line $D'C'B'A'$ of the polygon L' until it intersects the side $O'A'$ of the angle $A'O'D'$, and denote the segment obtained by $C'B'B_1'$. The triangles $\overline{A}BB_1$ and $A'B'B_1'$ are equal to one another. Hence $BB_1 = B'B_1'$. It follows that the polygons $L_1 = O_1B_1CD$ and $L_1' = O'B_1'C'D'$ satisfy the conditions of this lemma but have fewer links at the section B_1CD and $B_1'C'D'$ than L and L'. Repeating this construction for the polygons L_1 and L_1', we obtain the triangles O_2C_1D and $B_1'C'D'$, where $O_2 \in O_1D$ and DCC_1 and $D'C'C_1'$ are the links DC and $D'C'$ extended up to the intersection with the segments $O_2\overline{B}_1$ and $O'B_1'$ respectively. In these triangles $C_1D \geq C_1'D'$, and the corresponding angles at the sides C_1D and $C_1'D'$ are equal to one another. Hence $O_2D \geq O'D'$. Actually, $O_2D > O'D'$ since, by construction, $OD \geq O_1D \geq O_2D$. But by the condition of the lemma, one of the links of the polygonal line $ABCD$ is strictly larger than the corresponding link of $A'B'C'D'$ if their links satisfy the inequalities $AB \geq A'B'$, $BC > B'C'$, and $CD \geq C'D'$. The lemma is proved.

LEMMA 7. *Let the conditions of Lemma 6 be satisfied with one variation for the relation between the lengths of the corresponding links of the polygonal lines $A \dots B \dots CD$ and $A' \dots B' \dots C'D'$. Namely, the lengths of the links of the first polygonal line are not smaller than the lengths of the corresponding links of the second polygonal line at the sections $A \dots B$ and $A' \dots B'$. The opposite inequalities are satisfied at the sections $B \dots CD$ and $B' \dots C'D'$. For each pair of corresponding sections, let the inequality be strict for at least one pair of corresponding links. Then at least one of the inequalities $OA > O'A'$ or $OD < O'D'$ is satisfied for the polygons L and L'.*

PROOF. Suppose that the lemma is false, and $OA \leq O'A'$ and $OD \geq O'D'$. Let α and α' be straight lines serving as supporting lines to the polygonal lines $A \dots B \dots CD$ and $A' \dots B' \dots C'D'$ at the vertices B and B' and drawn in such a way that the angles formed by α with the links of the polygonal line $A \dots B \dots CD$ which diverge from B are equal to the angles formed by α' with the corresponding links of $A' \dots B' \dots C'D'$ at the vertex B'. Let β be a straight line passing through the point O and orthogonal to α, and let DE (where $E \in \beta$) be a straight line orthogonal to β. Let us construct the right triangle $O'E'D'$ with the right angle at the vertex E'. The angles formed by the side $D'E'$ with the straight lines $D'O'$ and $D'C'$ are equal to the angles between the side DE of the triangle OED and the straight lines DO and DC respectively. It follows from the conditions of the lemma and the construction of the straight lines that α' and $D'E'$ do not intersect, since the straight lines α and DE do not intersect (similar arguments were used in the proof of Lemma 5). Let us denote by T the point of intersection of α and β, by T' the point of intersection of α' and $O'E'$, by $h_0 = OT$ and h_0' the distances from the points O and O' to the straight lines α and α', respectively, by $h_1 \equiv TE$ and h_1' the distances between the straight lines α and DE, and α' and $D'E'$. respectively, and finally by $h_0'' = O'T'$ and $h_1'' = T'E'$ the lengths of the segments of the small side $O'E'$ of the right triangle $O'E'D'$. According to our assumption, we see from a comparison of the right triangles OED and $O'E'D'$ that $h_0 + h_1 \geq h_0'' + h_1''$. Moreover, for the distances between the points and the straight lines introduced just above, we get the inequality $h_0'' + h_1'' \geq h_0' + h_1'$. Hence $h_0 + h_1 \geq h_0' + h_1'$. It follows from Lemma 5 and from the condition of this lemma on the relations between the lengths of the links of the polygonal lines $B \dots CD$ and $B' \dots C'D'$ that $h_1 < h_1'$. Hence $h_0 > h_0'$. Interchanging in our reasoning the polygons L and L' and comparing the polygonal lines $A' \dots B'$ and $A \dots B$, we conclude that the opposite inequality $h_0' > h_0$ is valid. The resulting contradiction shows that our initial assumption is not true. The lemma is proved.

For completeness, we present the lemmas dual to Lemmas 6 and 7. They are used, together with Lemma 2 (the Cauchy lemma), to prove two more uniqueness theorems for combinatorially equivalent polyhedra. In these theorems, pairs of polyhedra are considered either in Euclidean, spherical and hyperbolic spaces, or in pseudo-Euclidean and de Sitter spaces. In the latter cases, the polyhedra in question are supposed to be space-like. First we formulate the theorems. Note that they are proved by the methods similar to those used to prove Theorems 1 and 2. However, the conditions of these theorems are not intrinsically geometric.

THEOREM 3. *Let P and P' be two closed convex combinatorially equivalent polyhedra with equal respective edges and equal respective dihedral angles. Then P and P' are equal and can be made to coincide by a motion.*

Interestingly, if k is the number of edges and f_n the number of n-gonal faces of P, the number of conditions that P and P' must satisfy in Theorem 3 is the same as in Theorem 1, i.e., $2k = \sum n f_n$.

THEOREM 4. *Let P and P' be two combinatorially equivalent caps with equal respective edges and equal respective dihedral angles, including the dihedral angles along the border of the caps, these angles being considered the inclinations to the plane of the edge of the boundary edges. Then P and P' are equal and can be made to coincide by a motion.*

Now we formulate the respective duality lemmas that refer to polygons in planes of the spherical, Euclidean, and hyperbolic geometries.

LEMMA 8. *Let AOD and $A'O'D'$ be isometric angles with $OA = O'A'$ and $OD = O'D'$. We inscribe the isometric convex polygonal lines $AB \dots CD$ and $A'B' \dots C'D'$ with $AB = A'B'$, ..., $CD = CD'$, into these angles. Finally, let the angles of the polygonal line $AB \dots C'D'$ at the internal vertices from the side of vertex O be no larger than the respective angles of the polygonal line $A'B' \dots C'D'$ and let at least one angle be strictly smaller. Then $\angle OAB > \angle O'A'B'$ and $\angle ODC > \angle O'D'C'$.*

PROOF. Suppose that the polygonal lines are both convex in relation to the vertices O and O', so that the polygons $OAB \dots CD$ and $O'A'B' \dots C'D'$ are convex–concave. The lemma is proved in the same way as Lemma 1. More precisely, if we assume that $\angle ODC < \angle O'D'C'$, we arrive at a contradiction: $AB < A'B'$. Suppose that the polygonal lines are both concave in relation to the vertices O and O', so that the polygons $OAB \dots CD$ and $O'A'B' \dots C'D'$ are convex. The lemma follows from Lemma 2. If the assertion of the lemma does not hold, then in comparing the angles of the polygons $OAB \dots CD$ and $O'A'B' \dots C'D'$ by ascribing signs, the number of sign changes is less than four as we go around the contour of each polygon. Suppose that one of the polygonal lines $AB \dots CD$ or $A'B' \dots C'D'$ is convex in relation to the vertices O and O' and the other is concave. We examine the intermediate triangle $O''A''B'' \dots C''D''$ with one side being the rectilinear segment $A''B'' \dots C''D''$, while the triangle itself is isometric to the polygons $OAB \dots CD$ and $O'A'B' \dots C'D'$, with $O''A'' = OA$, $A''B'' = AB$, ..., $C''D'' = CD$, and $D''O'' = DO$. The validity of the lemma is established by consecutively comparing the polygons $OA \dots CD$ and $O''A'' \dots C''D''$, $O''A'' \dots C''D''$ and $O'A' \dots C'D'$ and applying the results to the convex–concave and convex polygons $OAB \dots CD$ and $O'A'B' \dots C'D'$.

The lemma is proved.

Note that Lemma 1 of 1.5 is a particular case of Lemma 8 for polygonal lines in the Euclidean case. Our proof of Lemma 8 is much simpler. but cannot be applied to space-like double prisms; see also [**4**].

LEMMA 9. *Suppose that the assumptions of Lemma 8 hold with the following refinements for two isometric polygonal lines $AB_1 \dots BC \dots C_1D$ and $A'B'_1 \dots B'C' \dots C'_1D$. These polygonal lines are either both convex or both concave in relation to the vertices O and O'; on the sections $A \dots B$ and $A' \dots B'$ all the angles at the interior vertices of the first polygonal line with respect to vertex O are no larger than the corresponding angles of the second polygonal line, while on the sections $C \dots D$ and $C' \dots D'$ the opposite is true; for each pair of corresponding sections the inequality is strict for one pair of corresponding angles. Then at least one inequality, $\angle OAB_1 > \angle O'A'B'_1$ or $\angle ODC_1 < \angle O'D'C'_1$, holds.*

PROOF. Let both polygonal lines be convex in relation to O and O'. Suppose that the lemma is false: $\angle OAB_1 \leq \angle O'A'B'_1$ and $\angle ODC_1 \geq \angle O'D'C'_1$. Let $E \in BC$ and $E' \in B'C'$ be two interior points of isometrically corresponding links. Comparing the convex–concave polygons $OA \dots BEO$ and $O'A' \dots B'E'O'$ as we did when studying the first case in Lemma 8, we find that $OE < O'E'$. Similarly, comparing the polygons $O'D' \dots C'E'O'$ and $OD \dots CEO$, we find that $O'E' < OE$. We have arrived at a contradiction. Now suppose that both polygons are concave in relation to O and O'. Repeating the above arguments, we again arrive at the opposite inequalities $OE < O'E'$ and $O'E' < OE$; however, here the comparison of the respective polygons and the closing chords is done by Lemma 1 (the Cauchy lemma). Thus, in either case we arrive at a contradiction if we assume that the lemma is false.

The lemma is proved.

3. Stability

Now we examine how the stability of ordinary space-like caps in the class of convex prismatic polyhedra depends on variations of a certain integral characteristic of a cap.

3.1. Variations of integral curvature. Let V be a prism with its upper base a triangle, $\triangle ABC$. Through the edge AB we draw the space-like plane τ, with the fastest descent line in relation to the xy-plane being the same edge AB. We introduce a numerical characteristic of the prism at edge AB, which in absolute value is equal to the angle between the lines normal to τ and $\triangle ABC$ and is directed into the half-space $z > 0$. The characteristic is positive if $\triangle ABC$ lies above τ and negative if it lies below, and we call it the complementary angle at the edge AB of the prism. Now let $ACBD$ be a double prism with AB the inner edge of the upper base. We call the sum of the complementary angles at the edge AB of the prisms with upper bases $\triangle ACB$ and $\triangle ADB$ the complementary angle at the inner edge of the double prism. Obviously, for a Euclidean double prism the complementary angle is equal to the respective exterior angle at the edge AB. Expressions like "the complementary angle at a boundary edge" and "the complementary angle at the inner edge of the upper base of a prismatic cap" require no additional explanation. Let P be a convex prismatic cap with curvatures at the inner vertical edges h_i, prismatic curvatures equal to ω_i, and complementary angles at the boundary and inner edges l_j of the upper base equal to α_j. Then the quantity

$$H(P) = \sum \omega_i h_i + \sum \alpha_j l_j$$

is said to be the integral curvature of P. We now wish to prove assertions of the type of stability theorems, the full analogs of Lemma 1 of [**9**], which establish the extremal properties of $H(P)$ for a space-like cap.

Let $\{P\}$ be the set of convex prismatic caps with upper bases constructed in a similar manner from equal triangles and representing the same developed surface S both metrically and combinatorially. We say that the caps $\{P\}$ have the same triangulations. A Euclidean vector $\mathbf{h} = (h_0, h_1, \dots, h_n)$ determined by the values of the altitudes of all the inner vertices $A_0, A_1, \dots, A_n$ of a specific cap can be introduced as a parameter on the set $\{P\}$. Under such a parametrization the functionals $H(P)$ and $\omega_i(P)$ are infinitely differentiable, or even analytic, functions. Here we are speaking of space-like caps, while the Euclidean variant requires refinements, both here and in other cases, owing to the possibility of P having vertical boundary faces. The properties of the increments $\Delta H(P)$ and $\Delta\omega_i(P)$ are expressed by the following proposition.

LEMMA 1. *For an arbitrary cap,*

$$\Delta\omega_i(P) = \sum_j \frac{\partial\omega_i}{\partial h_i}\,\Delta h_j + o(|\Delta\mathbf{h}|), \qquad \Delta H(P) = \sum_i \omega_i \Delta h_i + o(|\Delta\mathbf{h}|),$$

where the terms $o(\cdot)$ admit the general estimate $|o(\cdot)| \le \varepsilon(|\Delta\mathbf{h}|)|\Delta\mathbf{h}|$, with the function $\varepsilon(\cdot)$ determined solely by the developed surface having a negative intrinsic curvature and tending to zero as $\Delta h \to 0$.

The existence of a general estimate for the quantities $o(\cdot)$ follows from the differentiability of the functionals ω_i and H, the compactness of the domain of definition of the cap set $\{P\}$, and the compactness of the cap set $\{P\}$ in this domain; see the proof of, and remark concerning, Lemma 2 in 1.3. As for the leading term in the expression for $\Delta H(P)$, its form is established in the same way as in the Euclidean case [**9**]. In short, we must prove that when a vertically elongated prism with a rigid upper base triangle ABC is subjected to an infinitesimal deformation, the following equation holds true:

$$h_A\,d\theta_A + h_B\,d\theta_B + h_C\,d\theta_C - AB\,d\alpha_{AB} - BC\,d\alpha_{BC} - CA\,d\alpha_{CA} = 0,$$

where the $\theta_A, \dots$ are the linear angles between the vertical faces at the vertical edges $h_A, \dots$, and the $\alpha_{AB}, \dots$ are the complementary angles at the edges $AB, \dots$ of the upper base of the prism. We partition the prism into three pyramids in the standard manner, introduce linear and appropriately defined (i.e., converging at an edge) time- and space-like faces and complementary angles at the pyramid edges, and examine small deformations of the pyramids with only one vertex moving along only one edge. The expression on the left-hand side of the above equation is represented by the sum of similar expressions for the three pyramids. For pyramids, as can be elementarily established, the respective expressions vanish; here the calculations are especially simple if we employ the representation used in [**9**] for the angular velocity of an infinitesimal rotation of an edge.

Application of Lemma 1 is useful in conjunction with the assertion characterizing the possible types of deformation of a convex cap.

LEMMA 2. *Let P be a convex cap, and let the curvatures at some of the vertical inner edges of P be negative. Then, while remaining in the class of convex caps whose upper bases are constructed in a similar manner from equal triangles, the cap can be continuously deformed by raising to the same height all the vertices of the cap with negative curvature. Suppose that the curvature at the vertical edge at the inner vertex A of the upper base of P is positive. Then the cap can be continuously deformed by lowering only vertex A, with the shape of the polyhedron changing in the same manner as above.*

The first assertion was actually proved by the lemma in 1.2. It guarantees, among other things, that under deformation a preset triangulation of the upper base of a prismatic cap is conserved. The second assertion follows from the first: in this case it is sufficient to raise all vertices of the cap, including the boundary vertices but excluding vertex A, by the same small quantity, and then to lower all vertices of the newly deformed cap by the same quantity. Clearly, the second assertion is also true when the cap has several type-A vertices; then the assertion must imply the existence of a deformation of a similar form in which any subset of inner type-A vertices is lowered by the same quantity.

In the Euclidean variant, Lemma 2 can be formulated in a similar manner, the only requirement being that the sign of the curvature is reversed.

Now we examine the main results, the theorems on the deviation of the caps from equality as a function of the difference in integral curvatures.

THEOREM 1. *Let P^0 and P^1 be convex space-like prismatic caps with isometric upper bases, and let P^0 be an ordinary cap. Then $H(P^1) - H(P^0) \geq 0$, with equality being possible only if P^0 and P^1 are equal.*

PROOF. Let $Z_0 \in P^0$ and $Z_1 \in P^1$ be arbitrary points of the upper bases of the two caps, with the one requirement that the points correspond to one another under the isometry. Naturally, the fact that the two caps are equal means that the altitudes of the points over the plane are equal, i.e., $h_{Z_0} = h_{Z_1}$. As Theorem 2 below shows, the equality interpreted in this way is equivalent to the equality of altitudes of the vertices of the upper bases of the caps only if the vertices correspond to one another under the isometry. Let us examine the complete set $\{P\}$ of convex caps with upper bases isometric to the upper base of P^0. These bases are not required to be triangulated in the same way. The compactness of the domain of definition of $\{P\}$, the compactness of the set of caps proper, and the continuous dependence of the integral curvature of a cap on the parameter h yield the following. On $\{P\}$, the functional $H(P)$ reaches its minimum value on a certain complex space-like cap P. For this cap, obviously,

$$H(P^1) - H(P^0) \geq H(P) - H(P^0).$$

We assume that the prismatic curvatures at all the vertical edges at the inner vertices of the upper bases of the caps P^0 and P are all equal to zero. Then, by the theorem in 2.1 on the unique definiteness of caps, these caps are equal. Suppose that cap P^0 is not equal to cap P and, hence, there are corresponding vertices at which the curvatures are not equal.

We take the cap P. Suppose that it has inner vertices of the upper base at which the curvatures are negative. Then, by Lemma 2, the cap can be deformed in such a way that the triangulation of the upper base is preserved while the altitudes at these vertices increase, with the cap remaining convex. By Lemma 1, for such a deformation the integral curvature of P decreases. Let us assume that the cap has an inner vertex A whose vertical edge is characterized by a positive curvature. Then, by Lemmas 1 and 2, the cap P can be deformed in the class of polyhedra $\{P\}$ with a decrease in the integral curvature. In any case we arrive at a contradiction with the extremal property of the integral curvature $H(P)$ and, hence, with our assumption about the inequality of the caps. The assumption is, therefore, incorrect. The lemma is proved.

When compared with the existence theorem in 1.2, this result marks a new approach to the problem of realizing a convex cap. We will not dwell on this aspect any further, however.

The next simple assertion, as well as its suggested proof, is of some interest for the Euclidean case, too. Here, as for ordinary Euclidean general convex caps, there is the well-known proof by Pogorelov [21]. The proof given below is especially well suited precisely in the pseudo-Euclidean case, since it avoids the important difficulties related to the triangle property.

THEOREM 2. *Let P^0 and P^1 be convex prismatic caps with isometric upper bases. It is not required that these bases be constructed in the same manner from equal faces or that they be triangulated in the same way. Then, if all the real inner vertices that correspond to one another under the isometry and belong to the upper bases of the caps have the same altitude above the xy-plane, the caps are equal.*

PROOF. Let A_1B_1 be an edge of the upper base of P^1, and let A_0B_0 be a polygonal line on P^0 corresponding to A_1B_1 under the isometry. If A_0B_0 is a rectilinear segment, the respective points on A_0B_0 and A_1B_1 are, obviously, equally distant from the xy-plane. Suppose that A_0B_0 is not a rectilinear segment. Then the polygonal line A_0B_0 passes only through the interior points of the faces and the interior points of the edges of the upper base of P^0, which means that locally A_0B_0 is a shortest line. Then, since P^0 is convex, as a result of projecting A_0B_0 by a time-like vertical cylinder the line A_0B_0 becomes a convex polygonal line on this cylinder—an easily proved variant of Liberman's lemma on a geodesic line on a convex surface. If C_1 is an arbitrary interior point of an arbitrarily chosen face of the upper base of P^1, we draw on this face a rectilinear segment $A_1C_1B_1$ that ends at edges of the cap. Repetition of the above method of comparison, which now refers to the new isometric lines $A_1C_1B_1$ and $A_0C_0B_0$, with a slight complication of the initial data, namely $h_{A_0} \geq h_{A_1}$ and $h_{B_0} \geq h_{B_1}$, again leads to the required result $h_{C_0} \geq h_{C_1}$. Thus, for any points Z belonging to the upper bases of the two caps we have $h_{Z_0} \geq h_{Z_1}$. The same method leads to the opposite inequality $h_{Z_1} \geq h_{Z_0}$. As a result we have the identity $h_{Z_0} \equiv h_{Z_1}$. The theorem is proved.

THEOREM 3. *Let P^0 and P^1 be convex prismatic caps with isometric upper bases, and P^0 an ordinary cap. Then $H(P^1) - H(P^0) \geq \max\{0, c(P^0)(\Delta h)^3\}$, where $\Delta h = \max_j (h_j^1 - h_j^0)$, the quantities h_j^0 and h_j^1 are the altitudes of the vertices of the upper bases of the caps (the vertices correspond to one another under the isometry), and $c(P^0)$ is a positive constant.*

PROOF. Only the case where the maximum of Δh is positive needs to be considered. Suppose that the maximum is attained at the vertices A^0 and A^1 of the corresponding caps, so that $H \equiv h_{A^1}$, $h^0 \equiv h_{A^0}$, and $\Delta h = H - h^0$. Let $\{P\}$ be the set of all convex caps c, not necessarily triangulated in the same way, with upper bases isometric to the upper bases of P^0 and P^1 and with altitudes at the vertices corresponding to A^0 and A^1 under the isometry and being no less than H. We select a cap P for which the functional $H(P)$ attains its minimum value on this set. Using Lemmas 1 and 2 as in the proof of Theorem 1, we establish the following result. First, for P the curvature ω_0 at the vertical edge h_{A_0} at the vertex A_0 corresponding, by isometry, to the vertices A^0 and A^1 is nonnegative, with $h_{A_0} \equiv H$. Second, at the other inner vertices of the upper base of P the curvatures at the vertical edges are zero. The curvature ω_0 must be nonzero, since otherwise, by the theorem in 2.1 on the unique definiteness of caps, P^0 must equal P. The set of convex caps with upper bases isometric to the upper base of P^0, with altitudes H at the vertices A_0 corresponding to the vertex A^0 under the isometry, and with zero curvatures at the vertical edges at the other inner vertices of the upper bases, is reduced solely to the cap P^0, since this cap proves to be the only one possessing these properties. This can be established in the same way as was done in proving the theorem on the unique definiteness for ordinary caps; see Lemma 1 in 3.2. Here we compare two prismatic convex polyhedra with isometric upper bases, with zero curvatures at the inner vertical edges, and with unconnected boundaries. At two corresponding boundary vertices the altitudes of the polyhedra are equal to H, while at the other boundary vertices the altitudes are equal to zero. In view of what was said earlier, we add to the properties of P the inequality $\omega_0 > 0$, which follows from Lemmas 1 and 2, and call such a cap a special cap, denoting it by P_H.

Let us consider the interval $h^0 < h < H$. Obviously, to each value of h in this interval there corresponds a well-defined and unique special convex cap P_h obtainable in the same manner as P_H; here we also refer to P^0 with $\omega_0 = 0$ as a special cap and denote it by P_{h^0}. Other notation is also introduced. The caps $\{P_h\}$ have isometric upper bases. We denote all the real inner vertices of the upper base of a specific cap by $A_0, A_1, \dots, A_n$, the corresponding heights by $h_0, h_1, \dots, h_n$, and the prismatic curvatures at the vertical edges at these vertices by $\omega_0, \omega_1, \dots, \omega_n$. When the cap undergoes variations, the quantities h_i and ω_i can be considered functions of h. The ω_i can also be interpreted as functions of the altitudes h_j, but this will always be linked to a specific, preserved triangulation of the variable cap. This last remark is important because a given cap can carry several triangulations "imprinted" on its upper base. Obviously, there can be only a finite number of such triangulations, a number no greater than that of internally geometric triangulations of the isometric upper bases S of the caps, the total number of such triangulations of the entire set $\{P_h\}$. Our immediate task is to complete the existing inequality

$$H(P^1) - H(P^0) \geq H(P_H) - H(P_{h^0})$$

by expressing the lower bound on the right-hand side in terms of the quantity $H - h^0$; difficulties emerge in proving the equation

$$H(P_H) - H(P_{h^0}) = \int \omega_0 \, dh,$$

whose Euclidean analog is accepted in [**9**] without substantiation. To this end we use the uniqueness and, hence, the continuity of the map $h \to P_h$ and properties that will be established in Lemmas 1 and 2 in 3.2 on bending special caps into each other: the strict monotonicity of curvature $\omega_0 = \omega_0(h)$, the inequality $|\Delta h_i| < |\Delta h_0|$ for $i \neq 0$, and the estimate for the increment $\Delta\omega_0(h)$ of the function $\omega_0(h)$.

Let δ be an arbitrary positive small number. We select a number $h \in [h^0, H]$ and introduce the δ_h-interval $[h - \delta_h, h + \delta_h]$, a closed neighborhood of h, with $\delta_h > 0$ and $\delta_h < \delta$. Obviously, we can always select a small value of δ_h at which each special cap with a label belonging to the δ_h-interval has a triangulation of the upper base coinciding with one of the triangulations "imprinted" on the upper base of P_h. By Lemma 1, the integral curvature of such a cap can be compared numerically to the integral curvature of P_h, with the additional requirement in selecting δ_h that the difference of the corresponding parameter h, i.e., $|\Delta \mathbf{h}|$, is less than δ. Here we also assume that δ_h is such that the conditions of Lemma 2 in 3.2 are met, which ensures a proper estimate of the curvature increment $\Delta\omega_0(h)$ for caps belonging to the δ_h-interval.

We cover each point h of the closed interval $[h^0, H]$ with a corresponding δ_h-interval whose properties have just been described. Out of the resulting covering of $[h^0, H]$ we select a finite subcovering Δ of this closed interval by δ_h-intervals. To simplify matters, we assume that the center h of each δ_h-interval belonging to Δ is covered in the Δ-system only by this interval. We also select all the points in the δ_h-interval whose multiplicity of covering in Δ is unity, stipulate that the intersection of their union and $[h^0, H]$ is an open interval $\delta'_h \subset [h^0, H]$, and introduce the system Δ' of closed δ'_h-intervals. Any two intervals belonging to Δ' either do not intersect or have a single common point, the endpoints of the intervals. We denote the endpoints of a δ'_h-interval and its point h by $h^{(\cdot)}$. The set of all points $h^{(\cdot)}$ belonging to the interval $[h^0, H]$ in the Δ'-system is naturally ordered; we represent this set in the form $\{h^0, h^1, \dots, h^k \equiv H\}$ for an integral value of k and introduce the notation $\Delta h^s = h^{s+1} - h^s$. Now we can proceed with estimates, first writing the equation

$$H(P_H) - H(P_{h^0}) = \sum \big[H\big(P_{h^{s+1}}\big) - H\big(P_{h^s}\big)\big].$$

By Lemma 1,

$$H\big(P_{h^{s+1}}\big) - H\big(P_{h^s}\big) = \omega_0(h^s)\Delta h^s + o_s(|\Delta\mathbf{h}|).$$

By Lemma 1 of 3.2, $|\Delta\mathbf{h}| < \sqrt{n+1}\,\Delta h^s$, and then, using Lemma 1 again, we find that $o_s|\Delta\mathbf{h}| \le \varepsilon\Delta h^s$ uniformly in s, where $\varepsilon \to 0$ as $\delta \to 0$; the reader can recall that $|\Delta\mathbf{h}| < \delta$ and that δ was selected beforehand. Thus,

$$\sum o_s(|\Delta\mathbf{h}|) \le \varepsilon(H - h^0).$$

Since $\omega_0(h)$ is continuous, we can send δ to zero and write the following equation:

$$H(P_H) - H(P_{h^0}) = \int \omega_0(h)\,dh.$$

Let us introduce the function $J(h) = \int_0^h \omega_0(h)\,dh$. Using the inequality for the curvature increment $\Delta\omega_0(h)$ established in Lemma 2 of 3.2, allowing for the strict monotonicity of $\omega_0(h)$, and applying the methods used in finding the integral representation of $H(P_H) - H(P_{h^0})$, we obtain

$$\begin{gathered} J(h) \le C\omega_0^{3/2}(h), \\ \omega_o(h) \ge CJ(h)^{3/2} \sim J'(h) \ge CJ(h)^{2/3}, \\ J(H) \ge C(H - h^0)^3. \end{gathered}$$

We have paid little attention to the form of the constants C; they depend solely on the maximum inclination to the xy-plane of the faces of the special caps. Thus,

$$H(P^1) - H(P^0) \ge C(H - h^0)^3.$$

The theorem is proved.

In proving the Euclidean variant of this theorem, Volkov **[9]** claimed that the following general estimate holds:

$$H(P^0) - H(P^1) \geq C \max_j \left|h_j^0 - h_j^1\right|^3.$$

A fairly simple calculation shows that, for an ordinary pyramid,

$$\Delta H \sim \left|h^0 - h^1\right|^2,$$

so an even better estimate holds for this polyhedron. However, obtaining a general estimate by carrying the method suggested in **[9]** over to the pseudo-Euclidean case in the present paper is impossible. Unremovable obstacles emerge, for instance, in proving Lemma 2 in 3.2, a situation partially discussed in 3.3 and requiring further investigations. An estimate for ΔH in the case where $h^0 - h^1 > 0$ could be useful, since actually we are dealing here with a specific approximation of an ordinary cap in the class of normal caps. In what follows, while examining special caps in 3.2, we assume that such caps have at least two real inner vertices.

3.2. Bending a special cap. Let us study the caps $\{P_h\}$ in greater detail. Here for special caps we establish the geometric properties "in the large." The approach to this problem differs from that adopted in **[9]** for the Euclidean case. All investigations are based largely on the study of finite deformations, or bendings, of upper bases of a cap P_h punctured at point A_0.

LEMMA 1. *To each $h \in [h^0, H]$ there corresponds only one special cap P_h; if $h' > h$, then the cap $P_{h'}$ lies above P_h, i.e., $h'_j > h_j$ for all inner vertices of the upper bases of the caps, with $h'_0 - h_0 > h'_j - h_j$ for all $j \neq 0$. The function $\omega_0(h)$ is strictly monotone: $\omega_0(h') > \omega_0(h)$ if $h' > h$. Let $h_0 \in [h^0, H]$ be chosen arbitrarily, A_k an inner vertex at the upper base of P_{h^0} connected with the inner vertex A_0 by a real edge, l_k the projection of this edge onto the xy-plane, and r_k the rectilinear segment in the lower base of the cap obtained by extending the segment l_k and connecting the point $\bar{A}_0$, the projection of A_0, with the base boundary. Then, if $P_{h'_0}$ is a cap close to P_{h_0}, i.e., if $\Delta h_0 = h'_0 - h_0$ is sufficiently small, we have*

$$\Delta h_0 - \Delta h_k \geq \Delta h_0 \frac{l_k}{r_k} + o_{h_0}(\Delta h_0).$$

PROOF. The first assertion of the lemma is proved in practically the same way as we proved the theorem on the unique definiteness of ordinary caps in Chapter 2. Here we are forced to compare isometric surfaces that are, if not ordinary, at least locally convex ordinary surfaces with equal altitudes at the respective generally unconnected boundary polygonal lines. The same remark applies to the proof of the second assertion of the lemma.

Let us consider the third assertion; we assume that the caps have the same triangulations. To this end we first demonstrate that for a cap P_h with any h the determinant $|\partial\omega_0/\partial h_j|$, where $i, j = 0, 1, 2, \ldots, n$, is nonzero. Let us suppose that the contrary is true and $|\cdot| = 0$ for a certain h. Then the system of equations

$$(*) \qquad \sum \frac{\partial \omega_i}{\partial h_j} \xi_j = 0$$

has a nonzero solution $(\xi_0, \xi_1, \ldots, \xi_n)$. We select a positive number λ for the small parameter, set a certain arbitrary triangulation of the upper base of P_h, and build a prismatic surface Φ_λ with an upper base isometric to P_h by specifying a deformation $\Delta h_k = h_k^\lambda - h_k$ of the altitudes at the vertices of the cap according to the following rule: $\Delta h_k = \lambda\xi_k$ at $k = 0, 1, \ldots, n$, and $\Delta h_k = 0$ at the boundary of P_h. This defines the isometric deformation function $\lambda\xi$ at each point of the upper base S of the cap. Since not all the ξ_k are zeros, the function $\lambda\xi$ reaches its maximum positive or minimum negative value on the upper base of P_h. We assume that it attains its positive maximum on the cap. Since on the boundary of S the function $\lambda\xi$ vanishes, S contains a region G in whose interior the function is strictly greater than a certain constant λc, with $c > 0$, and is equal to λc on the boundary Γ of G. By proper selection of c we can ensure that Γ does not pass through a single vertice of S. Since the map $P_h \to \Phi_\lambda$ is linear on the faces of S, it immediately follows that Γ is a polygonal line whose vertices are interior points of the edges of P_h. It is also obvious that the region G and the polygonal line Γ do not change under variations of λ and that G contains as an interior point at least one vertex of the upper base of P_h. By $\bar{G}$ and $\bar{\Gamma}$ we denote the projections of G and Γ onto the xy-plane, which are objects on the lower base of the cap, while G_λ, Γ_λ, $\bar{G}_\lambda$, and $\bar{\Gamma}_\lambda$ are the corresponding objects on the prismatic polyhedron Φ_λ: G and G_λ correspond to one another by isometry, and so do Γ and Γ_λ. Since the system of equations $(*)$ has a nonzero solution, by Lemma 1 of 3.1 the increment acquired by the total intrinsic curvature of the developed surface $\bar{G}$ is $o(\lambda)$ as a result of the $\bar{G} \to \bar{G}_\lambda$ transition.

Let us show that the polygonal line Γ does not intersect a single real edge of the cap P_h. If it does intersect at least one real edge RQ, then at the corresponding vertex $\bar{\Gamma}$ the increment of the rotation of the polygonal line from the side of $\bar{G}$ in the $\bar{G} \to \bar{G}_\lambda$ transition is $O(\lambda)$. The increment of the rotation at each vertex $\bar{\Gamma}$ is nonnegative, so that the increment of the total rotation of $\bar{\Gamma}$ is also equal to $O(\lambda)$. These conclusions can be derived from Lemma 1 in 1.5 if we apply it to an infinitesimal deformation of a double prism, and from the fact that the displacement of the vertex of edge RQ, which is an interior point of G, is equal to $\lambda\xi_k$ for a certain value of k, with $\xi_k > c > 0$. Writing the Gauss–Bonnet formula for regions $\bar{G}$ and $\bar{G}_\lambda$ and going over to increments of total intrinsic curvatures and rotations, for sufficiently small values of λ we arrive at a contradiction: $o(\lambda) + O(\lambda) = 0$. Thus, when λ is small, Γ does not intersect a single real edge of P_h.

Let us assume that G contains only one vertex. We cut the neighborhood of the vertical line $A_0\bar{A}_0$ in the prismatic polyhedron, cap P_h, by a half-plane whose boundary is the same vertical line, and develop a neighborhood of the cut into a multiple prism. Since Γ does not intersect the real edges of the cap, the upper base of this prism lies in a plane. Since $\omega_0(h) \leq 0$, the full angle at the vertex $\bar{A}_0$ of the prism's lower base does not exceed 2π, which means that the full angle at A_0 of the prism's upper base does not exceed 2π either. However, the latter must be greater than 2π because the curvature at each inner vertex of the upper base of P_h is negative. Thus, G must contain at least two inner vertices of the upper base of the cap.

Let G' be a region composed of the triangles of the upper base of P_h, including the triangles containing Γ, in which we can pass from triangle to triangle by sequentially traversing only the unreal edges of the cap. Let Γ' be the polygonal line that is the shortest in this region and homotopic to the polygonal line Γ. The line Γ' is nondegenerate since at least two vertices A are interior points of G. We assert that this is a local line passing strictly in the interior of G'. It is sufficient to show that Γ' does not pass through a singe vertex of the upper base of P_h, and this follows from Lemma 2 of 1.4. Now, if we write the Gauss–Bonnet formula for the region on S with the boundary Γ' including all the inner vertices of the cap belonging to G, we come to a contradiction, since the total intrinsic curvature of this large region is negative, and the total rotation of its boundary is zero. The contradiction shows that the assumption that there is a nontrivial solution $(\xi_0, \xi_1, \dots, \xi_n)$ and, hence, that the determinant $|\partial\omega_i/\partial h_j|$ vanishes, is invalid. Let $\|\Gamma^{js}\|$ be the reciprocal of the matrix $\|\partial\omega_i/\partial h_j\|$; both matrices are symmetric, which, e.g., follows from the representation $dH = \sum \omega_i\, dh_i$, but can also be easily verified by direct calculation. For convenience we will from now on simply write h instead of h_0 and h' instead of h'_0.

Lemma 1 of 3.1, the fact that $\Delta h_0 > \Delta h_k > 0$ for $k = 1, 2, \dots, n$, and the fact that the determinant is nonzero yield the following relationships for the cap P_{h_0}:

$$0 < \Delta\omega_0 = \left(\frac{\partial\omega_0}{\partial h_0} + \varepsilon_0\right)\Delta h_0 + \frac{\partial\omega_0}{\partial h_1}\Delta h_1 + \cdots + \frac{\partial\omega_0}{\partial h_n}\Delta h_n,$$

$$0 = \Delta\omega_1 = \left(\frac{\partial\omega_1}{\partial h_0} + \varepsilon_1\right)\Delta h_0 + \frac{\partial\omega_1}{\partial h_1}\Delta h_1 + \cdots + \frac{\partial\omega_1}{\partial h_n}\Delta h_n,$$

$$\dots\dots\dots\dots\dots\dots\dots\dots\dots\dots\dots\dots\dots\dots\dots\dots$$

$$0 = \Delta\omega_n = \left(\frac{\partial\omega_n}{\partial h_0} + \varepsilon_n\right)\Delta h_0 + \frac{\partial\omega_n}{\partial h_1}\Delta h_1 + \cdots + \frac{\partial\omega_n}{\partial h_n}\Delta h_n,$$

$$\Delta h_0 = \left(\Gamma^{00} + \varepsilon'_0\right)\Delta\omega_0,\ \Delta h_1 = \left(\Gamma^{01} + \varepsilon'_1\right)\Delta\omega_0,\ \dots,\ \Delta h_n = \left(\Gamma^{0n} + \varepsilon'_n\right)\Delta\omega_0,$$

where both ε and ε' are sufficiently small. These relationships imply that $\Gamma^{0k} \geq 0$ for all values of k, and $\Gamma^{00} \geq \Gamma^{0k}$ for all nonzero values of k. Since $\|\partial\omega_i/\partial h_j\|$ and $\|\Gamma^{js}\|$ are mutually reciprocal, with $\sum(\partial\omega_0/\partial h_k)\Gamma^{0k} = 1$, we find that $\Gamma^{00} > 0$.

Let us now find the necessary estimate. The deformation field Δh_k of the cap P_{h_0}, when linearly extended along each face, represents the complete field of the vertical components of the isometric deformation of the entire upper base of the cap in the $P_{h_0} \to P_{h'_0}$ transformation. It is this total field Δh that we will examine. The reader will recall that P_{h_0} and $P_{h'_0}$ have equal triangulations.

We extend the segment $l_k = \bar{A}_0\bar{A}_k$, which connects the projections of points A_0 and A_k in the lower base $\bar{S}$ of cap P_{h_0}, beyond point $\bar{A}_k$ to the intersection at a point C_k of the boundary $\bar{L}$ of the base. By $\bar{A}_0C$ we denote a variable ray in the base plane, the xy-plane, with the initial position at $\bar{A}_0C_k$, where the point C moves along $\bar{L}$. We carry out a special gluing procedure involving the prisms of P_{h_0}, which is, generally, "many-sheeted" on the cap proper since $\omega_0(h) \geq 0$. The purpose of such gluing is to obtain an ordinary convex surface more suitable for our study than the initial cap P_{h_0}. We paste together all the prisms constituting P_{h_0} whose bases in the developed surface $\bar{S}$ intersect the open ray $\bar{A}_0C_k$, the point $\bar{A}_0$ excluded. Before proceeding any further with the gluing process, we draw through the point C_k of the resulting prismatic polyhedron a straight line l that is the supporting line in relation to the corresponding section of boundary $\bar{L}$ of the cap's base, and through $\bar{A}_0$ we draw a line l_0 parallel to l. Now we continuously rotate the ray $\bar{A}_0C$ first in one direction and then in the other in relation to $\bar{A}_0C_k$, terminating the process every time when the moving ray finds itself on l_0. Here to the already obtained prismatic polyhedron we paste the prisms constituting P_{h_0} that intersect in the developed surface $\bar{S}$ with the open ray $\bar{A}_0C$ from which $\bar{A}_0$ is excluded. We use the vertical plane that contains the straight line l_0 to cut from the resulting polyhedron its main part, the part containing C_k. This main part is an ordinary solid convex polyhedron, to which we "attach" the old notation P_{h_0} and consider it to be a prismatic polyhedron. We still denote the upper base of this polyhedron by S; it appears natural to extend to the convex polyhedral surface S the displacement field Δh, which isometrically transforms S into the upper base S' of a new prismatic convex polyhedron, also denoted by the old symbol $P_{h_0'}$. At each point $X \in S$ we have a displacement $\Delta h(X) = \varphi_\Delta(X)\Delta\omega$, where on S the quantity $\varphi_\Delta(X)$ tends uniformly to a limit $\varphi(X)$ as $\Delta h_0 \to 0$. For one thing, at A_0 we have $\Delta h_0 = \left(\Gamma^{00} + \varepsilon_0'\right)\Delta\omega_0$, at A_k we have $\Delta h_k = \left(\Gamma^{0k} + \varepsilon_k'\right)\Delta\omega_0$, on the section $\bar{\Gamma}$ of the boundary of S we have $\Delta h = 0$, and on the part projected on the straight line l_0 we have $\Delta h \leq \Delta h_0$.

We select the number $\Delta\omega_0 > 0$ as a small parameter and construct a prismatic surface Φ_Δ whose upper surface is isometric to P_{h_0} by specifying the deformations of the altitudes $\Delta h(\bar{X})$ at points $X \in S$ (the deformations are referred to the projections of these points on the lower base of P_{h_0}) according to the following rule: on the straight line l_0 we have $\Delta h(\bar{X}) = \Delta h_0$, on the straight line l we have $\Delta h(\bar{X}) = 0$, and on the straight line l_p located between l_0 and l and at a distance p from l we have $\Delta h(\bar{X}) = p\Delta h_0/p_0$, where p_0 is the distance between l_0 and l. Interpreting $\Delta\omega_0$ as the time variable, we can assume that the field $p\Delta h_0/\Delta\omega_0 p_0 = p\left(\Gamma^{00} + \varepsilon_0'\right)/p_0$ defined at points X of the surface S represents the vertical components of the trivial field of an infinitesimal rotation of this surface about the straight line l. Such a rotation is simply an infinitesimal motion in which the curvature ω at the vertical edge at each inner vertex of the base S of the prismatic polyhedron P_{h_0} is time-independent, $d\omega = 0$, and we must assume that the rotation of S induces an infinitesimal deformation of the entire polyhedron. The surface Φ_Δ is the result of the deformation of P_{h_0} after the time $\Delta\omega_0$, with the time dependence of the increments of the vertical components of the surface being linear. Hence, when the corresponding prismatic polygon is deformed, the increments of the curvatures ω for this surface admit, to within quantities of order $(\Delta\omega_0)^2$, the representation $\Delta\omega = O\left((\Delta\omega_0)^2\right) = o(\Delta\omega_0)$; the equality $\Delta\omega = o(\Delta\omega_0)$ can also be derived from Lemma 1 in 3.1, since $d\omega = 0$.

To prove the inequality $|\partial\omega_i/\partial h_j| \neq 0$ we set up on the surface S a deformation field $\lambda\xi$ with a small parameter λ and established that the function ξ can reach neither its positive maximum nor its negative minimum. Actually, the proof consists in comparing two isometric surfaces, the upper base S of the convex polygon P_h and the upper base of another prismatic polyhedron obtained from S as a result of the deformation $\lambda\xi$, with the boundary curves of the surfaces being equidistant from the xy-plane (more precisely, simply lying in this plane). This procedure can be repeated by comparing two other isometric surfaces, the upper base S of the convex polyhedron $P_{h_0'}$, and the surface Φ_Δ. The deformation field that transforms $P_{h_0'}$ into Φ_Δ has the form $\xi\Delta\omega_0$ at the points of S; at the boundary of S the surface Φ_Δ lies no lower than S. In the given case it is impossible to prove that

$\xi(X) = 0$ at $X \in S$. But the same method allows us to establish that, for all points $X \in S$ and all boundary points, Φ_Δ lies almost above X. More precisely, since as $\Delta\omega_0$ tends to zero the function $\xi(X)$ acquires a certain limit value at each point X, the displacement $\xi\Delta\omega_0$ does not have a negative value of order $O(\Delta\omega_0)$ at a single inner vertex of S. Thus, we conclude that $\xi\Delta\omega_0 \geq o_{h_0}(\Delta\omega_0)$. It remains to write this inequality explicitely for the vertex A_k, allowing for the fact that for this point $p_0 - p = l_k$, with $p_0 = r_k$:

$$\Delta h_0 \frac{r_k - l_k}{r_k} - \Delta h_k \geq o_{h_0}(\Delta\omega_0) = o_{h_0}(\Delta h_0),$$

$$\Delta h_0 - \Delta h_k \geq \Delta h_0 \frac{l_k}{r_k} + o_{h_0}(\Delta h_0).$$

Let $h \in [h^0, H]$ be arbitrary, and let P_h be a special cap with a fixed triangulation of the upper base. We displace P_h vertically; naturally, the curvature ω at each vertex of the cap does not change. The fact that the curvature of P_h at the vertex A_0 is time-independent in an infinitesimal displacement of the cap by dh is expressed as follows:

$$0 = d\omega_0 = \sum_j \frac{\partial\omega_0}{\partial h_j} dh \sim \sum_j \frac{\partial\omega_0}{\partial h_j} = 0,$$

where the summation is over the boundary vertices of P_h. In the second sum, all terms in addition to $\partial\omega_0/\partial h_0$ corresponding to the vertices of the upper base that are not connected by edges with A_0 vanish, i.e., can be ignored. It is convenient to divide the remaining term into two parts, one corresponding to inner vertices of the cap and the other to the boundary vertices. In the following representation of the derivative $\partial\omega_0/\partial h_0$, which will be used in the proof of Lemma 2, this division is emphasized by the respective symbols "int" and "ext":

$$\frac{\partial\omega}{\partial h_0} = -\sum_{j\neq 0}^{\text{int}} \frac{\partial\omega_0}{\partial h_j} - \sum_{j\neq 0}^{\text{ext}} \frac{\partial\omega_0}{\partial h_j}.$$

LEMMA 2. *Let $h_0 \in [h^0, H]$ be arbitrary, and let the conditions of Lemma 1 be satisfied for P_{h_0}; for one thing, $P_{h_0'}$ has the same triangulation. Then*

$$\Delta\omega_0 = \omega_0(h_0') - \omega_0(h_0) \geq C\sqrt{\omega_0(h_0)}\,\Delta h_0 + o_{h_0}(\Delta h_0),$$

where C depends only on the developed surface of the upper base of P_{h_0}.

PROOF. Lemma 1 of 3.1 and the fact that $\Delta h_0 > \Delta h_j$ for $j = 1, 2, \ldots, n$ yield

$$\Delta\omega_0 = \sum \frac{\partial\omega_0}{\partial h_j} \Delta h_j + o(\Delta h_0).$$

Expressing $\partial\omega_0/\partial h_0$ in terms of the derivatives $\partial\omega_0/\partial h_j$, we find that

$$\Delta\omega_0 = -\sum_{j\neq 0}^{\text{int}} \frac{\partial\omega_0}{\partial h_j}(\Delta h_0 - \Delta h_j) - \Delta h_0 \sum_{j\neq 0}^{\text{ext}} \frac{\partial\omega_0}{\partial h_j} \frac{l_j}{l_j} + o(\Delta h_0).$$

The trivial factor $l_j/l_j \equiv 1$ is added to each term in the second sum so that when we discuss the lower bound on $\Delta\omega_0$ we can use the inequalities $r_j, l_j \leq d$, where d is the inside diameter of the developed surface of the lower base of P_h (d is uniformly bounded above in h). We find that since $\partial\omega_0/\partial h_i \leq 0$, $\Delta h_0 > 0$, and the derivatives $\partial\omega_0/\partial h_i$ are uniformly bounded above in the entire interval $[h^0, H]$, we have

$$\begin{aligned}\Delta\omega_0 &\geq -\frac{\Delta h_0}{\alpha}\left(\sum_{j\neq 0}^{\text{int}} \frac{\partial\omega_0}{\partial h_j} l_j + \sum_{j\neq 0}^{\text{ext}} \frac{\partial\omega_0}{\partial h_j} l_j\right) + o_{h_0}(\Delta h_0)\\ &= -\frac{\Delta h_0}{\alpha}\sum_{j\neq 0} \frac{\partial\omega_0}{\partial h_j} l_j + o_{h_0}(\Delta h_0).\end{aligned}$$

By Lemma 1 in 3.3,

$$\Delta\omega_0 \geq C\sqrt{\omega_0}\,\Delta h_0 + o_{h_0}(\Delta h_0).$$

The lemma is proved.

Several refinements are needed if we wish to apply the methods developed in this section to the Euclidean case. Generally, the interval $[h_0, H]$ must be partitioned into a finite set of subintervals, corresponding to the situation in which a fraction of the faces of the developed surface S gets transformed into boundary vertical faces as h is increased. By Liberman's principle in 1.3 and Lemma 1, these faces remain vertical as h is increased still further. In view of this, now in each subinterval we must consider not caps but prismatic convex polyhedra with a border not necessarily lying in the xy-plane and with isometric upper bases that are the same subdeveloped surface of S. As we move from a subinterval to the subsequent subinterval, the subdeveloped surface shrinks to a certain fraction. Of course, allowing the inequalities $\omega_0 \leq 0$ and $\partial\omega_0/\partial h_k \geq 0$, we change the statements and conclusions contained in the Euclidean analogs of Lemmas 1 and 2. In the Euclidean variant of the exposition, actually based on using infinitesimal bendings of punctured caps P_h, the following inequalities are established: $\Gamma^{00} < 0$, $\Gamma^{00} \leq \Gamma^{0k} \leq 0$, and $\Gamma^{0k} - \Gamma^{00} \geq -\Gamma^{00} l_k/r_k$. Lemma 8 in Volkov's paper [**9**], which states that $-d\omega_0 \geq C\sqrt{-\omega_0}\, dh_0$, is the analog of Lemma 2.

Let $r_h(X)$ be the radius vector of a varying point X in the upper base of a cap P_h. Following Cohn-Vossen [**13**], we consider the surface $r(X) = r_{h'}(X) + r_h(X)$ and the vector field $z(X) = r_{h'}(X) - r_h(X)$ specified at points X of the surface. Since the upper bases of P_h and $P_{h'}$ are isometric, z is the field of an infinitesimal bending of a surface r punctured at a point A_0. The vertical component of this field is $\Delta h(X) = h'(X) - h(X)$, and at a point A_i this component is equal to $(\Gamma^{0i} + \varepsilon_i')\Delta\omega_0$. Thus, if we assume that $\Delta\omega_0$ is the deformation parameter and send it to zero, we find that the quantities $(\Gamma^{00}, \Gamma^{01}, \dots, \Gamma^{0n})$ are the vertical components at the points $(A_0, A_1, \dots, A_n)$ of the field of an infinitesimal bending of the upper base of P_h punctured at A_0, with the border of the cap fixed in the xy-plane. Hence, the properties of the Γ^{0j} obtained earlier are the "global" properties of the bending field of the special cap P_h. In [**9**] these properties were established without detailed substantiation, which, as the proof of Lemma 1 clearly reveals, is fairly cumbersome; there they are derived from a peculiar maximum principle not related, however, to the notion of an infinitesimal bending.

To find the estimate for Γ^{00} used in the proof of the inequality $-d\omega_0 \geq C\sqrt{-\omega_0}\,\Delta h_0$, Volkov **[9]** discarded the term corresponding to the vertices A_j of the special cap that in the above proof of Lemma 2 are labeled by the symbol "ext". Without this term one cannot establish the validity of the given inequality. This means, for one thing, that Lemmas 8 and 1 of **[9]** do not encompass the case of caps with one inner vertex in the upper base, i.e., pyramids.

To establish the inequality for the difference $\Delta h_0 - \Delta h_k$, Lemma 1 studied the restriction of the deformation field Δh of the entire upper base of P_{h_0} to the part of the base cut off by the vertical plane containing the straight line l_0; this part can be increased if necessary, when $\Delta\omega_0 < \tau$, by a "multi-sheeted continuation". The method of proof is taken from **[9]**, where, however, it was used in a rather non-rigorous way. In the process of obtaining, in the infinitesimal bending variant, the inequality for the difference $\Gamma^{0k} - \Gamma^{00}$ and applying the corresponding maximum principle, the field Γ^{0j} and its partitioning into "internal" and "external" components are determined by the variations of the altitudes of only the inner vertices of the upper base of the cut-off part of the cap. These, however, are not the field Γ^{0j} and the quantities Γ^{0k} required for finding the required inequality. The quantities are determined by the diagonal submatrix of $\|\partial\omega_i/\partial h_j\|$, while the desired inequality is established for the quantities Γ defined by the matrix proper. The proof of Lemma 1 shows how rigor can be reinstated: it is sufficient to apply the field Γ corresponding to the entire set of the inner vertices of the upper base of the cap P_{h_0}.

3.3. Estimating auxiliary quantities. The following lemma is formulated with maximum generality in the variant required for obtaining Theorem 3 in 3.1 to the extent expected in **[9]**. However, in proving Theorem 3.1 it turns out that the lemma refers only to cases where $\omega > 0$ for pseudo-Euclidean space and $\omega < 0$ for Euclidean space.

LEMMA 1. *Let V be a convex prismatic pyramid with A_0 the inner vertex and $A_1, A_2, \dots, A_n$ the boundary vertices of the upper base and the full angle P at A_0 greater than 2π. Let ω be the curvature at the vertical inner edge of the pyramid, h_k the altitude of A_k above the xy-plane, and l_k the length of the projection of the edge A_0A_k onto the xy-plane. Then*

$$\left|\sum_{k\neq 0} l_k \frac{\partial\omega}{\partial h_k}\right| \geq C\sqrt{|\omega|},$$

where $C > 0$ depends only on the maximum inclination of the edges of the upper base of the pyramid to the xy-plane.

PROOF. We start with a Euclidean pyramid. The hypothesis of the lemma does not change except that now $p < 2\pi$. This makes it possible to give a pictorial geometric interpretation of the inequality we wish to prove and, at the same time, to introduce necessary notation. Let $\widetilde{S}$ be a sphere of unit radius with center at the origin, $\widetilde{O}$ the pole of the sphere with coordinate $z = 1$, and τ^* the plane tangent to $\widetilde{S}$ at $\widetilde{O}$. We assume, for the time being, that V is an ordinary pyramid: $\omega = 0$. In this case the upper base of the prismatic pyramid can be interpreted as a convex cone K with its vertex at A_0. Let us build the reciprocal convex cone $\widetilde{K}$ with its vertex at the center of $\widetilde{S}$ and formed by the exterior normals to the support planes of K that pass through the generatrices of this cone. By $\widetilde{L}$ and L^* we denote the convex polygonal lines on $\widetilde{S}$ and τ^* representing the intersections of $\widetilde{K}$ with $\widetilde{S}$ and τ^*. These polygonal lines are called, respectively, the spherical and normal images of the cone K. The vertices of these polygonal lines are the endpoints of the exterior normals to the faces of the cones, and the sections of the polygonal lines are the segments on $\widetilde{S}$ and τ^*, respectively, that connect the endpoints of normals to neighboring faces. We say that these vertices and links are the spherical and normal images of the faces and edges of K. The angle at the vertex of the polygonal line $\widetilde{L}$ on $\widetilde{S}$ is the complement to π of the angle at A_0 of the corresponding faces of the cone. The length of a section in $\widetilde{L}$ is the complement to π of the corresponding dihedral angle. Simple calculations show that the length of a section in L^* that is the normal image of face A_0A_k of K is equal to $l_k(\partial\omega/\partial h_k)$; in the pseudo-Euclidean case this length is $-l_k(\partial\omega/\partial h_k)$. For a Euclidean prism, $\partial\omega/\partial h_k$ is nonnegative for all positive values of k and vanishes only for the edges whose dihedral angles of prism V are equal to a straight angle; in the pseudo-Euclidean case $\partial\omega/\partial h_k$ is nonpositive for all positive values of k and vanishes under the same conditions as in the Euclidean case. Thus, the right-hand side in the analytical expression of the lemma is simply the length of the polygonal line L^*.

Now let us examine a Euclidean convex prismatic pyramid, for which we assume that $p < 2\pi$ and $\omega < 0$. In this pyramid, as well as in an ordinary pyramid, we can introduce spherical and normal images of edges and faces of its upper base. To this end we select the generatrix A_0Q of K that forms the largest angle with the xy-plane if the cone has boundary vertices for which $h_k > h_0$, or the one that forms the smallest angle with the xy-plane in the opposite case. In the first case we assume, changing the prism V slightly if necessary, that a ray A_0Q coincides with an edge of K, stipulating especially that the maximum inclination to a plane is chosen among generatrices, rays OQ, that lie entirely in the half-plane $z > 0$ and are not horizontal. We also assume that the edge A_0Q is the coinciding side of the upper base of a degenerate prism whose well-defined base plane is incorporated in the convex prismatic pyramid V. For an ordinary cone this additional assumption corresponds to drawing through the edge A_0Q the support plane of K, considered the plane of the degenerate face of the cone. We cut V with the vertical half-plane containing A_0Q. We glue two copies of the degenerate prism to the vertical faces of the component prisms of V that adjoin A_0Q, and then glue together all the other components appropriately. In this way V is transformed into a multiple convex prism with vertex at A_0 whose adjoining component prisms are degenerate and are obtained from each other via rotation through an angle $-\omega$ about the $A_0\bar{A}_0$ axis, followed by obligatory superimposition of the planes of the degenerate upper faces of these prisms. Now spherical and normal images of the prism V are obtained as a result of uniting the spherical and normal images of the faces, including the degenerate faces ($\widetilde{L}$ on $\widetilde{S}$ and L^* in τ^*), and the inner edges of the upper base of the multiple prism. By construction, the radii $\widetilde{O}D_1$ and $\widetilde{O}D_2$ drawn on $\widetilde{S}$ to the endpoints of the polygonal line $\widetilde{L}$ are perpendicular, respectively, to the final links of $\widetilde{L}$; the lengths of the radii are the same and equal to the angle ρ of inclination of A_0Q to the xy-plane. When $\widetilde{S}$ is rotated in itself about $\widetilde{O}$ through the angle $-\omega$, the radii coincide, the finite links of $\widetilde{L}$ become the segments that form, on the sphere, the complementary rays of a single straight line, and the sum of the lengths of the projections on τ^* of these segments from the center of $\widetilde{S}$ is equal to the length of the normal image of edge A_0Q.

Let us consider a more complicated case, where the ray A_0Q lies entirely in the half-plane $z > 0$ and is not horizontal. We wish to examine, in a qualitatively metric manner, the way the polygonal line L^* is positioned in the τ^*-plane with respect to the pole $\widetilde{O}$. We assume that $\mathcal{L}$ is a circumference of radius RB in the τ^*-plane, $C_1 \in \mathcal{L}$ and $C_2 \in \mathcal{L}$ are points symmetric with respect to the straight line RB, and $\mathcal{L}_1$ and $\mathcal{L}_2$ are circumferences with diameters RC_1 and RC_2. We assume that $\widetilde{O}$ lies on the open ray BR strictly outside the small circumferences; by $D_1^* \in \mathcal{L}_1$ and $D_2^* \in \mathcal{L}_2$ we denote the points closest to $\widetilde{O}$. The smooth curve $D_1^*C_1BC_2D_2^*$ consisting of arcs of the circumferences represents a qualitatively normal image of the generalized cone K. In a sense this representation is metrically exact if we assume that $\widetilde{O}D_1^* = \widetilde{O}D_2^* = \tan\rho$ and that the angle between the radii $\widetilde{O}D_1^*$ and $\widetilde{O}D_2^*$ on the side of B is equal to $-\omega$; we smoothly close the line L^* by an arc Λ of the circumference centered at $\widetilde{O}$ and having radius $\tan\rho$ corresponding to that angle. The central projections of the points D_1^* and D_2^* and the arc Λ on the sphere $\widetilde{S}$ from the sphere's center are the endpoints of the polygonal line $\widetilde{L}$, the points D_1 and D_2, and the arc $\widetilde{\Lambda}$ of a circumference, on $\widetilde{S}$, of radius ρ centered at $\widetilde{Q}$. Now we must deal with the real polygonal lines $\widetilde{L}$ and L^*. We introduce the following notation: $L' \equiv L^* \cup \Lambda$ and $\widetilde{L}' = \widetilde{L} \cup \widetilde{\Lambda}$. Note that here we are limiting our discussion to the case where $-2\pi \leq \omega < 0$; this case alone was considered by Volkov [**9**]. The general situation is discussed at the end of this proof. If $-\pi \leq \omega < 0$, then the line L' and, hence, $\widetilde{L}'$ bound in the τ^*-plane a simply connected region G^* and, respectively, on $\widetilde{S}$ a simply connected region $\widetilde{G}$. If $-2\pi \leq \omega < -\pi$, we can assume that these lines bound simply connected regions with possible self-overlapping; the same notation is used for these regions. It must be noted here that due to the extremal conditions of selecting the generatrix A_0Q on the sphere $\widetilde{S}$, the polygonal line $\widetilde{L}$ lies outside the circumference of radius ρ centered at $\widetilde{O}$.

Now we can proceed with the estimates. Let us agree to denote the regions and polygonal lines, and their areas and lengths, by the same symbols. Applying the isoperimetric inequality to G^*, we get $2L^* \geq L^* + \Lambda \geq \sqrt{4\pi G^*} \geq c\sqrt{\widetilde{G}}$, or $L^* \geq c\sqrt{\widetilde{G}}$; note that the polygonal line L^* encompasses the convex arc Λ from the outside. Using the Gauss–Bonnet formula for $\widetilde{G}$ on $\widetilde{S}$, we find that $\widetilde{G} = 2\pi - p - \omega\cos\rho$. Thus, $L^* \geq c\sqrt{-\omega}$; here and in what follows we denote various constants, whose exact values are inessential, by the same symbol.

Now we examine a space-like pyramid V. Here we must repeat all the constructions for Euclidean space related to the definitions of the normal (L^*) and spherical $(\widetilde{L})$ images. We assume that this has been done, and point out only the necessary changes. Now $\partial\omega/\partial h_k$ is nonpositive for all positive values of k, and we consider the case where $\omega > 0$; as follows from Lemma 2 formulated below, ω must obey the following inequalities: $0 < \omega < \pi$. For $\widetilde{S}$ we take the connected component of a sphere with an imaginary-unit radius, with the pole $\widetilde{O}$ of this sphere having the coordinate $z = -1$. The reader will recall that the intrinsic curvature of this sphere is -1 and that the sphere is isometric to a hyperbolic plane. In building the spherical and normal images of the upper base of V the interior normals to the faces are assumed the normals to the faces. In describing L^* in τ^*, for the circumferences $\mathcal{L}_1$ and $\mathcal{L}_2$ tangent to $\mathcal{L}$ and the points C_1 and C_2 we take equal circumferences with a radius greater than that of $\mathcal{L}$. The pole $\widetilde{O}$ lies on the ray BR strictly outside $\mathcal{L}_1$ and $\mathcal{L}_2$. The smooth line $L^* \equiv D_1^* C_1 B C_2 D_2^*$ has a single self-intersection point E, at points D_1^* and D_2^* is tangent to a circumference of radius $\tanh \rho$ centered at $\widetilde{O}$, and lies outside this circumference. The angle in τ^* between the radii $\widetilde{O}D_1^*$ and $\widetilde{O}D_2^*$ is ω.

The line $\widetilde{L}'$ defines two regions on $\widetilde{S}$: the region $\widetilde{G}_1$, bounded by the part $\widetilde{L}_1$ of the polygonal line $\widetilde{L}$ with endpoints at the self-intersection point $\widetilde{E}$, and the region $\widetilde{G}_2$, bounded by the remaining two parts of $\widetilde{L}$ and the arc $\widetilde{\Lambda}$. To simplify matters we assume that $\widetilde{E}$ is an interior point of two intersecting sections of $\widetilde{L}$ and α is the angle of intersection of these sections from the side of $\widetilde{G}_1$ and, hence, from the side of $\widetilde{G}_2$. Let τ_1 be a rotation on $\widetilde{S}$ of the part $\widetilde{L}_1$ of the polygonal line $\widetilde{L}$ from the side of $\widetilde{G}_1$, and τ_2 the total rotation of the remaining parts of the same polygonal line from the side of $\widetilde{G}_2$; obviously, $\tau_1 - \tau_2$ is equal to p, the rotation of the polygonal line $\widetilde{L}$ on $\widetilde{S}$ from the concave side. The rotation of Λ from the side of $\widetilde{G}_2$ is equal to $-\omega \cosh \rho$. Employing the isoperimetric inequality for the region G_1^* on τ^*, we obtain $L^* \geq L_1^* \geq \sqrt{4\pi G_1^*} \geq c\sqrt{\widetilde{G}_1}$. Applying the Gauss–Bonnet theorem to $\widetilde{G}_1$ and $\widetilde{G}_2$, we find that $\tau_1 + \pi - \alpha - \widetilde{G}_1 = 2\pi$ and $\tau_2 + \pi - \alpha - \omega \cosh \rho + 2\pi - \widetilde{G}_2 = 2\pi$, or $\widetilde{G}_1 = \widetilde{G}_2 + p - 2\pi + \omega \cosh \rho$, i.e., $\widetilde{G}_1 \geq \omega \cosh \rho$. Thus, $L^* \geq c\sqrt{\omega}$, where the constant c depends only on the maximum inclination of the generatrices of the upper base of the prism V to the xy-plane.

Now we must examine the case where $h_k < h_0$ for all positive values of k. To this end we project from the point $\widetilde{O}$ the polygonal lines L^* and $\widetilde{L}$ section-by-section on the τ^*-plane and the sphere $\widetilde{S}$, and glue the resulting regions along the outermost segments directed toward the endpoints of the polygonal lines. We denote the emerging developed surfaces by G^* and $\widetilde{G}$; at $\widetilde{O}$ they will have the same curvature ω. According to Aleksandrov's isoperimetric inequality [**2**], for G^* we have $L^* \geq \sqrt{2(2\pi - \omega)G^*} \geq c\sqrt{\widetilde{G}}$, because $\omega < \pi$. But according to the Gauss–Bonnet formula, for $\widetilde{G}$ we have $\widetilde{G} = p - 2\pi + \omega > \omega$. Hence, $L^* \geq c\sqrt{\omega}$. Note that the corresponding Euclidean cases use Strel′tsov's general isoperimetric inequality [**25**] of the form $L^2 \geq 2(2\pi - \omega^+)G$, where ω^+ is the positive part of the curvature of the developed surface; see also Reshetnyak's paper [**23**] and Busemann's monograph [**6**].

A remark concerning the Euclidean variant of the lemma is in order. In Volkov's case [**9**], from which we began our proof, it is assumed that $-2\pi < \omega < 0$. But the negative curvature ω can be arbitrary. It appears that the transition of ω to the interval $-2k\pi < \omega < -2(k-1)\pi$, $k > 1$, corresponds to a well-defined change in the curve $L' = D_1^* C_1 B C_2 D_2^* \cup \Lambda$, which we assumed to be the topological interpretation of the normal image, the polygonal line L^*, smoothly complemented by the arc Λ of a circumference. Now we give the corresponding representation of L^* for the case of a minimum transition to $-4\pi < \omega < -2\pi$ and derive the necessary estimate. While the transition involves a larger number of steps, the construction is simply duplicated. On the "old" L' we select in the interior of L^* a point E and draw in τ^* a circumference L_1 that is tangent from the inside in E to L^*. Then we cut

L' and L_1 at E and smoothly glue the resulting curves together at the cut points. We denote the resulting line by L_1'. The section of this line without the arc Λ serves as an interpretation of the "new" normal image L^* of the pyramid V for a specific value of ω. When going over to the real "new" normal image, we retain the old notation for its sections and assume that E is the point of intersection of two links of L^* at interior points. We keep the former notation for the "old" region G^*, but introduce G_1^* to denote the added "new" region bounded by L_1. This system of notation is also applied to the objects on $\widetilde{S}$. The "new" polygonal line L^* lies in the τ^*-plane outside a circumference of radius $\tan\rho$ centered at $\widetilde{O}$. Let τ and τ_1 be the rotations of the "new" $\widetilde{L}$ on $\widetilde{S}$ from the side of $\widetilde{G}$ and $\widetilde{G}_1$, α the angle of $\widetilde{L}$ at point $\widetilde{E}$ in these regions, and L_2 and L_1 the perimeters of G^* and G_1^*. For region $\widetilde{G}$, the angle at point $\widetilde{O}$ between the radii $\widetilde{O}D_1$ and $\widetilde{O}D_2$ from the side of point B is assumed equal to $\omega_0 = \omega + 2\pi$. Employing isoperimetric inequalities for G^* and G_1^*, we find that $L_2 \geq c\sqrt{G^*} \geq c\sqrt{\widetilde{G}}$ and $2L_1 \geq L_1 + \Lambda \geq c\sqrt{G_1^*} \geq c\sqrt{\widetilde{G}_1}$, or $L^* = L_1 + L_2 \geq c\big(\sqrt{\widetilde{G}_1} + \sqrt{\widetilde{G}}\big) \geq c\sqrt{\widetilde{G}_1 + \widetilde{G}}$. Applying the Gauss–Bonnet theorem to $\widetilde{G}$ and $\widetilde{G}_1$, we find that $\tau + \omega_0\cos\rho + \pi - \alpha + \widetilde{G} = 2\pi$ and $\tau_1 + \pi - (2\pi - \alpha) + \widetilde{G}_1 = 2\pi$, or $p + \omega_0\cos\rho + \widetilde{G}_1 + \widetilde{G} = 4\pi$ and $\widetilde{G} + \widetilde{G}_1 = 2\pi - p - \omega_0\cos\rho + 2\pi \geq -\omega_0\cos\rho + 2\pi\cos\rho = -\omega\cos\rho$. Thus, $L^* \geq c\sqrt{-\omega}$.

Note that the estimate on L^* in the remaining cases with $\omega = -2k\pi$, $k = 1, 2, \ldots$, can be derived from the above cases by passing to the limit.

It remains to study prisms V for which $\omega > 0$ in Euclidean space and $\omega < 0$ in pseudo-Euclidean space. Here it must be noted that the geometric interpretation of the lines L', polygonal lines L^* and $\widetilde{L}$, circumferences Λ and $\widetilde{\Lambda}$, and regions G^* and $\widetilde{G}$ is highly specific and exhibits a certain duality: in the pseudo-Euclidean case it coincides with the Euclidean interpretation with $\omega < 0$, and in the Euclidean case it coincides with the pseudo-Euclidean interpretation with $\omega > 0$. Hence, here we use the already existing constructions. No expected estimate of L^* is obtained, however. To verify this statement, it is sufficient to examine the relationships for the $\widetilde{G}$ on $\widetilde{S}$ established by the Gauss–Bonnet formula. In the Euclidean case we have $\omega > 0$ and $p < 2\pi$. Examining $\widetilde{G}_1$ and $\widetilde{G}_2$, we find that $\tau_1 + \pi - \alpha + \widetilde{G}_1 = 2\pi$ and $\tau_2 + \pi - \alpha - \omega\cos\rho + 2\pi + \widetilde{G}_2 = 2\pi$, or $\widetilde{G}_1 - \widetilde{G}_2 = 2\pi - \omega\cos\rho - p$. To obtain an inequality in the last relationship we must exclude the term $2\pi - p > 0$. This necessarily leads only to $\widetilde{G}_1 \geq -\omega\cos\rho$. In the pseudo-Euclidean case we have $\omega < 0$ and $p > 2\pi$. Examining $\widetilde{G}$, we find that $\widetilde{G} = p - 2\pi + \omega\cosh\rho$, hence $\widetilde{G} \geq \omega\cosh\rho$. In each of these cases the lower bounds obtained for the areas are meaningless, since areas are estimated by negative quantities. Thus, it can be said that our method makes it possible to prove the lemma only for $\omega \leq 0$ in Euclidean space and $\omega \geq 0$ in pseudo-Euclidean. The lemma is proved.

In conclusion we formulate a lemma that was actually established in the proof of Lemma 2 in 1.4 and is a generalization of Lemma 2 in 1.5.

LEMMA 2. *Let $AC\ldots BD$ be a multiple convex prism in which the union of the vertical boundary faces $DA\bar{A}$ and $CA\bar{A}$ either constitutes a straight space-like angle or exceeds such an angle. Let the full angle at the vertex $\bar{A}$ of the lower base* *of the prism be smaller than π. Then the full angle at the vertex A of the upper base of the prism is also smaller than π.*

This lemma remains true in the Euclidean case with appropriate minor changes in the hypothesis.

References

1. A. D. Aleksandrov, *Convex polyhedra*, GITTL, Moscow, 1948; German transl., Akademie-Verl50, Berlin, 1958.
2. ———, *Inner geometry of convex surfaces*, GITTL, Moscow, 1948; German transl., Akademie-Verlag, Berlin, 1955.
3. A. D. Aleksandrov and E. P. Sen′kin, *On the inflexibility of convex surfaces*, Vestnik Leningrad. Univ. **1955**, no. 4, 3–13. (Russian)
4. I. Ya. Bakel′man, A. L. Verner, and B. E. Kantor, *Introduction to differential geometry "in the large"*, "Nauka", Moscow, 1973. (Russian)
5. M. Berger, *Géométrie*, Cedic, Paris, 1977.
6. H. Busemann, *Convex surfaces*, Wiley, New York, 1958.
7. Yu. D. Burago and V. A. Zalgaller, *Realization of developed surfaces in the form of polyhedra*, Vestnik Leningrad. Univ. **1960**, no. 7, 60–80. (Russian)
8. Yu. A. Volkov, *Existence of a convex polyhedron with a given developed surface*. 1, Vestnik Leningrad. Univ. **1960**, no. 19, 75–86. (Russian)
9. ———, *Estimating the deformation of a convex surface as a function of variation of its intrinsic metric*, Ukrain. Geometr. Sb. **5–6** (1968), 44–49. (Russian)
10. V. L. Gurevich, *Convex surfaces in a pseudo-Euclidean space*, Dokl. Akad. Nauk SSSR **240** (1978), 512–514; English transl in Soviet Math. Dokl. **19** (1978).
11. N. V. Efimov, *Qualitative aspects of the theory of surface deformation*, Uspekhi Mat. Nauk **3** (1948), no. 2, 47–158; English transl. in Amer. Math. Soc. Transl. (1) **6** (1962).
12. V. A. Zalgaller, *On deformations of a polygon on a sphere*, Uspekhi Mat. Nauk **11** (1956), no. 5, 177–178. (Russian)
13. S. Cohn-Vossen, *Flexibility of surfaces "in the large"*, Uspekhi Mat. Nauk (1936), no. 1, 33–76. (Russian)
14. J. Liberman, *Geodesic lines on convex surfaces*, C. R. (Dokl.) Acad. Sci. URSS **32** (1941), 310–313.
15. A. D. Milka, *What is geometry "in the large"?*, "Znanie", Moscow, 1986. (Russian)
16. ———, *Convex hypersurfaces in a pseudo-Euclidean space*, Dokl. Akad. Nauk SSSR **284** (1985), 1314–1316; English transl in Soviet Math. Dokl. **32** (1985).
17. ———, *On continuous bending of convex surfaces*, Ukrain. Geometr. Sb. **13** (1973), 129–141. (Russian)
18. ———, *Unique definiteness of general closed convex surfaces in Lobachevskiĭ's space*, Ukrain. Geometr. Sb. **23** (1980), 99–107. (Russian)
19. ———, *On the unique definiteness of convex surfaces in a pseudo-Riemannian spherical space*, Ukrain. Geometr. Sb. **29** (1986), 113–118. (Russian)
20. S. P. Olovyanishnikov, *Generalization of Cauchy's theorem on convex polyhedra*, Mat. Sb. **18 (60)** (1946), 441–446. (Russian)
21. A. V. Pogorelov, *Extrinsic geometry of convex surfaces*, Amer. Math. Soc., Providence, RI, 1973.
22. ———, *Quasigeodesic lines on a convex surface*, Mat. Sb. **25 (67)** (1949), 275–306; English transl. in Amer. Math. Soc. Transl. (1) **6** (1962).
23. Yu. G. Reshetnyak, *On the isoperimetric property of manifolds whose curvature is not greater than K*, Vestnik Leningrad. Univ. **1961**, no. 19, 58–76. (Russian)
24. D. D. Sokolov, V. G. Gaĭdalovich, and U. Il′khamov, *Closed convex surfaces in a pseudo-Euclidean space*, Trudy Inst. Mat. Sibirsk. Otd. Akad. Nauk SSSR **9** (1987), 154–159. (Russian)
25. V. V. Strel′tsov, *Estimating the length of a polygonal line on a polyhedron*, Izv. Akad. Nauk Kazakh. SSR No. **116** (1952) (Ser. Astr. Fiz. Mat. Mekh. no. 1 (6)) 3–36. (Russian)
26. L. A. Shor, *Applicable convex surfaces*, Ukrain. Geometr. Sb. **13** (1973), 179–183. (Russian)

27. A. Cauchy, *IIe mémoire sur les polygones et les polyèdres*, J. École Polytech. **9** (1813), 87–98.
28. H. Minkowski, *Volumen und Oberfläche*, Math. Ann. **57** (1903), 447–495.
29. R. Connelly, *A flexible sphere*, Math. Intelligencer **1** (1978), 130–131.

Amer. Math. Soc. Transl.
(2) Vol. **176**, 1996

Small Parameters in the Theory of Isometric Imbeddings of Two-dimensional Riemannian Manifolds in Euclidean Spaces

È. G. Poznyak and E. V. Shikin

Introduction

The introduction of a small parameter in problems of regular isometric imbeddings of two-dimensional Riemannian manifolds in Euclidean spaces made it possible to connect simple geometrical ideas with effective analytical methods. This method was found to be extremely fruitful and led to the solution of a whole range of interesting problems. The first results concerned nonlocal imbeddings of Riemannian manifolds of negative curvature [**1–7**], but the range of applications of a small parameter in problems of isometric imbeddings was extended considerably to cover wide classes of manifolds with nonnegative, nonpositive, and sign-alternating curvatures [**8–15**]. The transition from a small numerical parameter to a small functional parameter made it possible to solve many new problems [**16–20**]. Significant advances were made in the application of a small parameter to solve problems of regular isometric imbeddings in Euclidean spaces of dimensions higher than three [**21**], and to study the properties of solutions of the sine-Gordon equation [**22, 23**].

Let us begin with an illustrative example.

Isometric imbedding of rotational metrics. Suppose that a rotational metric

$$ds^2 = dx^2 + B^2(x)dy^2 \tag{1}$$

is defined on a plane, with the function $B(x) \in C^1$ on the Ox-axis. It is well known that the parametric equations of the rotational surface in three-dimensional Euclidean space E^3 with an intrinsic metric (1) have the form

$$X = B(x)\cos y, \quad Y = B(x)\sin y, \quad Z = \int_0^x \sqrt{1 - B'(t)^2}dt. \tag{2}$$

1991 *Mathematics Subject Classification.* Primary 53A07, 53C42.

While analyzing this formula, one could naturally suggest that every metric (1) can be imbedded isometrically (at least locally) into the space E^3 in the form of a rotational surface. After all, if $|B'| > 1$, the coordinate Z becomes imaginary according to the third formula in (2) and there is no real imbedding. However, it turns out that the problem of isometric imbedding of the metric (1) into the space E^3 can be solved by introducing a small numerical parameter. Moreover, the statement about imbedding *in the large* can be proved by simple means.

Let

$$X = X(x, y), \quad Y = Y(x, y), \quad Z = Z(x, y)$$

be parametric equations of the required surface in the space E^3 which realizes the metric defined in (1). Then the functions $X(x, y)$, $Y(x, y)$, and $Z(x, y)$ satisfy the following differential equation:

$$dx^2 + B^2 dy^2 = dX^2 + dY^2 + dZ^2. \tag{3}$$

We introduce a small parameter ε into this equation and go over to new variables $r(x, y)$ and $Z(x, y)$:

$$X = \varepsilon r(x, y) \cos \frac{y}{\varepsilon}, \qquad Y = \varepsilon r(x, y) \sin \frac{y}{\varepsilon}, \qquad Z = Z(x, y). \tag{4}$$

From (3) and (4) we obtain

$$dx^2 + B^2(x) dy^2 = \varepsilon^2 dr^2 + r^2 dy^2 + dZ^2. \tag{5}$$

Formula (5) is a compact notation of the imbedding equations. Thus a small parameter is introduced in the imbedding equations.

It can easily be seen that the functions

$$r(x, y) = B(x), \quad Z(x, y) = \int_0^x \sqrt{1 - \varepsilon^2 B'^2(t)}dt$$

are solutions of equation (5). Hence equations (4) of a rotational surface with the intrinsic metric (1) have the form

$$X = \varepsilon B(x) \cos \frac{y}{\varepsilon}, \qquad Y = \varepsilon B(x) \sin \frac{y}{\varepsilon}, \qquad Z = \int_0^x \sqrt{1 - \varepsilon^2 B'^2(t)}dt. \tag{6}$$

By an appropriate choice of the small parameter ε in the above formulas, we can always ensure the integrand in the last formula in (6) is positive on any given interval of the variable t. Moreover, we have obtained an interesting statement: a necessary and sufficient condition for the metric (1) to be an isometric imbedding *in the large* into the space E^3 in the form of a rotational surface of class C^1 is that the derivative $B'(x)$ be bounded on the axis Ox.

Note that the rotational surface on which the metric (1) is realized lies in a cylinder of radius $R = \varepsilon \sup B$ with Oz as axis and is a twisted surface resembling a roll of paper which degenerates into the axis OZ as $\varepsilon \to 0$ (the cylinder radius tends to zero).

Locally, any metric (1) is isometrically imbedded into the space E^3 as a rotation surface. However, no metric of type (1) with a negative curvature K which differs from zero through a negative constant can be imbedded *in the large* into the space E^3 in the form of a rotational surface. Indeed, the equality $K = -B''/B$ implies an exponential rise in the function B.

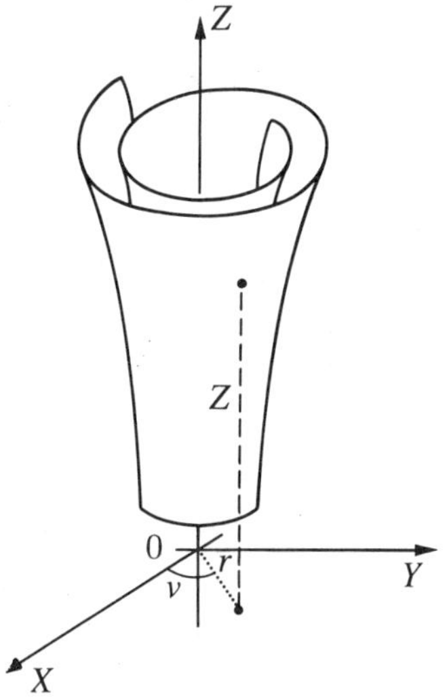

FIGURE 1

Asymptotic regular realization of two-dimensional metrics. Suppose that a regular two-dimensional metric W is defined in a certain two-dimensional region and a family of regular two-dimensional metrics W_ε (ε is a parameter), which are realized in the space E^3 through regular surfaces S_ε, is defined on the same set.

We say that *the surfaces S_ε asymptotically realize the metric W* if the metrics W_ε converge to the metric W (in a certain class of regularity) as $\varepsilon \to 0$. The convergence of the surfaces S_ε to any surface is not assumed.

Let us now consider a method of asymptotically generating a metric W (with an accuracy $O(\varepsilon^{2n})$, n being a natural number), which is defined in the strip

$$\pi = \{(x,y)|0 \leq x \leq a, -\infty < y < \infty\}$$

by the linear element

$$ds^2 = dx^2 + B^2(x,y)dy^2 \tag{7}$$

with an additional requirement of boundedness of all functions $B(x,y)$ in π.

In cylindrical coordinates (Z, v, r) (Figure 1), we consider a family of surfaces S_ε defined by the parametric equations

$$\overset{1}{Z} = x, \quad \overset{1}{v} = y/\varepsilon, \quad \overset{1}{r} = \varepsilon B(x,y). \tag{8}$$

The coefficients $\overset{1}{E}$, $\overset{1}{F}$, and $\overset{1}{G}$ of the first quadratic forms of the surfaces $\overset{1}{S}_\varepsilon$ have the form

$$\overset{1}{E} = 1 + \varepsilon^2 B_x^2, \quad \overset{1}{F} = \varepsilon^2 B_x B_y, \quad \overset{1}{G} = B^2 + \varepsilon^2 B_y^2. \tag{9}$$

Comparing the first quadratic form (7) with the first quadratic forms

$$d\overset{1}{s}{}_\varepsilon^2 = \overset{1}{E}dx^2 + 2\overset{1}{F}dxdy + \overset{1}{G}dy^2 \tag{10}$$

of the surfaces $\overset{1}{S}_\varepsilon$, we find that these surfaces asymptotically realize the metric W to within $O(\varepsilon^2)$.

The asymptotic realization of the metric W to within $O(\varepsilon^2)$ can be carried out in steps. The asymptotic construction of the metric with an accuracy $O(\varepsilon^2)$ is the *first step*.

The *second step* involves a transition from the family $\overset{1}{S}_\varepsilon$ to the family $\overset{2}{S}_\varepsilon$, and realizes the metric W with an accuracy $O(\varepsilon^4)$.

The basic idea of the transition from $\overset{1}{S}_\varepsilon$ to $\overset{2}{S}_\varepsilon$ can be described as follows. We add increments $\Delta\overset{1}{Z}$, $\Delta\overset{1}{v}$, and $\Delta\overset{1}{r}$ to the coordinates $\overset{1}{Z}$, $\overset{1}{v}$, and $\overset{1}{r}$. This results in a change in the coefficients $\overset{1}{E}$, $\overset{1}{F}$, and $\overset{1}{G}$ of the first quadratic forms of the surfaces $\overset{1}{S}_\varepsilon$. Choosing the increments in an appropriate manner, we can obtain the required family $\overset{2}{S}_\varepsilon$.

The choice of the increments is made in several steps. First of all, we choose $\Delta\overset{1}{Z}$, and put $\Delta\overset{1}{v}$ and $\Delta\overset{1}{r}$ equal to zero, thereby constructing an intermediate family of surfaces $\overset{11}{S}_\varepsilon$ defined by the equations

$$\overset{11}{Z} = x + \Delta\overset{1}{Z}, \qquad \overset{11}{v} = \overset{1}{v} = y/\varepsilon, \qquad \overset{11}{r} = \overset{1}{r} = \varepsilon B. \tag{11}$$

The coefficients of the first quadratic forms of the surfaces $\overset{11}{S}_\varepsilon$ have the following form:

$$\begin{aligned} \overset{11}{E} &= 1 + \varepsilon^2 B_x^2 + 2\Delta\overset{1}{Z}_x + \Delta\overset{1}{Z}{}_x^2, \\ \overset{11}{F} &= \varepsilon^2 B_x B_y + \Delta\overset{1}{Z}_y + \Delta\overset{1}{Z}_x \Delta\overset{1}{Z}_y, \\ \overset{11}{G} &= B^2 + \varepsilon^2 B_y^2 + \Delta\overset{1}{Z}{}_y^2. \end{aligned} \tag{12}$$

Next, we choose $\Delta\overset{1}{Z}$ in such a way that the sum $\varepsilon^2 B_x^2 + 2\Delta\overset{1}{Z}_x$ in the expression for the coefficient $\overset{11}{E}$ is equal to zero. In other words, we put

$$\Delta\overset{1}{Z} = -\frac{\varepsilon^2}{2}\int_0^x B_x^2\,dx.$$

For such a choice of $\Delta\overset{1}{Z}$, we have

$$\begin{gathered} \overset{11}{E} = 1 + \frac{\varepsilon^2}{4}\left[\int_0^x B_x^2\,dx\right]^2 = 1 + O(\varepsilon^4), \\ \overset{11}{F} = O(\varepsilon^2), \qquad \overset{11}{G} = B^2 + O(\varepsilon^2). \end{gathered} \tag{13}$$

The expressions $O(\varepsilon^2)$ in the last two formulas are polynomials in ε^2. This follows from (12) and the choice of $\Delta\overset{1}{Z}$.

Without changing $\overset{11}{Z} = x + \triangle\overset{1}{Z}$ and $\overset{11}{r} = \varepsilon B$, we now choose the increment $\triangle\overset{1}{v}$ in such a way that the coefficients $\overset{12}{E}$, $\overset{12}{F}$, and $\overset{12}{G}$ of the first quadratic forms in the new family $\overset{12}{S}_\varepsilon$ have the following structure:

$$\overset{12}{E} = O(\varepsilon^4), \qquad \overset{12}{F} = O(\varepsilon^4), \qquad \overset{12}{G} = B^2 + O(\varepsilon^2).$$

For this purpose, we turn to the explicit expressions $\overset{12}{E}$, $\overset{12}{F}$, and $\overset{12}{G}$. Using (13), we obtain

$$\begin{aligned} \overset{12}{E} &= \overset{11}{E} + \varepsilon^2 B^2 \triangle \overset{1}{v}{}_x^2 = 1 + O(\varepsilon^4) + \varepsilon^2 \overset{1}{v}{}_x^2, \\ \overset{12}{F} &= \overset{11}{F} + \varepsilon^2 B^2 \left(\frac{1}{\varepsilon} \triangle \overset{1}{v}_x + \triangle\overset{1}{v}_x \triangle \overset{1}{v}_y \right) \\ &= O(\varepsilon^2) + \varepsilon B^2 \triangle \overset{1}{v}_x + \varepsilon^2 B^2 \triangle \overset{1}{v}_x \triangle \overset{1}{v}_y, \\ \overset{12}{G} &= \overset{11}{G} + \varepsilon^2 B^2 \left(\frac{2}{\varepsilon} \triangle \overset{1}{v}_y + \triangle\overset{1}{v}{}_y^2 \right) \\ &= B^2 + O(\varepsilon^2) + 2\varepsilon B^2 \triangle \overset{1}{v}_y + \varepsilon^2 B^2 \triangle \overset{1}{v}{}_y^2. \end{aligned} \tag{14}$$

Here $O(\varepsilon^4)$ and $O(\varepsilon^2)$ are known polynomials in ε^2.

The increment $\triangle\overset{1}{v}$ is chosen in such a way that the sum $O(\varepsilon^2) + \varepsilon B^2 \triangle \overset{1}{v}_x$ in the final expression for the coefficient $\overset{12}{F}$ vanishes. Obviously, $\triangle\overset{1}{v}$ is a polynomial in ε, and $\triangle\overset{1}{v} = O(\varepsilon)$. For such a choice of $\triangle\overset{1}{v}$, the coefficients $\overset{12}{E}$, $\overset{12}{F}$, and $\overset{12}{G}$ have the form

$$\overset{12}{E} = 1 + O(\varepsilon^4), \qquad \overset{12}{F} = O(\varepsilon^4), \qquad \overset{12}{G} = B^2 + O(\varepsilon^2). \tag{15}$$

The quantities $O(\varepsilon^4)$ and $O(\varepsilon^2)$ are known polynomials in ε^2.

Without changing $\overset{12}{Z} = x + \triangle\overset{1}{Z}$, $\overset{12}{v} = y/\varepsilon + \triangle\overset{1}{v}$, we choose $\triangle\overset{1}{r}$ in such a way that the coefficients of the first quadratic forms obtained from the surfaces $\overset{2}{S}_\varepsilon$ have the form

$$\overset{12}{E} = 1 + O(\varepsilon^4), \qquad \overset{2}{F} = O(\varepsilon^4), \qquad \overset{2}{G} = B^2 + O(\varepsilon^4).$$

For this purpose, we turn to an explicit expression for the coefficients $\overset{2}{E}$, $\overset{2}{F}$, and $\overset{2}{G}$, and assume $\triangle\overset{1}{r}$ to be arbitrary. Using expression (15) for $\overset{12}{E}$, $\overset{12}{F}$, and $\overset{12}{G}$, we obtain

$$\begin{aligned} \overset{2}{E} =& 1 + O(\varepsilon^4) + 2\varepsilon B_x \triangle \overset{1}{r}{}_x^2 + \left(2\varepsilon B \triangle \overset{1}{r} + \triangle\overset{1}{r}{}^2 \right) \triangle \overset{1}{v}{}_x^2, \\ \overset{2}{F} =& O(\varepsilon^4) + \varepsilon B_x \triangle \overset{1}{r}_y + \varepsilon B_y \triangle \overset{1}{r}_y + \triangle\overset{1}{r}_x \triangle \overset{1}{r}_y \\ &+ \left(2\varepsilon B \triangle \overset{1}{r} + \triangle\overset{1}{r}{}^2 \right) \triangle \overset{1}{v}_x \left(1/\varepsilon + \triangle\overset{1}{v}_y \right), \\ \overset{2}{G} =& B^2 + O(\varepsilon^2) + 2\varepsilon B_y \triangle \overset{1}{r}_y + \triangle\overset{1}{r}{}_y^2 \\ &+ \left(2\varepsilon B \triangle \overset{1}{r} + \triangle\overset{1}{r}{}^2 \right) \left(1/\varepsilon + \triangle\overset{1}{v}_y \right)^2. \end{aligned} \tag{16}$$

The term $O(\varepsilon^2)$ in the last of the above formulas is a polynomial in ε^2. Let $\varepsilon^2 \triangle G$ be the first term in the expansion of $O(\varepsilon^2)$ in powers of ε^2, $O(\varepsilon^2) = \varepsilon^2 \triangle G + O(\varepsilon^4)$. Using this relation and allowing for the fact that $\triangle \overset{1}{v}_y = O(\varepsilon)$, we write the following expression for the coefficient $\overset{2}{G}$:

$$\overset{2}{G} = B^2 + O(\varepsilon^4) + \varepsilon^2 \triangle G + 2\frac{B}{\varepsilon} \triangle \overset{1}{r} + O(\varepsilon) \triangle \overset{1}{r} + \left[\frac{1}{\varepsilon^2} + O(1)\right] \triangle \overset{1}{r}{}^2 + O(\varepsilon) \triangle \overset{1}{r}_y + \triangle \overset{1}{r}{}^2_y.$$

We put

$$\triangle \overset{1}{r} = -\frac{1}{2}\frac{\triangle G}{B^2}\varepsilon^3 = O(\varepsilon^3).$$

For such a choice of $\triangle \overset{1}{r}$, taking into account the equality $\triangle \overset{1}{v} = O(\varepsilon)$, we obtain for the coefficients $\overset{2}{E}$, $\overset{2}{F}$, and $\overset{2}{G}$ the following expressions:

$$\overset{2}{E} = 1 + O(\varepsilon^4), \qquad \overset{2}{F} = O(\varepsilon^4), \qquad \overset{2}{G} = B^2 + O(\varepsilon^4).$$

Thus we have constructed the family of surfaces $\overset{2}{S}_\varepsilon$ that asymptotically generate the metric W with an error not exceeding $O(\varepsilon^4)$. This completes the second step.

The parametric equations for the surfaces belonging to the family $\overset{2}{S}_\varepsilon$ have the following form:

$$\overset{2}{Z} = x - \varepsilon^2 \int_0^x B_x^2 dx, \qquad \overset{2}{v} = \frac{y}{\varepsilon} - \varepsilon A,$$
$$\overset{2}{r} = \varepsilon B - \varepsilon^3 \frac{B_y^2 - 2BA_y}{B},$$

where

$$A = \int_0^x \frac{B_x B_y - \int_0^x B_x B_y dx}{B^2} dx.$$

The families $\overset{3}{S}_\varepsilon, \overset{4}{S}_\varepsilon \dots$ are constructed in the same manner.

The possibility of asymptotic generation of the metric W with an accuracy $O(\varepsilon^{2n})$ can easily be proved by induction.

In the following analysis, we shall require some auxiliary formulas in order to solve the problem of regular isometric imbedding *in the large* of two-dimensional metrics W^- with a negative curvature $K = -k^2$, defined in the strip $\pi = \{(x, y) | 0 \le x \le a, -\infty < y < \infty\}$ by the linear element (7).

We use the asymptotic generation of such metric to within $O(\varepsilon^2)$. The surfaces from the family $\overset{1}{S}_\varepsilon$ responsible for such a realization are defined by the parametric equations (8). Hence the angular coefficients r_ε and s_ε of the images of asymptotic lines of surfaces from the family S_ε on the plane of parameters (x, y) have the following form:

(17) $$r_\varepsilon = \varepsilon k + \varepsilon^2 R(x, y, \varepsilon), \qquad s_\varepsilon = -\varepsilon k + \varepsilon^2 S(x, y, \varepsilon).$$

This can easily be verified by simple direct computations. The functions $R(x, y, \varepsilon)$ and $S(x, y, \varepsilon)$ belong to the class of $C^{2,1}$-bounded functions (if the function $B(x, y)$

is $C^{4,1}$-bounded in the strip π) on the set

$$\{(x, y, \varepsilon) | 0 \leq x \leq a, -\infty < y < \infty, -\infty < \varepsilon < \infty\}.$$

1. A numerical parameter in isometric imbeddings of negative-curvature manifolds in the space E^3

Briefly, the method of solving imbedding problems by using a small parameter can be broken into the following main stages:

1. Choice of imbedding equations.
2. Introduction of a small parameter into the imbedding equations and the corresponding transformations.
3. Proof of the existence of solutions of an auxiliary system of partial differential equations (as a rule, this is a complicated problem from the technical point of view).
4. Geometrical conclusions.

Convex sets on negative-curvature manifolds. Let W^- be a complete Riemannian manifold of negative curvature homeomorphic to a two-dimensional plane. We shall describe certain convex subsets belonging to this manifold.

1. *Geodesic disk.* Let O be a point in W^- and let R be a positive number. On each geodesic emanating from the point O, we lay off an arc of length R. The set of endpoints of these arcs (having a common origin O) is called a geodesic circle, while the region bounded by this circle is called a geodesic disk of radius R.

2. *Equidistant strip.* Let Γ_0 be a complete geodesic of the manifold W^- and d a positive number. Arcs of length d are laid off along the geodesic $\Gamma_\perp$ on both sides of the point M at which this geodesic intersects the given geodesic Γ_0. The set of the tips of these arcs consists of two curves that are equidistant from Γ_0. The region between these curves is called a strip of constant width or an equidistant strip.

3. *Horodisk.* Let O be a point on the manifold W^- and Γ_0^+ a geodesic ray starting at this point. We shall consider a family of geodesic circles with centers $M(s)$ on the ray Γ_0^+ that pass through the point O. For an infinite separation between the centers ($s \to +\infty$), these circles tend to the limiting cycle, called a horocycle. The convex (in the metric of the manifold) with the limiting cycle as the boundary is called a horodisk.

4. *Expanding strip.* Let Γ_0 be a geodesic of the manifold W^-, O a point on the geodesic Γ_0, and s a natural parameter (length of the arc) on the geodesic Γ_0 measured from the point O. Let $d = d(s)$ be a continuous function satisfying the condition

$$\lim_{|s| \to +\infty} d(s) = +\infty.$$

Let $M(s)$ be the point on Γ corresponding to s, and $\Gamma_\perp$ the geodesic intersecting Γ_0 perpendicularly at this point. An arc of length $d(s)$ is laid out on $\Gamma_\perp$ at each side of $M(S)$. The set of all the tips of such arcs consists of two diverging curves. The region between these curves is called an infinitely expanding strip.

5. *Infinite polygon.* Any complete geodesic Γ of the manifold W^- breaks it into two regions (half-planes) with the common boundary Γ. Any nonempty intersection

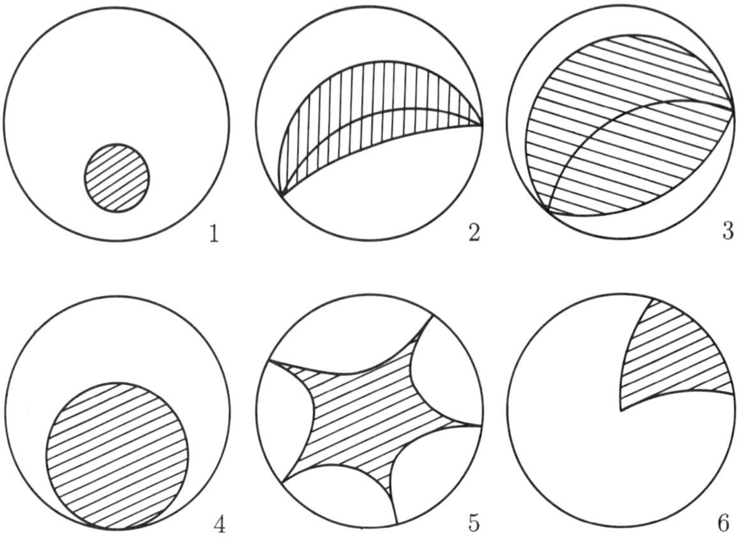

FIGURE 2

of a finite or countable set of closed half-planes whose boundaries have no common points is called an infinite polygon.

6. *Angle.* The region of the manifold W^- enclosed between two geodesic rays starting at the same point is called an angle.

Figure 2 shows all the above convex regions for the case when the curvature K of the manifold is constant, i.e., $K = -1$ (the Lobachevsky plane). Here, we have used the conformal interpretation of the Lobachevsky plane in a disk.

7. *Semigeodesic coordinates.* Semigeodesic coordinates can be introduced immediately on a manifold of negative curvature homeomorphic to a plane. We show how to construct semigeodesic coordinate systems on a manifold taking into account the structure of each of the convex regions described above.

A. Let Γ_0 be a complete geodesic of the manifold W^-. Choosing the geodesics $\Gamma_\perp$ orthogonal to this geodesic (lines x) and curves equidistant from this geodesic (lines y) as the coordinate lines, we define the metric of W^- by the linear element

$$ds^2 = dx^2 + B^2(x,y)dy^2. \tag{1}$$

For an infinite strip of the manifold W^-, the domain of variation of the coordinates x and y on the parametric plane is the infinite strip

$$\pi_d = \{(x,y) \mid |x| \leq d, -\infty < y < \infty\}.$$

B. Let O be a point on the manifold W^-. Taking the geodesics emerging from O as the first family (lines x) and the orthogonal trajectories as the second family (lines y), we can reduce the metric of W^- to the form (1), also. The domain of variation of the coordinates x and y on the plane of parameters for the angle is the set

$$\{(x,y)|0 \leq x \leq \infty, |y| \leq d\}.$$

C. Let Ω be a horocycle of W^-. For the lines y, we choose horocycles equidistant from Ω. The orthogonal trajectories to these horocycles are taken as the lines x. The metric of W^- has the form (1), while the domain of variation of coordinates x and y corresponding to the horodisk encompassed by the horocycle Ω is defined as

$$\{(x,y) \mid 0 \leq x \leq \infty, -\infty < y < \infty\}.$$

D. Let Γ_0 be one of the geodesics confining an infinite polygon. Choosing the geodesics $\Gamma_\perp$ orthogonal to this geodesic as the lines x, and the geodesics equidistant from Γ_0 as the lines y, we can reduce the metric of the manifold to the form (1). The coordinate system obtained in this way will be convenient below (in particular, for analyzing infinite polygons with a finite number of sides).

The facts that the metric of the manifold W^- has the form

$$ds^2 = dx^2 + B^2(x,y)dy^2,$$

and that the semigeodesic coordinate system (x,y) can be naturally connected with each of the convex regions introduced above are important for constructing the required isometric imbeddings.

System of Rozhdestvenskiĭ–Poznyak equations. The classical approach based on the Gauss–Peterson–Codazzi equations and Bonnet's theorem, when used for solving the problem of isometric imbedding of negative-curvature manifolds, leads to a system of two quasilinear equations for Riemannian invariants.

Suppose that a metric is defined on the plane of parameters (x,y) by a linear element of form (1)

$$\begin{cases} LN - M^2 = B^2K, \\ M_x - L_y = -\dfrac{B_x}{B}M, \\ N_x - M_y = (N + B^2L)\dfrac{B_x}{B} - M\dfrac{B_y}{B}. \end{cases} \tag{2}$$

The Gauss–Peterson–Codazzi equations for the coefficients of the second quadratic form can be transformed, by using the substitutions

$$r = \frac{-M - Bk}{N}, \qquad s = \frac{-M + Bk}{N}, \tag{3}$$

where $k - \sqrt{-K}$, to a system of weakly nonlinear equations of the following form (see [**24**]):

$$\begin{cases} \dfrac{\partial r}{\partial x} + s\dfrac{\partial r}{\partial y} = A_1 r + A_2 s + A_3 r^2 + A_4 rs + A_5 r^2 s, \\ \dfrac{\partial s}{\partial x} + r\dfrac{\partial s}{\partial y} = A_1 s + A_2 r + A_3 s^2 + A_4 sr + A_5 s^2 r. \end{cases} \tag{4}$$

The coefficients A_i of (4) are defined af follows:

$$A_1 = \frac{kx}{2k} - \frac{B_x}{B}, \qquad A_2 = -\frac{kx}{2k} - \frac{B_x}{B},$$
$$A_3 = \frac{ky}{2k}, \qquad A_4 = -\frac{ky}{2k} - \frac{B_y}{B}, \qquad A_5 = -BB_x.$$

For brevity, we define

$$F_1(x, y, r, s) = A_1 r + A_2 s + A_3 r^2 + A_4 rs + A_5 r^2 s.$$

For a linear element in the general form

$$ds^2 = E dx^2 + 2F dx dy + G dy^2$$

the system of equations (4) has a similar structure [**2**].

Under the condition $r \neq s$, we see that the coefficients L, M, N defined by

$$L = \frac{2krs}{s-r} B, \qquad M = -\frac{k(s+r)}{s-r} B, \qquad N = \frac{2k}{s-r} B$$

are the solution of system (2).

The system (4) is a system of quasilinear hyperbolic equations. Its characteristics are the integral curves defined by the differential equations

$$dy/dx = r(x, y), \quad dy/dx = s(x, y).$$

The required functions $r(x, y)$ and $s(x, y)$ are called *Riemannian invariants.*

The left-hand sides of the equations in (4) are simply total derivatives of the functions $r(x, y)$ and $s(x, y)$ along the characteristics of the system.

Integral transformation. Let us consider in the strip π_h the Cauchy problem for the system (4) with the following initial conditions on the axis Oy:

$$r(0, y) = r_0(y), \quad s(0, y) = s_0(y). \tag{5}$$

We examine the linear system

$$\begin{aligned} r_x + \sigma r_y &= F_1(x, y, \rho, \sigma), \\ s_x + \rho s_y &= F_1(x, y, \sigma, \rho), \end{aligned} \tag{6}$$

where $\rho(x, y)$ and $\sigma(x, y)$ are known functions and $\rho(x, y) \neq \sigma(x, y)$.

The characteristics of the system (6) are the integral curves of the system of ordinary differential equations

$$\frac{dy}{dx} = \sigma(x, y), \qquad \frac{dy}{dx} = \rho(x, y). \tag{7}$$

We assume that the couple $\{r, s\}$ is obtained as a result of transformations of the couple $\{\rho, \sigma\}$ if it is a solution of the Cauchy problem for the system (6) with the initial conditions (5).

Let (ξ, η) be an arbitrary fixed point in the strip π_h and let

$$y = y_I(x; \xi, \eta), \qquad -y = y_{II}(x; \xi, \eta)$$

be the characteristics of the system (6) passing through the point (ξ, η), i.e., the solution of the system (7) defined by the initial conditions

$$y_I(\xi; \xi, \eta) = \eta, \qquad y_{II}(\xi; \xi, \eta) = \eta,$$

Integrating equations (6) from 0 to ξ along the corresponding characteristics, we obtain the integral transformation in which we are interested:

$$(8)\qquad \begin{aligned} r(\xi,\eta) =& r_0\left(y_I(0;\xi,\eta)\right) \\ &+ \int_0^\xi F_1\left(x, y_I(x;\xi,\eta), \rho\left(x, y_I(x;\xi,\eta)\right), \sigma\left(x, y_I(x;\xi,\eta)\right)\right) dx, \\ s(\xi,\eta) =& s_0\left(y_{II}(0;\xi,\eta)\right) \\ &+ \int_0^\xi F_1\left(x, y_{II}(x;\xi,\eta), \sigma\left(x, y_{II}(x;\xi,\eta)\right), \rho\left(x, y_{II}(x;\xi,\eta)\right)\right) dx. \end{aligned}$$

The solution $\{r(x,y), s(x,y)\}$ of the Cauchy problem for system (4) with the initial conditions (5) is a stationary element of the integral transformation constructed in (8). The quest for this stationary element is carried out through different modifications of the method of successive approximations. A considerable part of the proof of the existence theorem involves the proof of sufficient smoothness of the stationary element obtained.

Isometric imbedding of equidistant strips. The following statement is valid.

THEOREM [**1,2**]. *Suppose that a metric is defined on the plane of the parameters by a linear element of type* (1). *If the curvature K of this metric is confined between two negative constants and is a $C^{2,1}$-bounded function in the strip*

$$\pi_a = \{(x,y) \| x| \le a, -\infty < y < \infty\}$$

of width $2a$ (a is an arbitrary positive number), then the given metric in the strip π_a can be imbedded into E^3 as a surface of class $C^{3,1}$.

The proof of this theorem is carried out by a fine analysis of the solution of the system of imbedding equations and can be broken down into the following main steps.

First step. Asymptotic realization. We consider the surface S_ε defined in cylindrical coordinates by the parametric equations

$$(9)\qquad Z = x, \quad \theta = \frac{y}{\varepsilon}, \quad \rho = \varepsilon B(x,y).$$

The first quadratic form of the surface S_ε is

$$(10)\qquad ds_\varepsilon^2 = (1+\varepsilon^2 B_x^2)dx^2 + 2\varepsilon^2 B_x B_y dx dy + (B^2 + \varepsilon^2 B_y^2)dy^2,$$

while the angular coefficients r_ε and s_ε of the inverse images of the asymptotic lines of the surface S_ε on the plane of the parameters (x,y) are computed from the formulas

$$r_\varepsilon = \varepsilon k(x,y) + \varepsilon^2 R(x,y,\varepsilon), \qquad s_\varepsilon = -\varepsilon k(x,y) + \varepsilon^2 S(x,y,\varepsilon),$$

where $R(x, y\varepsilon)$ and $S(x,y,\varepsilon)$ are $C^{2,1}$-bounded functions in the strip π_a.

Second step. Well-posedness of the solution of the system of imbedding equations and its structure. Direct computations show that the coefficients of the right-hand sides of the system of imbedding equations, calculated for metrics (1) and (10) respectively, differ in magnitude by not more than $\varepsilon^2 T(x,y,\varepsilon)$, where $T(x,y,\varepsilon)$ is a $C^{1,1}$-bounded function in the strip π_a.

By appropriately choosing the parameter ε, we can ensure that *small solutions of the system of imbedding equations are well-posed.* This property can be described as follows:

There exists $\delta > 0$ such that all systems whose right-hand side coefficients differ from those of the given system in the class $C^{1,1}$ by less than δ, have a solution in the strip π_a for any initial data differing from zero by less than δ.

This property of well-posedness of small solutions leads to the existence of the solution $r(x,y)$, $s(x,y)$ of a system of equations in the strip π_a with the following initial conditions:

$$r_0(y) = r_\varepsilon(0,y), \quad s_0(y) = s_\varepsilon(0,y).$$

The solution has the form

$$\begin{aligned} r(x,y) &= \varepsilon k(x,y) + \varepsilon^2 O(1), \\ s(x,y) &= -\varepsilon k(x,y) + \varepsilon^2 O(1). \end{aligned} \tag{11}$$

Third step. Geometrical conclusions. For all sufficiently small values of the parameter ε, the condition

$$r(x,y) \neq s(x,y),$$

which guarantees the existence of the required regular realization, is satisfied at each point in the strip π_a.

In conclusion, it can be stated that although the parameter ε is not introduced explicitly into the system of imbedding equations, it plays a key role at all stages of the proof and, among other things, in finding the structure (11) of the solution.

Taking this structure into consideration, we can obtain the above result in a different manner, i.e., by introducing the small parameter ε into the system of imbedding equations directly [**4**]. In this case, it is sufficient to require that the function $B(x,y)$ be C^4-bounded in the strip π_a.

First step. Using

$$r = \varepsilon k + \varepsilon^2 \rho, \quad s = -\varepsilon k - \varepsilon^2 \sigma \tag{12}$$

as substitutions for the unknown functions, we can transform the system of imbedding equations to

$$\begin{cases} \rho_x - (\varepsilon k + \varepsilon^2 \sigma)\rho_y = \Phi_1(\varepsilon, x, y, \rho, \sigma), \\ \sigma_x + (\varepsilon k + \varepsilon^2 \rho)\sigma_y = \Phi_2(\varepsilon, x, y, \rho, \sigma). \end{cases} \tag{13}$$

Second step. The method of successive approximations is used to prove the existence of a smooth bounded solution of this system in the strip π_a under the zero initial conditions $\rho(0,y) = 0$, $\sigma(0,y) = 0$.

It should be interesting to study the problem of determining the minimal conditions imposed on the regularity of the metric under which this metric can be realized in E^3.

Considerable progress can be made in this direction by using Darboux's classical approach [**25**] for the case when the function $B(x,y)$ belongs to the class C^3.

For a linear element of type (1), Darboux's imbedding equation assumes the form

$$\begin{aligned} Z_{xx}Z_{yy} - Z_{xy}^2 = K\left(B^2(1 - Z_x^2) - Z_y^2\right) + \frac{B_x^2}{B_2} Z_y^2 \\ - \left(BB_x Z_x - \frac{B_y}{B} Z_y\right) Z_{xx} - 2\frac{B_x}{B} Z_y X_{xy}. \end{aligned} \tag{14}$$

If the manifold has negative curvature, equation (14) can be reduced to a system of five quasilinear hyperbolic equations [**26**] with the unknown functions

$$x, y, Z, \quad p = Z_x, \quad q = Z_y$$

whose arguments are the characteristic variables u and v:

$$\begin{cases} x_{uv} = f_1(u, v, x, y, x_u, x_v, y_u, y_v), \\ y_{uv} = f_2(u, v, x, y, x_u, x_v, y_u, y_v), \end{cases} \tag{15}$$

$$\begin{cases} z_{uv} = f_3(u, v, x, y, x_u, x_v, y_u, y_v, p, q), \\ p_{uv} = f_4(u, v, x, y, x_u, x_v, y_u, y_v, p, q), \\ q_{uv} = f_5(u, v, x, y, x_u, x_v, y_u, y_v, p, q). \end{cases} \tag{16}$$

If the solution of the system (15)–(16) belongs to the class C^*, then, under the condition

$$\begin{vmatrix} x_u & x_v \\ y_u & y_v \end{vmatrix} \neq 0,$$

this solution can be used to reconstruct the C^2-smooth solution of the Darboux equation and to construct a C^2-smooth imbedding of (1).

The inclusion of a small parameter ε in the initial data of the system (15)–(16) enables us to construct its solution with the above properties by the method of successive approximations [**7**].

Let π^* be a strip of arbitrary width between the straight lines l and l_+ equidistant from the straight line l_0, with equation $u + v = 0$. In this case, for all ε that are sufficiently small, there exists in π^* a solution of class C^* for the system (15) that satisfies the initial conditions

$$\begin{aligned} x|_{l_0} &= 0, \quad y|_{l_0} = 2\varepsilon u, \\ x_u|_{l_0} &= x_v|_{l_0} = \frac{1}{k(0, 2\varepsilon u)}, \\ y_u|_{l_0} &= \varepsilon, \quad y_v|_{l_0} = -\varepsilon. \end{aligned}$$

In this case, the derivatives of the solution with respect to the variables u and v in π^* have the form

$$\begin{aligned} x_u &= 1/k + \varepsilon O(1), \quad x_v = 1/k + \varepsilon O(1), \\ y_u &= \varepsilon + \varepsilon^2 O(1), \quad y_v = -\varepsilon + \varepsilon^2 O(1). \end{aligned}$$

In proving the existence of such a solution, it is essential to note that the unknown functions z, p, and q do not appear in (15). It is interesting to note that such an effect is also observed for a linear element of a general type. If we turn to a Monge–Ampère hyperbolic equation (equation (14) is an equation of this type), all

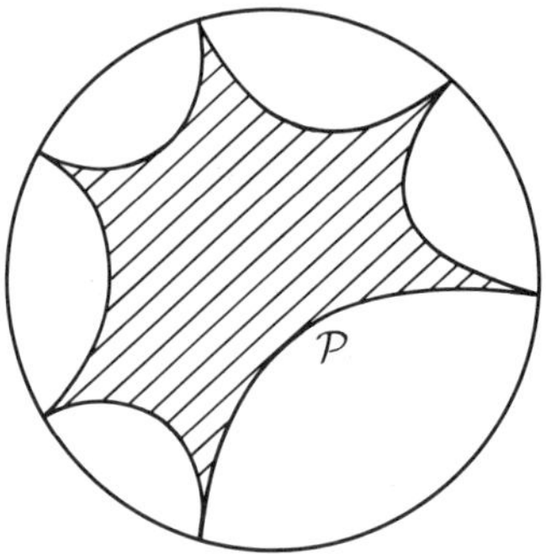

FIGURE 3

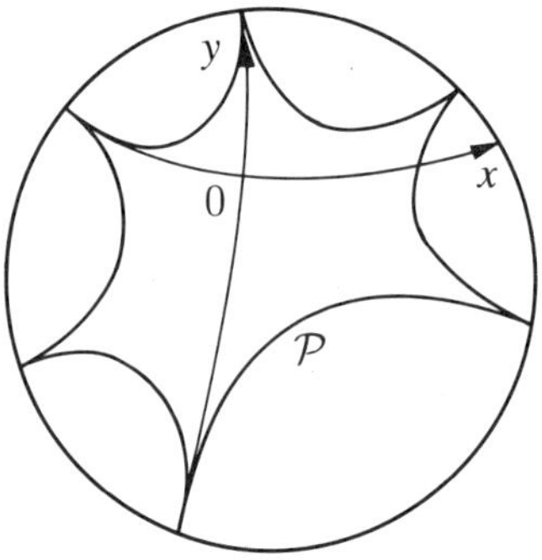

FIGURE 4

that has been stated above with the exception of the imbedding equation is valid under the condition that all the coefficients of the Monge–Ampère equation depend only on x and y and are independent of the function Z and its derivatives Z_x and Z_y [**27**].

Isometric imbedding of infinite polygons in the Lobachevsky plane. Let $\mathcal{P}$ be an infinite polygon in the Lobachevsky plane, with no half-planes and with a finite number of sides (Figure 3). We consider a complete geodesic in this polygon and construct in the Lobachevsky plane a semigeodesic system of coordinates (Figure 4) by the technique described earlier.

The linear element in the Lobachevsky plane in coordinates x and y can be reduced to the form

$$ds^2 = dx^2 + \cos h^2 x \, dy^2, \tag{17}$$

while the set P (Figure 5) is the range of variation of the variables x and y corresponding to the infinite polygon $\mathcal{P}$ (Figure 5).

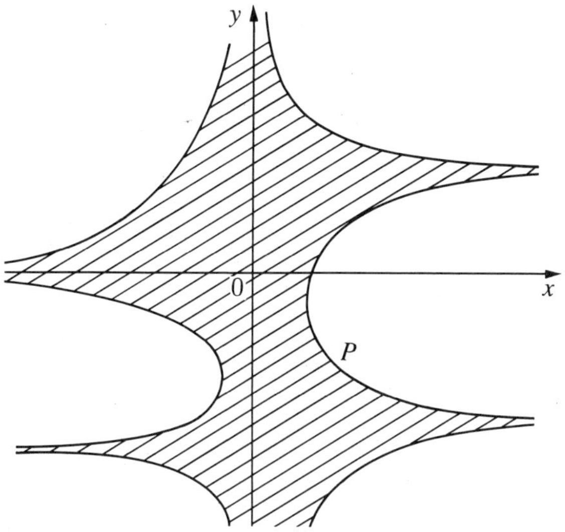

FIGURE 5

For the linear element (17), the Rozhdestvenskiĭ–Poznyak system of equations assumes the following form:

$$\left\{\begin{aligned} r_x + s r_y &= -\tanh x(r+s) - \cosh x \sinh x r^2 s, \\ s_x + r s_y &= -\tanh x(r+s) - \cosh x \sinh x r s^2. \end{aligned}\right. \tag{18}$$

Any Euclidean strip can be realized in E^3 in a class of analytical surfaces [**28**] as a universal covering of a rotational surface.

Stipulating for the system of equations (18) constant initial conditions of the type

$$r(0,y) = \varepsilon, \quad s(0,y) = -\varepsilon \tag{19}$$

on the axis Oy, we obtain its solution in the form

$$r(x,y) = \frac{\varepsilon}{\sqrt{1-\varepsilon^2\sinh^2 x}}, \quad s(x,y) = -\frac{\varepsilon}{\sqrt{1-\varepsilon^2\sinh^2 x}}.$$

For a given ε, we can easily indicate the maximum possible width of the strip beyond which the solution of system (18) with initial conditions (19) cannot be continued without singularities. From the equation $1-\varepsilon^2\sinh^2 x = 0$ we obtain

$$l = \ln\frac{1+\sqrt{1+\varepsilon^4}}{\varepsilon^2}$$

(Figure 6).

As we move away from the axis Oy toward the straight lines $x = \pm l$, the angular coefficients r and s of the inverse images of the asymptotic lines on the plane of the parameters (x,y) increase in absolute value and become infinitely large in the vicinity of the straight lines $x = \pm l$ (Figure 7). Hence the angles formed by these

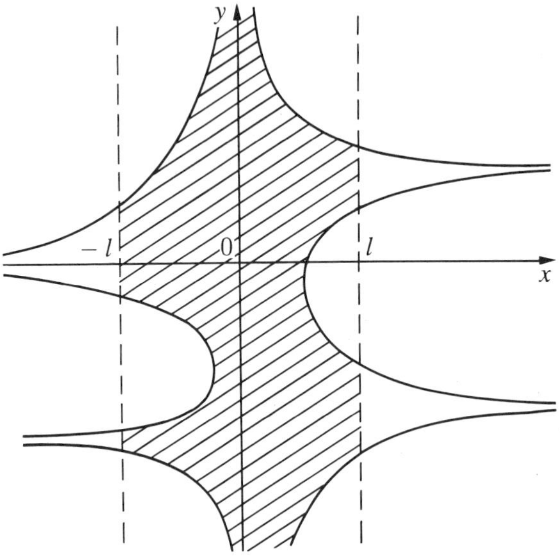

FIGURE 6

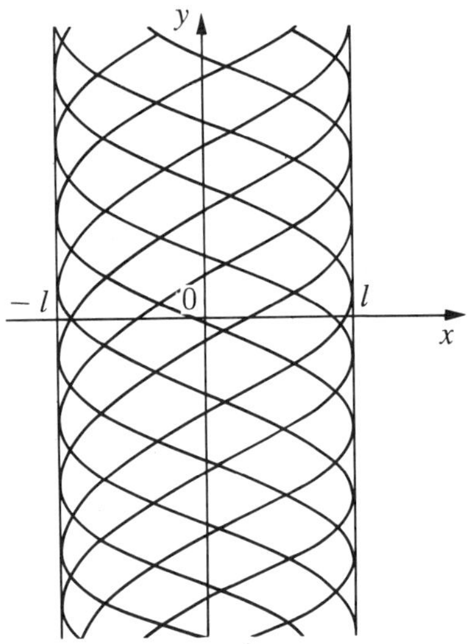

FIGURE 7

curves with the axis Ox are close to right angles in the vicinity of the points $(\pm l, 0)$ (Figure 8).

A transition from the strip π_l to an adjacent strip $\pi_{l-\delta}$ (δ is a small positive number) makes it possible to depart from the singular lines while retaining large

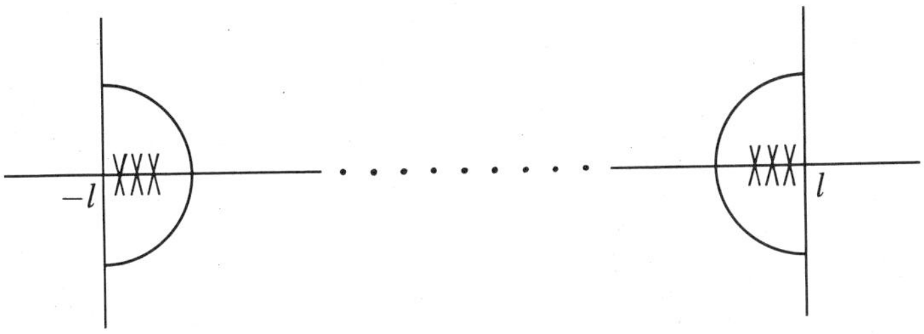

FIGURE 8

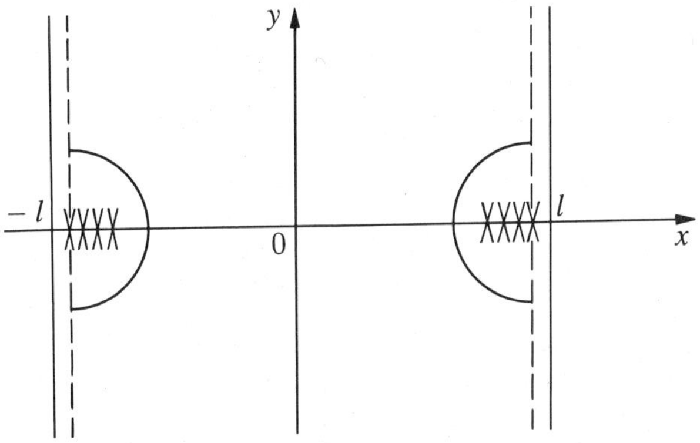

FIGURE 9

values of the functions $r(x, y)$ and $s(x, y)$ near the straight lines $x = \pm(l - \delta)$ by appropriately choosing the value of δ (Figure 9).

Obviously, the width of the imbedded strip $\pi_{l-\delta}$ can be made as large as possible by an appropriate choice of the parameter ε. Hence it is possible to decrease indefinitely the distance between the points M_- and M_+ where the equidistant curve $x = -l + \delta$ (see Figure 6) intersects adjacent sides of the polygon $\mathcal{P}$ (Figure 10).

These observations make it possible to construct imbeddings of the infinite parts of the polygon under consideration that cannot be contained within the strip $\pi_{l-\delta}$ so that the imbeddings are in conformity with the existing imbedding of the strip.

We construct in the Lobachevsky plane a new coordinate system (ξ, η), taking the axis Ox as the basic geodesic (Figure 11).

The system of imbedding equations for the linear element

$$ds^2 = d\xi^2 + \cosh \xi^2 d\eta^2$$

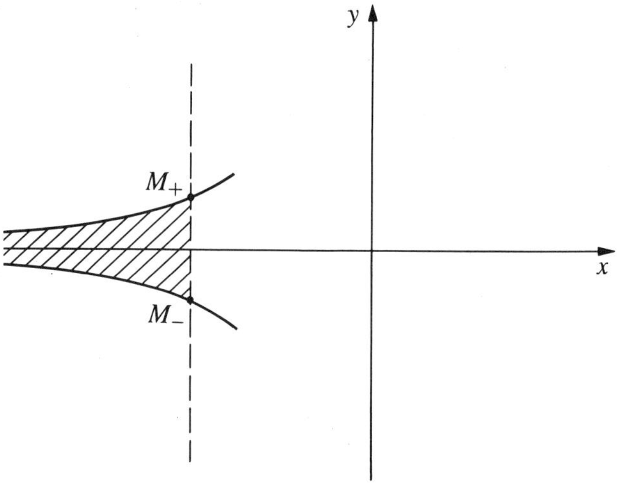

FIGURE 10

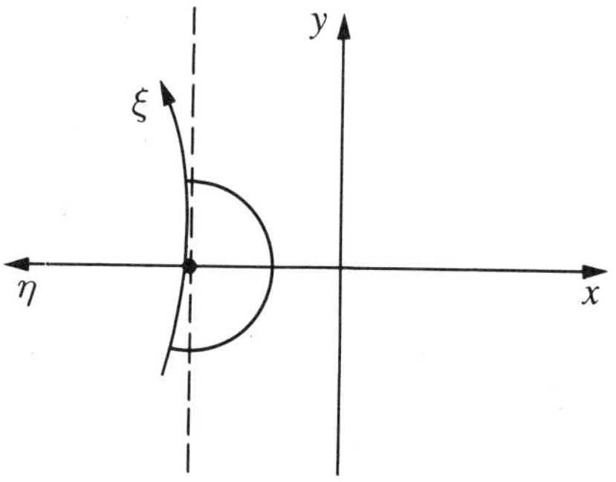

FIGURE 11

has the form

$$R_\xi + SR_\eta = -\tanh\xi(R+S) - \cosh\xi\sinh\xi R^2 S,$$
$$S_\xi + RS_\eta = -\tanh\xi(R+S) - \cosh\xi\sinh\xi RS^2. \tag{20}$$

The initial conditions for the system (20) are constructed by using the smallness of the quantities

$$R_0(\eta) = \frac{1}{r(x,0)} > 0, \quad S_0(\eta) = \frac{1}{s(x,0)} < 0$$

near the point $\eta = 0$.

Continuing the functions $R_0(\eta)$ and $S_0(\eta)$ along the entire positive semiaxis η while preserving the signs and the above-mentioned smallness, and using the functions obtained here as the initial data for the system (20), we arrive at the

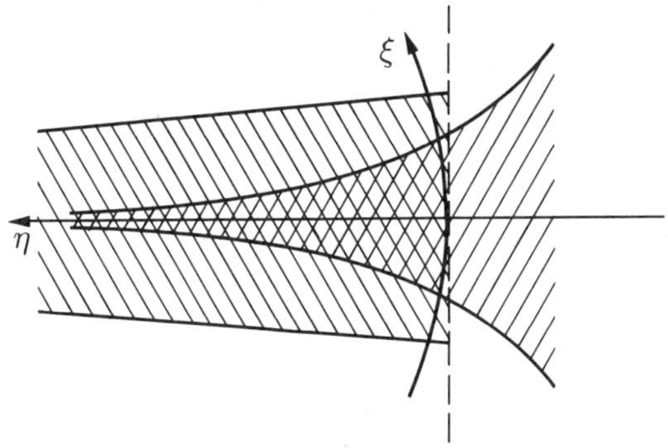

FIGURE 12

possibility of "gluing" an equidistant half-strip to the immersed region, including the region which we are interested in (Figure 12).

The possibility of "gluing" all other infinite tails is established similarly.

Such a method can be used to prove the existence of a regular isometric imbedding of infinite polygons with an infinite number of sides in E^3 [**6**].

Isometric imbedding of horodisks. Let us define on a plane a complete metric W^- whose curvature K satisfies the following conditions.

1. In the vicinity of each point, there is a semigeodesic system of coordinates (u, v) such that the curvature as a function of u and v belongs to the class C^3.
2. There are positive constants k_1 and k_2 $(k_1 \leq k_2)$ such that
$$-k_2^2 \leq K \leq -k_1^2.$$
3. The absolute values of the derivatives
$$\frac{\partial^p K}{\partial g_\perp^i \partial g^j}, \quad p = 1, 2, 3; \quad i = 0, 1; \quad j = p - i$$
(here $\partial/\partial g$ is the differentiation along the arc of the geodesic Γ, and $\partial/\partial g_\perp$ is the differentiation along the arc of the geodesic $\Gamma^\perp$ orthogonal to Γ), do not exceed a constant that is independent of the choice of the point and the geodesic passing through this point.
4. There exists a constant $d > 1$ such that
$$|K(M) - K_0| \leq C(M_0)/r^\alpha(M_0, M),$$
where M_0 is an arbitrary fixed point and $K_0 = \text{const}$.

THEOREM [**11**]. *Suppose that the conditions* (1)–(4) *are satisfied for the metric* W^-. *In this case, any horodisk in this metric can be imbedded isometrically in the space* E^3 *as a surface of class* C^3.

First step. Constructing horocyclic coordinates. Let Ω be an arbitrary horoidisk and Ω^* the horocycle encompassing it. On the entire manifold determined by the metric W^- defined on a plane, we construct a special system of semigeodesic coordinates, one family of which is formed by geodesic curves orthogonal to Ω^* and the other by the orthogonal trajectories (equidistant horocycles). In these coordinates, the linear element can be written in the form

$$ds^2 = dx^2 + B^2(x,y)dy^2.$$

We shall prove that the conditions imposed on the curvature K ensure the smoothness of the coordinate system thus constructed, which will be required for further analysis, and discuss the behavior of the function B and its derivatives.

Second step. Basic substitution. The system (4) can be transformed by introducing new unknown functions R and S as follows:

$$r = \varepsilon k + \varepsilon^2 \frac{R}{B^2}, \qquad s = -\varepsilon k - \varepsilon^2 \frac{S}{B^2}. \tag{21}$$

This transformation makes it possible to

(1) isolate the main part of the solution of (4) in order to satisfy the condition $r \neq s$;
(2) introduce a numerical parameter ε into the coefficients of the obtained system; and
(3) use the significant role of the functional parameter B while constructing the solution of the transformed system.

Third step. Constructing a solution in the half-plane of parameters. By imposing conditions on the behavior of the required functions R and S, namely,

$$\lim_{x\to+\infty} R = 0, \qquad \lim_{x\to+\infty} S = 0,$$

we shall prove the existence and uniqueness of the smooth solution of the transformed system in the half-plane $x > 0$.

Geometric remark. While considering the problem of imbedding a specific horodisk, the conditions of the theorem can be made less stringent in the following way.

The continuity of the curvature K of the metric W^- and condition (2) ensure the existence in the large on the plane of a one-parameter family of horocyclic coordinate systems. Among other things, there exists a coordinate system whose base is the horocycle Ω_0^* confining a given horodisk Ω_0. In this case, one of the families is formed by geodesics orthogonal to Ω_0^* and the other by horocycles equidistant from Ω_0^*.

We require that condition (1) be satisfied, together with the following condition:

(3*) the absolute values of the derivatives

$$\frac{\partial^p K(M)}{\partial g_\perp^i \partial s^j}, \quad p = 1,2,3; \quad i = 0,1; \quad j = p - i$$

(where $\partial/\partial s$ and $\partial/\partial g_\perp$ are the operators of partial differentiation along the arcs of coordinate horocycles and along the arcs of orthogonal geodesics, respectively), do not exceed a constant.

In this case, the horocyclic coordinate system (with the base Ω_0^*) will be regular. Among other things, the geodesic curvatures κ of the coordinate horocycles will be regular.

We impose additional conditions on the behavior of κ and K. Namely, we assume that there exists an horodisk $\Omega \subset \Omega_0$ satisfying the following conditions:

(4*)

$$\left| \iint_D \frac{\partial \kappa}{\partial s} d\sigma \right| \leq \text{const}$$

(here $D \subset \Omega$ is an arbitrary coordinate quadrangle in the constructed horocyclic coordinate system);

(5*)

$$\left| \frac{\partial \ln \sqrt{-K}}{\partial g_{\perp}} + \kappa \right| \geq \triangle = \text{const} > 0.$$

THEOREM. *If the conditions* (1)–(2), (3*)–(5*) *are satisfied, the horodisk* Ω_0 *admits a* C^3*-regular isometric imbedding in the space* E^3.

Isometric imbedding of arbitrary angles. Let us formulate conditions to be imposed on a metric so that if they are satisfied, we can use a line of reasoning similar to the above to prove the possibility of regular realization of an arbitrary angle apart of the plane lying between two geodesic rays emanating from the same point in this metric.

Let us define on a plane a complete metric W_∞^- of bounded negative curvature K, for which the following conditions are satisfied:

1. Each point has a neighborhood in which we can introduce semigeodesic coordinates (u, v) such that $K \in C^3(u, v)$. There exist a constant $r_0 \geq 1$ and a point M_0 such that the following inequalities are satisfied outside the disk $\omega(M_0, r_0)$ with radius r_0 and center at M_0:

2.

$$k(M) \leq C_0/r^q, \quad q = \text{const} > 2.$$

3.

$$\left| \frac{\partial^p Q(M)}{\partial g_{\perp}^i \partial g^j} \right| \leq \frac{C_p}{r^p}, \quad p = 1, 2, 3; \quad i = 0, 1; \quad j = p - i,$$

where $k = \sqrt{-K}$, $Q = \frac{1}{2} \ln k$, $r = r(M_0, M)$, and $C_0, C_1 < 1$ while C_2 and C_3 are positive constants.

THEOREM [**11**]. *Suppose that the conditions* (1)–(3) *are satisfied for the metric* W_∞^-. *In this case, any angle defined by two different geodesic beams in the metric* W_∞^- *can be imbedded isometrically in the space* E^3 *through a surface of class* C^3.

2. A small functional parameter in problems of imbedding negative-curvature manifolds

The transition from a small numerical parameter to a small functional parameter enables us to enlarge significantly the class of regions in Riemannian manifolds of negative curvature that possess a regular isometric imbedding into E^3. In this section we discuss the results concerning the isometric imbedding of infinitely expanding strips.

The first result in this area of research was obtained for manifolds of constant negative curvature, namely, for the Lobachevsky plane [**16**].

Let l be an arbitrary straight line in the Lobachevsky plane. As noted earlier, a strip of constant width is specified by two equidistant lines each separated by a distance d from l. But if d varies and increases without limit as we move along l, the strip determined by the resulting curves expands infinitely.

For the linear element of the Lobachevsky plane

$$ds^2 = dx^2 + \cosh^2 x\, dy^2$$

the system of imbedding equations simplifies considerably:

$$\begin{aligned} r_x + s r_y &= -\tanh x\,(r+s) - r^2 s \cosh x \sinh x, \\ s_x + r s_y &= -\tanh x\,(r+s) - r s^2 \cosh x \sinh x. \end{aligned}$$

Let us examine a function $\varepsilon(y)$ with the following properties.

1. It is defined for all values of the variable y, and is positive.
2. It has an infinite number of sections in which it remains constant.
3. As $|y| \to +\infty$, the limit relations

$$\varepsilon(y) \to 0, \qquad \frac{\varepsilon'(y)}{\varepsilon(y)} \to 0, \qquad \frac{\varepsilon''(y)}{\varepsilon(y)} \to 0$$

hold. (Such a function was constructed in [**16**].)

The introduction of the small functional parameter $\varepsilon(y)$ into the system of imbedding equations via transformation of the unknown functions r and s,

$$r = \frac{\varepsilon(y)(1+\rho)}{\sqrt{1-\varepsilon^2(y)\sinh^2 x}}, \qquad s = -\frac{\varepsilon(y)(1+\sigma)}{\sqrt{1-\varepsilon^2(y)\sinh^2 x}},$$

made it possible to build the solution of this system in a strip of the form

$$\{(x,y) \mid |x| \le -\ln \varepsilon(y)\}.$$

Although the rate at which this strip expands is extremely low, the fact that ε has large sections in which it remains constant has made it possibly to apply the method developed in [**6**] and to construct a regular isometric imbedding of another new class of infinite convex polygons of the Lobachevsky plane into E^3.

Later the conditions on the rate of variation of strip width were made considerably less stringent [**17, 18**], and the results were generalized to incorporate manifolds of variable negative curvature [**20**]. The notion of an expanding strip can be transformed in a natural manner to apply to this class of manifolds (the whole geodesic serves as the base).

Below we give the main steps in the proof for L-type manifolds (complete Riemannian manifolds that are homeomorphic to a plane and have a curvature whose value lies between two negative constants).

First step. Introducing a small functional parameter into the imbedding equation. By the formulas

$$r = \xi(x,y)(1+\rho), \qquad s = \xi(x,y)(1+\sigma),$$

where

$$\xi(x,y) = \frac{\varepsilon(y)k(x,y)}{\sqrt{1-\varepsilon^2(y)k^2(x,y)B^2(x,y)}}, \qquad \varepsilon(y) = \exp\{-\lambda\sqrt{y^2+a^2}\},$$

the new unknown functions ρ and σ are introduced into the system of imbedding equations. The transformed system is

$$\rho_x - \xi(x,y)(1+\sigma)\rho_y = \Psi_1(x,y,\rho,\sigma),$$
$$\sigma_x + \xi(x,y)(1+\rho)\sigma_y = \Psi_2(x,y,\rho,\sigma).$$

Second step. Constructing a bounded solution of the transformed system in an expanding strip. Selecting the zero initial data

$$\rho(0,y) = 0, \qquad \sigma(0,y) = 0,$$

and arbitrarily choosing the positive numbers m and n, we find parameters λ and α such that in the expanding strip

$$\pi = \{(x,y) \mid |x| \leq m|y| + n\}$$

the transformed system has a smooth solution $\{\rho(x,y), \sigma(x,y)\}$ satisfying the condition

$$|\rho(x,y)| < 1, \qquad |\sigma(x,y)| < 1.$$

The existence of a solution with such properties is proved by the well-known method of successive approximations.

Third step. Geometric conclusions. The above inequalities readily yield the relation $r(x,y) \neq s(x,y)$, which implies that there is regular realization into E^3 of the given metric in a strip π.

There is also a somewhat different way of introducing a small parameter into the system of imbedding equations, one that allows us to prove the possibility of regular isometric imbedding into E^3 of any expanding strip of a manifold of negative curvature [**19**].

First step. Introducing a small functional parameter. Consider the transformation

$$r = \varepsilon(y)\big(k(x,y) + g(x,y)\big) + \varepsilon^2(y)\rho,$$
$$s = -\varepsilon(y)\big(k(x,y) - g(x,y)\big) + \varepsilon^2(y)\sigma,$$

where $\varepsilon(y)$ is a positive function tending to zero as $|y| \to \infty$ at a rate determined by how fast the imbedded strip expands, and $g(x,y)$ is a smooth solution of the Riccati equation

$$g_x = -\varepsilon' g^2 - 2\frac{B_x}{B} + \varepsilon' k^2;$$

this transformation introduces the new unknown functions ρ and σ into the system of imbedding equations.

Second step. Constructing and investigating the auxiliary system. The transformation

$$X = \frac{x}{w(y)}, \qquad Y = y$$

maps the strip $\pi_w = \{(x,y) \mid |x| \leq w(y)\}$ in the plane (x,y) into the constant-width strip

$$\pi = \{(X,Y) \mid |X| \leq 1\}.$$

The system of equations for ρ and σ also undergoes certain changes. Then the method of successive approximations is used to construct in the strip π a smooth solution of the transformed system of equations, which in the final analysis ensures the validity of the geometrically justifiable inequality $r(x, y) \neq s(x, y)$.

In conclusion we specify conditions on the coefficient $B(x, y)$ of the metric W^- under which all arguments of this section are valid.

THE SMOOTHNESS CONDITIONS. The function $B(x, y)$ belongs to class C^4 in the complete plane of parameters (x, y), and the imbedded surface is C^3-regular.

Note that the existence of a regular realization in E^3 of an arbitrarily expanding strip of a manifold of negative curvature implies the possibility of a regular isometric imbedding into this space of any horodisk, of a convex hull of any pair of horodisks, and of a convex region having a tangency of any order with the absolute. The possibility of using a small numerical parameter to imbed horodisks can be proved if extremely stringent conditions are applied to the behavior of the curvature [**10**–**12**], but the application of a small functional parameter yields stronger results with no limiting conditions on the curvature, and considerably broadens the class of regions that admit regular realization in E^3. However, the extrinsic properties of the existing imbeddings of horodisks are different: with a numerical parameter the twisting is "along" parallel geodesics, while with a functional parameter the twisting is "across" such a bundle. In both cases, however, the small parameter, either numerical or functional, plays an important role.

3. A small parameter in the sine-Gordon equation

The sine-Gordon equation. The equation

$$z_{xy} = \sin z \tag{1}$$

is known as the sine-Gordon equation. Initially it was related to geometrical studies [**29, 30**], in which it described the behavior of the angle $z(x, y)$ of a Chebyshev array in the Lobachevsky plane. Employing this equation in the theory of surfaces of constant negative curvature made it possible to prove a number of interesting and important results [**30**–**36**]. Today equation (1) is one of the basic equations of mathematical physics. One reason is that it has solutions in the form of solitary waves, or solitons.

Chebyshev arrays. The method of successive approximations for the sine-Gordon equation. Arrays on surfaces (on two-dimensional Riemannian manifolds) are called Chebyshev arrays if in each array quadrangle the opposite sides are equal [**37**].

If the coordinate array (x, y) on a surface is of the Chebyshev type, the linear element of this surface has the form

$$ds^2 = dx^2 + 2\cos z \, dx \, dy + dy^2, \tag{2}$$

where $z(x, y)$ is the array angle, and x and y are the lengths of the arcs of the coordinate lines.

Let $K = K(x, y)$ be the curvature of the linear element (2). Then we have

$$z_{xy} = -K \sin z. \tag{3}$$

For $K = -1$ equation (3) assumes the form

$$z_{xy} = \sin z,$$

which is the sine-Gordon equation. Note that the linear element (2) with curvature $K = -1$ is the linear element of the Lobachevsky plane.

If $z(x, y)$ is the solution of equation (1) satisfying the natural geometric condition for the array angle

$$0 < z < \pi, \tag{4}$$

then in a certain section of the Lobachevsky plane we can construct a Chebyshev array with this array angle, since the geodesic curvatures of the lines x and y calculated in the metric (2) are, respectively, $-z_x$ and z_y. Then the geodesic curvatures can be used to reconstruct the curves of the array. Note that this section of the Lobachevsky plane can be regularly and isometrically imbedded into E^3 in such a way that the lines x and y on the resulting surfaces are asymptotic. This follows from the fact that the Gauss–Peterson–Codazzi equations for the desired surface written in asymptotic coordinates x and y can be reduced to equation (1).

One way to construct solutions for the sine-Gordon equation is by successive approximations, first applied to equation (1) by L. Bianchi.

THEOREM [**33**]. *Let $\varphi(x) \in C^1$ and $\psi(y) \in C^1$ be functions on Ox and Oy respectively, with $\varphi(0) = \psi(0)$. Then in the plane (x, y) there is a unique solution $z(x, y) \in C^*$ of equation* (1) *satisfying the conditions*

$$z(x, 0) = \varphi(x), \qquad z(0, y) = \psi(y), \qquad \varphi(0) = \psi(0). \tag{5}$$

The proof is carried out by applying the method of successive approximations to the integral equation

$$z(x, y) = \varphi(x) + \psi(y) \quad \varphi(0) + \int_0^x\!\!\int_0^y \sin z \, dx \, dy \tag{6}$$

equivalent to problem (1), (5). For the initial approximation we take $z_0 \equiv 0$, and the approximations z_{n+1} and z_n are related by the formula

$$z_{n+1} = \varphi(x) + \psi(y) - \varphi(0) + \int_0^x\!\!\int_0^y \sin z_n \, dx \, dy. \tag{7}$$

The convergence of the approximations z_n to a solution of equation (1) in the entire plane follows from the estimate

$$|z_{n+1} - z_n| \leq \frac{|xy|^n}{(n!)^2},$$

which yields the following estimate for the desired solution $z(x, y)$:

$$|z(x, y)| \leq c \exp\{|xy|\}, \quad c = \sup\{|\varphi(x)| + |\psi(y)| + \varphi(0)\}. \tag{8}$$

Amsler [**34**] used Bianchi's result to construct in E^3 an analytic surface of constant negative curvature containing two complete intersecting geodesics, the asymptotic lines (straight lines on this surface). He proved that the surface's natural boundary (the collection of its limit points as $z \to 0$ and $z \to \pi$) for $0 < z < \pi$ consists of four analytic arcs, each homeomorphic to a straight line. A generalization of this result based on special asymptotic representations of small solutions of the sine-Gordon equation is examined in this section.

Small solutions of the sine-Gordon equation. Asymptotic representation of small solutions. Let us consider the Darboux problem for equation (1),

$$\begin{gathered} z_{xy} = \sin z, \\ z(x,0) = \varepsilon\varphi(x), \qquad z(0,y) = \varepsilon\psi(y), \qquad \varphi(0) = \psi(0), \end{gathered} \tag{9}$$

where ε is a numerical parameter, $\varphi(x) \in C^1$ on the axis Ox, and $\psi(y) \in C^1$ on the axis Oy.

For given functions $\varphi(x)$ and $\psi(x)$ a small solution of problem (9) is a solution of this problem for all sufficiently small values of ε (at $\varepsilon = 0$ the solution of problem (9) is identically zero).

Let $z(x, y, \varepsilon)$ be the solution of problem (9) in the entire plane with the additional requirement that the functions $\varphi(x)$ and $\psi(y)$ and their first derivatives be bounded. Then for $x \geq 0$ and $y \geq 0$ the following asymptotic representation holds true:

$$\begin{aligned} z(x,y,\varepsilon) = \varepsilon\Bigg[\varphi(0)\sum_{n=0}^{\infty}\frac{x^n y^n}{(n!)^2} &+ \int_0^x \varphi'(\xi)\sum_{n=0}^{\infty}\frac{(\xi - x)^n(-y)^n}{(n!)^2}\,d\xi \\ &+ \int_0^y \psi'(\eta)\sum_{n=0}^{\infty}\frac{(-x)^n(\eta - y)^n}{(n!)^2}\,d\eta\Bigg] + O(\varepsilon^3). \end{aligned} \tag{10}$$

In absolute value the term $O(\varepsilon^3)$ does not exceed $\frac{1}{6}\varepsilon^3 c_0^3 \exp\{4|xy|\}$, where $c_0 = \sup\{|\varphi(x)| + \psi(y) + |\varphi(0)|\}$.

The validity of this representation can be established by solving the Darboux problem for the telegraph equation:

$$\begin{gathered} z_{xy} = z + \varepsilon^2 \mathring{f}(z,\varepsilon), \\ z(x,0) = \varphi(x), \qquad z(0,y) = \psi(y), \end{gathered}$$

where

$$\mathring{f}(z,\varepsilon) = \begin{cases} \dfrac{\sin\varepsilon\mathring{z} - \varepsilon\mathring{z}}{\varepsilon^3} & \text{if } \varepsilon \neq 0, \\ -\dfrac{1}{6}\left[\mathring{z}(x,y,0)\right]^3 & \text{if } \varepsilon = 0, \end{cases}$$

with the functions $\mathring{z}(x, y, \varepsilon)$ and $z(x, y, \varepsilon)$ are related as follows:

$$z(x,y,\varepsilon) = \varepsilon\mathring{z}(x,y,\varepsilon).$$

Geometric applications. Let Γ be a line that lies in the Lobachevsky plane and is homeomorphic to a straight line. We take the part of the Lobachevsky plane containing the whole line imbedded regularly and isometrically into E^3 in such a way that Γ is entirely asymptotic only if one of the following conditions is satisfied.

1. The geodesic curve $\rho(x)$ of Γ, where x is arc length, retains its sign, and

$$0 \leq \left| \int_{-\infty}^{+\infty} \rho(x)\,dx \right| < \pi.$$

2. We have

$$0 \leq \left| \int_{a}^{b} \rho(x)\,dx \right| < \frac{\pi}{2} - \alpha, \quad 0 < \alpha < \frac{\pi}{2},$$

where a and b are arbitrary numbers.

We introduce the following notation. Let $\varphi_0 \neq 0$ be the angle between the complete lines Γ_1 and Γ_2 at the single point of their intersection, x and y the lengths of arcs of Γ_1 and Γ_2 measured from point P, and $\rho_1(x) \geq 0$ and $\rho_2(y) \geq 0$ the geodesic curvatures of these lines. We put

$$\mu_1 = \int_{-\infty}^{+\infty} \rho_1(x)\,dx, \qquad \mu_2 = \int_{-\infty}^{+\infty} \rho_2(y)\,dy, \qquad \mu = \mu_1 + \mu_2. \tag{11}$$

If

$$\begin{aligned} &\mu_1 < \pi, \qquad \int_{-\infty}^{0} \rho_1(x)\,dx < \varphi_0 < \pi - \int_{0}^{\infty} \rho_1 x)\,dx, \\ &\mu_2 < \pi, \qquad -\int_{-\infty}^{0} \rho_2(y)\,dy < \varphi_0 < \pi - \int_{0}^{\varphi} \rho_2(y)\,dy, \end{aligned} \tag{12}$$

then in a region Ω of the Lobachevsky plane containing Γ_1 and Γ_2, a Chebyshev array can be introduced in such a way that these lines are the base lines for this array, x and y are coordinates in Ω, and the array angle $z(x, y)$ satisfies the condition

$$0 < z(x, y) < \pi. \tag{13}$$

Using this assertion and the asymptotic representation of the small solutions of the sine-Gordon equation obtained earlier (see equation (10)), we can estimate the size of Ω.

THEOREM [**23**]. *Let Ω be the maximum part of the Lobachevsky plane in which a regular Chebyshev array can be introduced with base lines Γ_1 and Γ_2 and coordinates x and y, and for which the conditions* (12) *and* (13) *are met. Then, if φ_0 and μ are related as*

$$(3\varphi + 0 + \mu)^3 e^{4\delta} < 6\varphi_0, \qquad (\varphi_0 + \mu)e^{\delta} + \varphi_0 < \pi, \tag{14}$$

the region Ω^* defined by the inequality $|xy| < \delta$ is contained in Ω.

Among the values of φ_0, μ, and δ that satisfy inequalities (14) are, for instance, those obeying the following inequalities:

$$0 < \varphi_0 < \frac{\sqrt{3}}{8\sqrt{2}}, \qquad 0 \leq \mu \leq \varphi_0, \qquad 0 < \delta \leq \frac{1}{2}\ln 2.$$

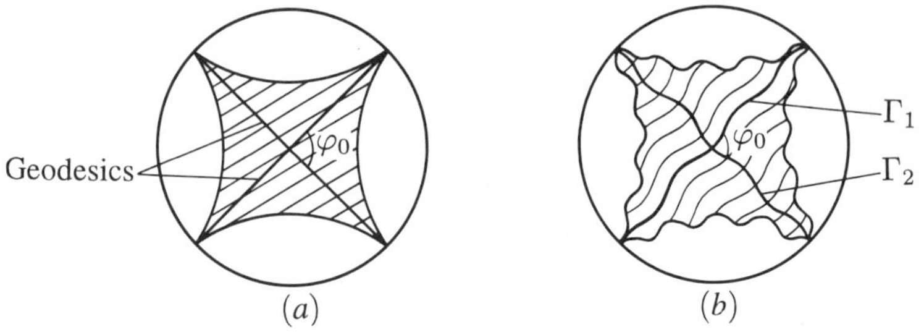

FIGURE 13

The theorem just proved is a generalization of the Bianchi–Amsler result. Figure 13a shows the region of the Lobachevsky plane studied by Bianchi and Amsler, and Figure 13b shows the region Ω^* defined by the hypothesis of the theorem.

4. Isometric imbedding of two-dimensional manifolds into four-dimensional Euclidean space

The problem of regular isometric imbeddings of two-dimensional manifolds into E^3 can be broken down into the local problem and the problem in the large. The local problem for manifolds with curvature of fixed sign has been solved completely: locally, manifolds with positive or negative curvature can be imbedded both regularly and isometrically into E^3 (see [**38, 39**]). The local problem of regular isometric imbeddings of manifolds with sign-alternating curvature has yet to be solved in full. The complexity of the local problem and the difficulties encountered in its solution are illustrated by the following example of Pogorelov [**40**]. A special surface of class $C^{1,1}$ with an intrinsic metric of class $C^{2,1}$ is constructed, with the surface curvature changing its sign in any neighborhood of a given point P. It is then proved that there can be no surface of class C^2 whose intrinsic metric in any neighborhood of a point coincides with the intrinsic metric of the initial surface in the neighborhood of P. In other words, the metric of the initial surface in any neighborhood of P cannot be realized in E^3 on a surface of class C^2 but can be realized on a surface of class $C^{1,1}$. Lin's results show that as the regularity of a metric with sign-alternating curvature distributed in a certain way in the neighborhood of a point P increases, it becomes possibile to find a local imbedding of this metric into E^3 both regularly and isometrically.

One might get the impression that any compact part of a complete manifold that has a sign-alternating curvature and is homeomorphic to a plane can be imbedded both regularly and isometrically into E^3. However, this is not the case. In [**41**] there is an example of a metric of class C^∞ defined in a disk (it can be continued without difficulty into the entire plane with the regularity retained) that cannot be imbedded into E^3 in the form of a surface of class $C^{1,1}$. Other examples can be found, for instance, in [**42**]. For this reason it is natural to study the possibility of isometrically imbedding arbitrary compact parts of complete manifolds homeomorphic to a plane into E^4.

This possibility can be implemented [**21**]. And again the small parameter idea is used.

Suppose that an arbitrary metric of class $C^{3,\alpha}$ is defined on a plane. We take a simply connected compact region Q in this plane. The linear element in Q can be assumed to be defined in terms of isometric coordinates x and y (see [**43**]):

$$ds^2 = \lambda^2(dx^2 + dy^2), \tag{1}$$

where $\lambda(x, y)$ is a function of class $C^{3,\alpha}$ in the domain of variation of x and y.

THEOREM [**21**]. *Let a metric of class $C^{3,\alpha}$ be specified in a simply connected compact domain of variation of the coordinates x and y by a linear element of type* (1). *Then this metric is realized in E^4 as a surface of class $C^{2,\alpha}$.*

Below we give a schematic description of the proof. In the space E^4 with the Cartesian coordinates X_1, X_2, X_3, and X_4 we introduce a set of coordinates ρ, σ, u, and v through the following relations:

$$X_1 = \varepsilon\rho\cos\frac{u}{\varepsilon}, \quad X_2 = \varepsilon\rho\sin\frac{u}{\varepsilon}, \quad X_3 = \varepsilon\sigma\cos\frac{v}{\varepsilon}, \quad X_4 = \varepsilon\sigma\sin\frac{v}{\varepsilon}, \tag{2}$$

where ε is a positive number. Equations (2) yield the following expression for the linear element in E^4 in terms of the coordinates ρ, σ, u, and v:

$$\sum_i dX_i^2 = \rho^2\, du^2 + \sigma^2\, dv^2 + \varepsilon^2(d\rho^2 + d\sigma^2). \tag{3}$$

Let the metric (1) be isometrically imbedded into E^4, with the imbedding specified by parametric equations of the form

$$\rho = \sigma = \exp\{w(x, y)\}, \qquad u = u(x, y), \qquad v = v(x, y). \tag{4}$$

Combining equations (1), (3), and (4), we see that in the domain of variation of the parameters x and y the following equation holds:

$$\lambda^2 \exp\{-2w\}(dx^2 + dy^2) - 2\varepsilon^2\, dw^2 = du^2 + dv^2. \tag{5}$$

Since $du^2 + dv^2$ is the linear element of the Euclidean plane, equation (5) implies that the linear element

$$\lambda^2 \exp\{-2w\}(dx^2 + dy^2) - 2\varepsilon^2\, dw^2 \tag{6}$$

has zero curvature. Calculating this curvature and equating it to zero, we arrive at the following expression for the function $w(x, y)$:

$$\Delta w + \frac{1}{\lambda^2}\left[(\operatorname{grad}\lambda)^2 - \lambda\,\Delta\lambda\right] + \varepsilon\,\frac{1}{\lambda^6}\,\Phi(\varepsilon, \lambda, \lambda', \lambda'', w, w', w'') = 0. \tag{7}$$

Here Φ is a polynomial in ε, λ, w, e^{-w}, and derivatives of λ and w up to the second order.

Equation (7) is the equation for imbedding the metric (1) into E^4. Indeed, if $w(x, y)$ is the solution of (7) for which the linear element (6) is positive definite in the entire domain of variation of x and y, equation (5) is used to find the functions $u(x, y)$ and $v(x, y)$, and then equations (4) are used to construct the imbedding of the metric (1) into E^4.

Let us examine the solution w_0 of equation (7) at $\varepsilon = 0$ with the boundary conditions $w_{\partial\Omega} \equiv 0$. We put $U = w - w_0$. Clearly, the function U satisfies an equation of the form

$$\Delta U = \varepsilon^2 f(\varepsilon, \lambda, \lambda', \lambda'', U, U', U''). \tag{8}$$

If in solving equation (8) we use *a priori* estimates for the solutions at zero boundary conditions $U_{\partial\Omega} \equiv 0$ and estimates of the norm $\|f\|_\alpha$, we can clearly see that the iterative sequence $\{U_n\}$ in which U_n is defined as the solution of a Dirichlet problem in Ω, namely

$$\Delta U_n = \varepsilon^2 f(\varepsilon, \lambda, \lambda', \lambda'', U_{n-1}, U'_{n-1}, U''_{n-1}),$$
$$U_n|_{\partial\Omega} \equiv 0, \qquad U_0 \equiv 0,$$

converges, for sufficiently small values of parameter ε, in the region Ω, and in class $C^{2,\alpha}$, to a function $U(x,y,\varepsilon) \in C^{2,\alpha}$ that is the solution of equation (8). Using the fact that $U(x,y,\varepsilon)$ is uniformly bounded in ε, we can easily verify that for all sufficiently small values of ε the discriminant of the quadratic form (6) for the function $w = U(x,y,\varepsilon) + w_0$ is positive.

5. A small parameter in the problem of isometric imbeddings of analytic metrics

Introductory remarks. Local imbedding of two-dimensional analytic manifolds into E^3 is done by applying the local Cauchy–Kovalevskaya theorem to the Darboux equation, an equation of Monge–Ampère type for one of the coordinates $X(x,y)$, $Y(x,y)$, and $Z(x,y)$ of the points of the required surface with a given analytic intrinsic metric

$$ds^2 = E(x,y)\,dx^2 + 2F(x,y)\,dx\,dy + G(x,y)\,dy^2. \tag{1}$$

The problem of isometrically imbedding such manifolds in the large into E^3 is more complicated. The solution is acquired by introducing a small parameter into the Gauss–Peterson–Codazzi equations and applying a modified variant of the Cauchy–Kovalevskaya theorem to the resulting equations.

Formulation of the fundamental theorem. A special form of imbedding equations. We consider complete analytic manifolds without conjugate points and homeomorphic to a plane (any two geodesics can intersect at no more than one point). In the entire plane (x,y) we can select a system of coordinates such that an arbitrarily chosen geodesic disk corresponds to a rectangle

$$\pi = \{(x,y) \mid 0 \le x \le a,\ 0 \le y \le b\}$$

and the linear element of the metric in π has the form

$$ds^2 = dx^2 + B^2(x,y)\,dy^2, \tag{2}$$

where $B(x,y)$ is a function analytic in π.

FUNDAMENTAL THEOREM [8]. *The analytic metric specified in the rectangle* π *by the linear element* (2) *can be isometrically imbedded into* E^3. *The surface on which the metric* (2) *is realized is analytic.*

Thus, any geodesic disk of a complete analytic manifold that has no conjugate points and is homeomorphic to a plane can be isometrically imbedded into E^3.

The proof of the theorem requires using a system of first-order partial differential equations containing a small parameter and obtained as a result of a special transformation of the main equations of the theory of surfaces, i.e., the Gauss–Peterson–Codazzi equations for the linear element (1) and for the reduced coefficients l, m, and n of the second quadratic form of the desired surface, which are the ordinary coefficients L, M, and N of the second quadratic form divided by $\sqrt{EG-F^2}$. The equations are

$$\begin{aligned} l_y - m_x &= -\Gamma^2_{22} l + 2\Gamma^2_{12} m - \Gamma^2_{11} n, \\ n_x - m_y &= -\Gamma^1_{22} l + 2\Gamma^1_{12} m - \Gamma^1_{11} n, \\ ln - m^2 &= K, \end{aligned} \tag{3}$$

where Γ^i_{jk} are the Christoffel symbols calculated for metric (1), and K is the curvature of this metric.

The substitutions

$$u = B^2 m, \qquad v = \frac{1}{\varepsilon n}, \qquad 0 < \varepsilon < 1,$$

reduce system (3) to

$$\begin{aligned} u_x &= \varepsilon\left[\frac{2}{B^2}\, uvu_y + \left(B^2K + \frac{1}{B^2}\, u^2\right)v_y + B(BK)_y v - \frac{2B_y}{B}\, u^2 v\right], \\ v_x &= \varepsilon\left[-\frac{1}{B^2}\, v^2 u_y + \frac{2B_y}{B^3}\, uv^2 - \varepsilon^2 B_x BK v^3 - \varepsilon\, \frac{B_x}{B}\, u^2 v^3\right]. \end{aligned} \tag{4}$$

Note that if in π the solution $u(x,y)$, $v(x,y)$ of (4) satisfying $v \neq 0$ is found, the reduced coefficients l, m, and n can be expressed algebraically in terms of the functions u and v.

Generalization of the Cauchy–Kovalevskaya theorem. As is known, for a function $f(x,y)$ analytic in the rectangle π one can specify numbers M and r in such a way that the following estimates hold for the derivatives of this function:

$$\left|\frac{\partial^{\nu+\mu} f}{\partial x^\nu\, \partial y^\mu}\right| < \nu!\,\mu!\,\frac{M}{r^{\nu+\mu}}. \tag{5}$$

An analytic function $f(x,y)$ is called a *majorant at a point* $P(x_0, y_0)$ of an analytic function $g(x,y)$ if all the coefficients in the expansion of $f(x,y)$ in a power series in $x - x_0$ and $y - y_0$ with center at P are nonnegative and no smaller than the absolute values of the corresponding coefficients in the expansion of $g(x,y)$ in a power series in $x - x_0$ and $y - y_0$ with center at P. An analytic function $f(x,y)$ is called a *majorant of* $g(x,y)$ *in a region* Q if it is a majorant of $g(x,y)$ at each point of Q.

Below we formulate the conditions for a function $f(x,y)$ analytic in a region Q to be the majorant of a function $g(x,y)$ in a simply connected subregion $Q^* \subset Q$.

Suppose that the point $P(x_0, y_0)$ belongs to the boundary of Q^* and that any point in Q^* can be connected with P by a smooth curve with nonnegative slope and lying entirely in Q^*. We also assume that for a point $(x, y) \in Q^*$ the differences $x - x_0$ and $y - y_0$ are nonnegative.

LEMMA. *Let a function $f(x, y)$ that is analytic in a region Q be the majorant at a point $P(x_0, y_0)$ of a function $g(x, y)$ that is analytic in the neighborhood of P. Then $g(x, y)$ can be analytically continued into Q^*, with $f(x, y)$ being the majorant of $g(x, y)$ in Q^*.*

Usually the proof of the Cauchy–Kovalevskaya theorem is carried out by the method of majorant equations, the concept of which is introduced by analogy with majorant analytic functions (the right-hand sides of the equations are analytic functions of all the variables: the independent variables, the desired functions, and the y-derivatives of the desired functions). Estimate (5) supports the notion that there are constants M and r such that the equation

$$z_x = \varepsilon M c^6 (z_y + 1)\left(1 - \frac{x - x_0}{r}\right)^{-1}\left(1 - \frac{y - y_0}{r}\right)^{-1}\left(1 - \frac{z}{c}\right)^{-1}, \tag{6}$$

with c a fixed constant not less than unity, is the majorant equation for system (4).

To prove the generalized Cauchy–Kovalevskaya existence theorem for systems of type (4) we will need several additional properties of equation (6).

Let us examine equation (6) in the region

$$-r + \delta \le x \le r - \delta, \qquad -r + \delta \le y \le r - \delta, \qquad |z| < c,$$

with $c \ge 1$, $0 < r < 1$, and $0 < \delta < \frac{1}{2}r$. Since the properties we need are independent of the values of x_0 and y_0, we assume that $x_0 = 0$ and $y_0 = 0$.

Let Q be a region defined in the following way:

$$-r + \delta \le x \le r - \delta, \qquad -r + \delta \le y \le r - \delta.$$

We solve the Cauchy problem for equation (6) with the initial data

$$x = 0, \qquad y = Y, \qquad z = \omega(Y), \qquad 0 \le Y \le Y_0, \qquad Y_0 < r - \delta. \tag{7}$$

Let Q^* be a fixed isosceles trapezoid whose larger base is the segment $[0, Y_0]$ of the axis Oy and whose smaller base lies on the straight line $x = r - \delta$. Then, applying the method of characteristic curves to equation (6) with the initial data (7), we can see that for every positive ε not exceeding some ε_0 $(0 < \varepsilon_0 < 1)$ and for a certain $c \ge 1$, the Cauchy problem (6), (7) has a solution in the closed region Q^*. If $P(x, y)$ is a point of the trapezoid (7) lying on the projection L^* onto the plane Oxy of a characteristic curve of equation (6), where L^* starts from the point Y of the segment $[0, Y_0]$, and z is the z-coordinate of point P, then the following estimates hold:

$$\begin{aligned} |z| &\le |\omega(Y)| + \varepsilon \frac{M r^2 c^7}{\delta^2} |x|, \\ |z_y| &\le |\omega'(Y)| + \varepsilon \frac{2 M r^2 c^8}{\delta^2} \left[3 + |\omega(Y)| + \frac{1}{\delta} + \frac{|\omega'(Y)|}{\delta}\right]^2 |x|. \end{aligned} \tag{8}$$

Let us assume that the coefficients in the right-hand sides of the equations in (4) are specified in a rectangle π' containing π.

EXISTENCE THEOREM. *Let arbitrary analytic initial data for system* (4) *be specified on the left vertical side of the rectangle* π'. *Then for all sufficiently small values of the parameter* ε *the system* (4) *has an analytic solution in the rectangle* π. *Both the solution and its* y*-derivatives remain uniformly bounded as* $\varepsilon \to 0$.

Below we give a schematic description of the proof. We take the system (4) in π' and build an isosceles trapezoid Ω with a small slope k of the lateral sides. The larger base of the trapezoid is the left vertical side of π', on which we specify the arbitrary analytic initial data $u_0(y)$ and $v_0(y)$ for (4). Let

$$\omega(y) = \mu\left(1 - \frac{y - y_0}{\delta}\right)^{-1}, \qquad \mu, \rho = \text{const},$$

be the majorant function for $u_0(y)$, $v_0(y)$ at any point y_0 of the left vertical side of π'. From the above properties of the solutions of equation (6) we conclude that for sufficiently small ε and k, equation (6) with the initial data $\omega(y)$ on the segment $[y_0 - \rho^*, y_0 + \rho^*]$, $\rho^* < \rho$, has a solution $z(x, y)$ in the isosceles trapezoid Q^*, with the larger base lying on this segment, the altitude being δ (the choice of δ is specified above), and the slope of the lateral sides being k. Applying the lemma formulated earlier, we obtain a solution of (4) with the initial data $u_0(y_0)$, $v_0(y)$, in the trapezoid Ω_1, which is the intersection of Ω and the vertical strip

$$\{(x, y) | 0 \leq x \leq \delta, \ -\infty < y < +\infty\}.$$

Here the solution $z(\delta, y)$ of (6) is a majorant on the right base of Ω_1 of the solution we have just obtained. Selecting these solutions on the right base of Ω_1 as the initial data for system (4) and the solution $z(\delta, y)$ of (6) as the majorant of our solutions, we continue the solution of (4) into the trapezoid Ω_2, the intersection of Ω with the strip

$$\{(x, y) | \delta \leq x \leq 2\delta, \ -\infty < y < +\infty\}.$$

Thus, taking sequential steps, i.e., going from a trapezoid Ω_i to a trapezoid Ω_{i+1}, we build the solution $u(x, y)$, $v(x, y)$ of system (4) in the trapezoid Ω, which for k sufficiently small contains the rectangle π.

The second assertion of the theorem follows from the estimates (8) for the majorant equation. Since the solution of the majorant equation and the derivatives of this solution majorize the solution of (4) with its derivatives, by applying the estimates (8) sequentially in the transition from Ω_i to Ω_{i+1}, we arrive at the following estimates for the solution $u(x, y)$, $v(x, y)$ in Ω:

$$\begin{gathered} \max_{\Omega}\{|u(x, y)|, |v(x, y)|\} \leq |\omega| + O(\varepsilon), \\ \max_{\Omega}\{|u_y(x, y)|, |v_y(x, y)|\} \leq |\omega'| + O(\varepsilon). \end{gathered} \tag{9}$$

PROOF OF THE FUNDAMENTAL THEOREM. Let the initial data $u_0(y), v_0(y)$ be bounded from below by a positive constant. We want to construct the solution $u(x, y)$, $v(x, y)$ of the Cauchy problem for system (4) in the rectangle π. By the existence theorem, this solution and its y-derivatives remain uniformly bounded as $\varepsilon \to 0$. Hence, system (4) implies that both u_x and v_x are functions of the form $O(\varepsilon)$. But then system (4) yields the following representation of the functions $u(x,y)$ and $v(x,y)$ in the rectangle π:

$$u(x,y) = u_0(y) + O(\varepsilon), \qquad v(x,y) = v_0(y) + O(\varepsilon).$$

Combining this with the fact that $v_0(y)$ is bounded below by a positive constant, we conclude that for all sufficiently small ε the function $v(x,y)$ is nonzero at all points of the rectangle π. With these conditions imposed on the solution $u(x,y)$, $v(x,y)$ of system (4), well-known methods can be used to find a surface with the intrinsic metric (2).

6. Applying a small parameter for imbedding two-dimensional manifolds with sign-alternating curvature

It is of considerable interest to use a small parameter in proving the existence of regular isometric imbeddings into three-dimensional Euclidean space E^3 of two-dimensional Riemannian manifolds whose curvature is not of fixed sign, i.e., is nonnegative or nonpositive or sign-alternating. Results have been achieved in each of these three areas. More about this in the next subsection.

Local isometric imbedding of manifolds with nonnegative curvature. The following assertion holds true.

THEOREM [**13**]. *Suppose that the metric*

$$ds^2 = E\,dx^2 + 2F\,dx\,dy + G\,dy^2, \tag{1}$$

whose curvature K is nonnegative, with $K(0,0) = 0$, is specified in a neighborhood of the point $P(0,0)$ in the plane of parameters u and v. If the coefficients E, F, and G of the metric (1) *belong to the class C^s, where $s \geq 10$, then in a certain* (*smaller*) *neighborhood of P it admits regular realization in E^3 in the form of a surface of the class C^{s-6}.*

The proof is based on a smooth solution of the Monge–Ampère equation with the use of a modified Nash–Moser–Hörmander scheme [**44**] and can be broken down into several steps.

First step. Deriving the imbedding equations. This is done by a method suggested by Darboux [**25**].

To solve the local problem of imbedding the metric (1) it is sufficient to prove the existence in a neighborhood of P of smooth functions $X(u,v)$, $Y(u,v)$, and $Z(u,v)$ for which

$$dX^2 + dY^2 + dZ^2 = E\,du^2 + 2F\,du\,dv + G\,dv^2.$$

This is equivalent to the metric

$$E\,du^2 + 2F\,du\,dv + G\,dv^2 - dZ^2$$

having zero curvature, i.e.,

$$\left(Z_{11} - \Gamma_{11}^i Z_i\right)\left(Z_{22} - \Gamma_{22}^i Z_i\right) - \left(Z_{12} - \Gamma_{12}^i Z_i\right)^2 = \mathcal{K}(u,v,\nabla Z), \tag{2}$$

where

$$Z_1 = Z_u, \qquad Z_2 = Z_v, \qquad Z_{11} = Z_{uu}, \qquad Z_{12} = Z_{uv}, \qquad Z_{22} = Z_{vv},$$
$$\mathcal{K}(u, v, \nabla Z) = K\big(EG - F^2 - EZ_2^2 + 2FZ_1Z_2 - GZ_1^2\big),$$

and Γ^i_{jk} are the Christoffel symbols calculated for the metric (1). With the solution $Z(u, v)$ of equation (2) known, the other two functions, $X(u, v)$ and $Y(u, v)$, can be found in terms of integrals.

A key issue in proving the existence of a smooth solution of equation (2) is the preliminary transformations related to introducing into this equation a small parameter and building the main operators.

Second step. Introducing a small parameter. If in equation (2) we introduce new variables using the substitutions

$$u = \varepsilon^2 x, \qquad v = \varepsilon^2 y, \qquad Z = \tfrac{1}{2}v^2 + \varepsilon^5 w,$$

where ε is a small numerical parameter, after simple transformations we arrive at the following equation:

$$w_{xx} + \varepsilon\big\{F_1(\varepsilon, x, y, \nabla w, \nabla^2 w) + F_2(\varepsilon, x, y, \nabla w)\big\} = 0, \tag{3}$$

where

$$\begin{aligned} F_1 =&\big(w_{xx} - \varepsilon\Gamma^2_{11}y - \varepsilon^2\Gamma^1_{11}w_x - \varepsilon^2\Gamma^2_{11}w_y\big)\big(w_{yy} - \varepsilon\Gamma^2_{22}y - \varepsilon^2\Gamma^1_{22}w_x - \varepsilon^2\Gamma^2_{22}w_y\big) \\ &- \big(w_{xy} - \varepsilon\Gamma^2_{12}y - \varepsilon^2\Gamma^1_{12}w_x - \varepsilon^2\Gamma^2_{12}w_y\big)^2 - \frac{1}{\varepsilon^2}\mathcal{K}(\varepsilon^2 x, \varepsilon^2 y, \varepsilon^3\nabla w), \\ F_2 =& -\Gamma^2_{11}y - \varepsilon\Gamma^1_{11}w_x - \varepsilon\Gamma^2_{11}w_y. \end{aligned}$$

Clearly, F_1 and F_2 are smooth functions that remain bounded as $\varepsilon \to 0$.

Third step. Statement of the boundary value problem. In the rectangle

$$D = \big\{(x, y) \,\big|\, |x| \le x_0,\ |y| \le y_0\big\},$$

where x_0 and y_0 are positive numbers, let us consider two nonnegative cut-off functions $\chi_1(x, y)$ and $\chi_2(x, y)$ belonging to class C^∞ and satisfying the following conditions:

$$\chi_1(x, y) = \begin{cases} 1 & \text{if } |y| \le \frac{1}{2}y_0, \\ 0 & \text{if } |y| \ge \frac{3}{4}y_0, \end{cases} \qquad \chi_2(x, y) = \begin{cases} 1 & \text{if } |y| \le \frac{3}{4}y_0, \\ 0 & \text{if } |y| \ge \frac{7}{8}y_0. \end{cases}$$

Using these cut-off functions, we can construct a differential operator

$$G(w) = w_{xx} + \varepsilon F(\varepsilon, x, y, \nabla w, \nabla^2 w), \tag{4}$$

called the smoothing operator of equation (2). The function $F = \chi_1 F_1 + \chi_2 F_2$ belongs to the class C^{s-3}, coincides with the sum $F_1 + F_2$ near the initial point $(0, 0)$, and vanishes near the horizontal segments of the boundary of D, i.e., $y = \pm y_0$. Note that in the region

$$D' = \big\{(x, y) \,\big|\, |x| \le x_0,\ |y| \le \tfrac{1}{2}y_0\big\}$$

the equation

$$G(w) = 0 \tag{5}$$

and equation (2) coincide. In addition, the structure of the smoothing operator $G(w)$ makes it possible to stipulate the boundary conditions only for a part of the boundary of D, the vertical line segments $x = \pm x_0$.

Clearly, by building the smooth solution of the problem

$$\begin{gathered} G(w) = 0, \\ w(x_0, y) = w(-x_0, y) = 0, \end{gathered} \tag{6}$$

we at the same time prove the existence of the solution of equation (2), and with it the solution of the entire problem.

Fourth step. Solving the auxiliary problem. Using the operator $L(w)$, we linearize the operator (4) and construct the singular second-order elliptic operator

$$L_\Theta(w) = L(w) + \Theta\chi_1 \frac{\partial^2}{\partial y^2},$$

where Θ is a positive constant. In the rectangle we examine the boundary value problem

$$\begin{gathered} L_\Theta(w)\rho = g, \\ \rho(x_0, y) = \rho(-x_0, y) = 0, \end{gathered} \tag{7}$$

where $\Theta = |G(w)|_{L^\infty(D)}$.

Assuming that $\|w\|_{H^6} \leq 1$ and that ε and Θ are sufficiently small parameters and using a regularization method, we can prove that there is a number $s_0 = s_0(\varepsilon, \Theta)$ such that if $g \in H^s$, $0 \leq s \leq s_0$, problem (7) has exactly one solution $\rho \in H^s$, for which the following important estimate holds:

$$\|\rho\|_{H^s} \leq C_s\big(\|g\|_{H^s} + \|w\|_{H^{s+4}} + \|\rho\|_{H^z}\big). \tag{8}$$

Here H^s is a Sobolev space with the norm

$$\|u\|_{H^s} = \Big(\sum_{|\alpha|\leq s} \|D^\alpha u\|^2_{L^2}\Big)^{1/2},$$

and C_s is a constant.

Fifth step. Solution of the main problem. The boundary value problem (6) can be solved in D according to the Nash–Moser–Hörmander scheme. First, smoothing operators S_n are used to construct a sequence of functions w_n such that

$$w_{n+1} = w_n + \rho_n,$$

where ρ_n is the solution of a boundary value problem of type (7):

$$\begin{gathered} L_{\Theta_n}(S_n w_n)\rho_n = g_n, \\ \rho_n(x_0, y) = \rho_n(-x_0, y) = 0, \end{gathered}$$

with $\Theta = |G(S_n w_n)|_{L^\infty(D)}$.

Then, using the properties of the solution of the auxiliary problem and the appropriate estimate of type (8), we can prove the convergence of this sequence to the solution w of problem (6) in H^{s-4}. The fact that H^{s-4} is imbedded into C^{s-6} completes the proof.

In conclusion we note that no restrictions are imposed on the number of zeros of the curvature K of the metric (1).

Local isometric imbedding of manifolds with sign-alternating curvature that have a single parabolic line. The following assertion holds:

THEOREM [**14**]. *Suppose that in a neighborhood of the point $P(0,0)$ in the plane of parameters (u,v) we define a metric of type (1) with curvature K at P satisfying the following conditions*:

$$K(0,0)=0, \qquad \nabla K(0,0)\neq 0.$$

If the curvature K of the metric belongs to class C^s, where $s\geq 6$, then, in a neighborhood of P, the metric (1) admits a C^{s-3}-regular isometric imbedding into E^3.

The proof is also done by reduction to the Monge–Ampère equation (2) and subsequent study of a perturbed Tricomi equation.

Omitting intricate technical details, we focus on the main steps in the proof.

First step. Introducing a small parameter in the imbedding equation. Without loss of generality we can assume that

$$\Gamma^i_{jk}(0,0)=0,$$
$$\mathcal{K}(u,v,\nabla z)=v+O(|u|^2+|v|^2+|\nabla z|^2)$$

in a neighborhood of P.

If in (2) we put

$$u=\varepsilon^2\xi, \qquad v=\varepsilon^2\eta, \qquad z=\tfrac{1}{2}u^2+\tfrac{1}{6}v^3+\varepsilon^7 w, \tag{9}$$

and then perform fairly simple transformations, we arrive at the perturbed Tricomi equation

$$\eta w_{\xi\xi}+w_{\eta\eta}+\varepsilon\Phi(\varepsilon,\xi,\eta,\nabla w,\nabla^2 w)=0, \tag{10}$$

where Φ is a smooth function of its arguments and remains bounded as $\varepsilon\to 0$.

REMARK. Note that the suggested transformation (9) has a purely geometric origin: the graph of the function

$$z=\tfrac{1}{2}u^2+\tfrac{1}{6}v^3,$$

which is the principal part in the equation for z in (9), is the surface of translation of the parabola $z=\frac{1}{2}u^2$ along the cubic parabola $z=\frac{1}{6}v^3$ (Figure 14). The parabola

$$z=\tfrac{1}{2}u^2, \quad v=0$$

lying on this surface is a zero-curvature line that divides the surface into elliptic ($v>0$) and hyperbolic ($v<0$) parts. In the previous case the principal part specifies a parabolic cylinder.

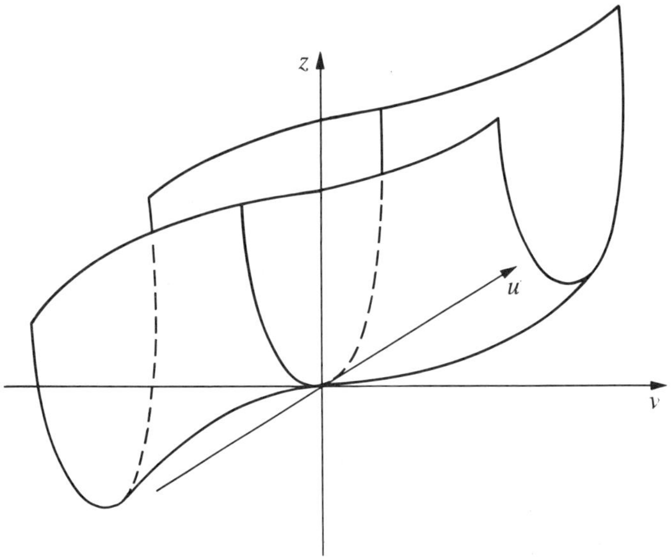

FIGURE 14

Second step. Statement of the boundary value problem. In the plane (ξ, η) we define a region Ω with a boundary consisting of four parts (Figure 15): three straight line segments

$$\xi = \pm\xi_0, \qquad \eta = \eta_0$$

and one slightly curved segment γ specified by the equation

$$\eta = \varphi(\xi).$$

The slight nonlinearity of γ means that there exists a small positive δ such that

$$|\varphi'|_{C^2} < \delta.$$

This region Ω cuts out a segment on the axis ξ that divides Ω into two parts: the one where equation (10) is elliptic $(\eta > 0)$, and the one where it is hyperbolic $(\eta < 0)$.

The solution w of equation (10) is sought in Ω with the following conditions specified on the boundary of Ω:

$$\begin{aligned} &w = w_\xi = w_\eta = 0 && \text{at } \xi = \xi_0, \\ &w = 0 && \text{on } \gamma, \\ &w_\xi + w_\eta = 0 && \text{at } \eta = \eta_0. \end{aligned} \tag{11}$$

Third step. Transition to a quasilinear system. The transformation

$$x = \xi, \qquad y = \psi(\xi, \eta),$$

where $\psi(\xi, \eta)$ is a fairly smooth function coinciding with $\eta - \varphi(\xi, \eta)$ near γ and with η near η_0, maps Ω into the rectangle

$$D = \{(x, y) \mid |x| \le \xi_0, \ \eta_0 \le y \le 0\}.$$

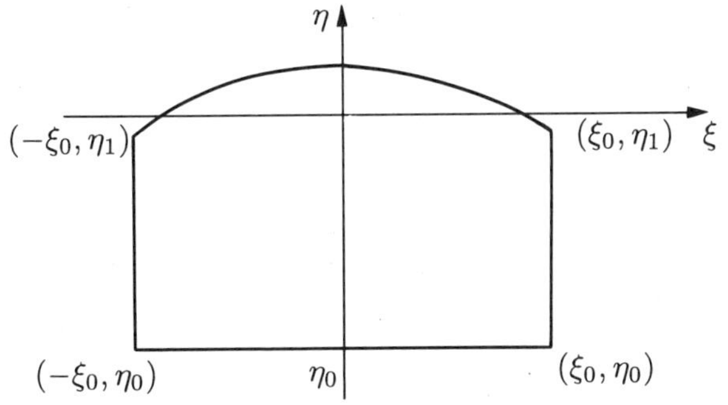

FIGURE 15

Equation (10) is transformed accordingly:

$$\eta w_{xx} + 2\psi_\xi \eta w_{xy} + (\eta\psi_\xi^2 + \psi_y^2)w_{yy} + (\eta\psi_{\xi\xi} + \psi_{\eta\eta})w_y + \varepsilon\Phi = 0.$$

Simple operations, including differentiation with respect to the variable x and several algebraic manipulations, reduce this equation to an equivalent system of first-order equations, which in matrix notation has the following form:

$$2A^1 u_x + 2A^2 u_y + Bu = \varepsilon g. \tag{12}$$

The components u_1, u_2, u_3, and u_4 of the vector-valued function u are related to the function w as

$$\begin{pmatrix} u_1 \\ u_2 \\ u_3 \\ u_4 \end{pmatrix} = e^{\lambda x} \begin{pmatrix} w_x \\ w_y \\ (w_x)_x \\ (w_y)_x \end{pmatrix},$$

and, therefore, must obey the following boundary conditions:

$$\begin{aligned} &u_1 = u_2 = u_3 = u_4 = 0 && \text{at } x = \xi_0, \\ &u_1 = u_3 = 0 && \text{at } y = 0, \\ &u_1 + u_2 = u_3 + u_4 = 0 && \text{at } y = \eta_0. \end{aligned} \tag{13}$$

For sufficiently small ε and δ, the quasilinear system (12) proves to be symmetrically positive [**45**]:

- the matrices A^1 and A^2 are symmetric;
- the matrix $\frac{1}{2}(B + B^*) - A_x^1 - A_y^2$ is positive definite and the boundary conditions are admissible; and
- the function $A^1 n_x + A^2 n_y$, with (n_x, n_y) an external normal to the rectangle D, is nonnegative only on the set of solutions of system (13).

Fourth step. Proof of the main result. Developing the Friedrichs method and applying the fixed-point theorem to problem (12), (13) enables us, using highly nontrivial arguments, to prove the existence of its solution in the Sobolev space H^{s-2}. From this it immediately follows that the isometric imbedding of the metric (1) defined by this solution belongs to the class C^{s-3}.

Isometric imbedding in the large of manifolds of nonnegative curvature. Let the complete metric W be specified in a plane by a linear element of the form

$$ds^2 = d\xi^2 + B^2(\xi, \eta)\, d\eta^2, \tag{14}$$

whose curvature K is nonpositive and satisfies the following conditions:

$$k(\xi,\eta) = \begin{cases} \left(g_1(\xi_1(\eta) - \xi)\right)^{\alpha_1(\eta)} f_1(\xi,\eta) & \text{for } \xi < \xi_1(\eta), \\ 0 & \text{for } \xi_1(\eta) \le \xi \le \xi_2(\eta), \\ \left(g_2(\xi - \xi_2(\eta))\right)^{\alpha_2(\eta)} f_2(\xi,\eta) & \text{for } \xi > \xi_2(\eta), \end{cases}$$

with $k = \sqrt{-K}$. Let us also assume that the above functions satisfy the following conditions.

1. B and the f_i are positive and twice continuously differentiable in the entire plane.
2. The α_i are positive and twice continuously differentiable, and the ξ_i are thrice continuously differentiable, on the axis η.
3. The $g_i(x)$ are continuously differentiable for $x > 0$, the $g_i'(x)$ are positive, and $\lim_{x\to+0} g_i(x) = 0$.
4. The $\left(g_i(x)\right)^{\alpha_i(y)} \ln^2\{g_i(x)\}/g_i'(x)$ are uniformly continuous on bounded subsets of the half-plane $x > 0$.

THEOREM [**9**]. *Suppose that the above conditions are satisfied and $\gamma_1(\eta)$ and $\gamma_2(\eta)$ are arbitrary continuous functions satisfying the following inequalities*:

$$\gamma_1(\eta) < \xi_1(\eta) \le \xi_2(\eta) < \gamma_2(\eta).$$

Then the metric (14) *in the region*

$$\pi(\gamma_1, \gamma_2) = \{(\xi,\eta) | \gamma_1(\eta) < \xi < \gamma_2(\eta)\}$$

can be isometrically imbedded into E^3 in the form of a surface of class C^s.

The method used in the proof is a development of the method used in [**19**].

REMARK. The result remains true at $\gamma_1 = \xi_1 = -\infty$ or at $\xi_2 = \gamma_2 = +\infty$. In the first case the manifold W has zero curvature for $\xi \le \xi_2(\eta)$, and the region that admits isometric imbedding consists of points at which $\xi < \gamma_2(\eta)$. In the second case the curvature vanishes for $\xi \ge \xi_1(\eta)$, and the imbedded region is specified by the inequality $\xi > \gamma_1(\eta)$.

References

1. È. G. Poznyak, *On regular realization in the large of two-dimensional metrics with negative curvature*, Dokl. Akad. Nauk SSSR **170** (1966), 786–789; English transl. in Soviet Math. Dokl. **7** (1966).
2. ———, *On regular realization in the large of two-dimensional metrics with negative curvature*, Ukrain. Geom. Sb. **3** (1966), 78–92. (Russian)
3. E. V. Shikin, *On regular realization in the large in E^3 of two-dimensional metrics of class C^2 with negative curvature of class C^1*, Dokl. Akad. Nauk SSSR **188** (1969), 1014–1016; English transl. in Soviet Math. Dokl. **10** (1969).

4. ______, *On regular embedding in the large in R^3 of metrics of class C^4 with negative curvature*, Mat. Zametki **14** (1973), 261–266; English transl. in Math. Notes **14** (1973).
5. ______, *On the existence of solutions of the Peterson–Codazzi–Gauss system of equations*, Mat. Zametki **17** (1975), 765–781; English transl. in Math. Notes **17** (1975).
6. È. G. Poznyak, *Isometric embedding in E^3 of some noncompact parts of the Lobachevsky plane*, Mat. Sb. **102** (1977), 3–12; English transl. in Math. USSR Sb. **31** (1977).
7. E. V. Shikin, *On isometric embedding of two-dimensional manifolds with negative curvature by the Darboux method*, Mat. Zametki **27** (1980), 779–794; English transl. in Math. Notes **27** (1980).
8. È. G. Poznyak, *Realization in the large of two-dimensional analytic metrics with alternating-sign curvature*, Ukrain. Geom. Sb. **7** (1970), 89–97. (Russian)
9. R. Ts. Musaelyan, *On regular embedding in the large in E^3 of some complete metrics with alternating-sign curvature*, Akad. Nauk Armyan. SSR Dokl. **73** (1981), 17–23. (Russian)
10. E. V. Shikin, *On isometric embedding in the large in R^3 of some metrics with nonpositive curvature*, Dokl. Akad. Nauk SSSR **215** (1974), 61–63; English transl. in Soviet Math. Dokl. **15** (1974).
11. ______, *Isometric embedding in E^3 of noncompact regions with nonpositive curvature*, Itogi Nauki i Tekhniki: Problemy Geometrii, vol. 7, VINITI, Moscow, 1975, pp. 249–266. (Russian)
12. ______, *Isometric embedding in E^3 of noncompact regions with nonpositive curvature*, Mat. Zametki **25** (1979), 785–797; English transl. in Math. Notes **25** (1979).
13. Lin Chang-Shou, *The local isometric embedding in R^3 of 2-dimensional Riemannian manifolds with nonnegative curvature*, J. Differential Geom. **21** (1985), 213–230.
14. ______, *The local isometric embedding in R^3 of two-dimensional Riemannian manifolds with Gaussian curvature changing sign cleanly*, Comm. Pure Appl. Math. **39** (1986), 867–887.
15. Gen Nakamura, *Local isometric embedding of 2-dimensional Riemannian manifolds into R^3 with nonpositive Gaussian curvature*, Proc. Japan Acad. Ser. A Math. Sci. **61** (1985), 211–212.
16. Zh. Kaĭdasov, *On regular isometric embedding in E^3 of an expanding strip of the Lobachevsky plane*, Studies in the Theory of Surfaces in Riemannian Spaces, Leningrad. Gos. Ped. Inst., Leningrad, 1984, pp. 119–129. (Russian)
17. Zh. Kaĭdasov and E. V. Shikin, *On isometric embedding in E^3 of a convex region of the Lobachevsky plane containing two horodisks*, Mat. Zametki **39** (1986), 612–617; English transl. in Math. Notes **39** (1986).
18. ______, *On isometric embedding in E^3 of a convex region of the Lobachevsky plane containing a horodisk*, Vestnik Moskov. Univ. Ser. I Mat. Mekh. **1986**, no. 5, 79–81, English transl. in Moscow Univ. Math. Bull. **41** (1986).
19. D. V. Tunitskiĭ, *On regular isometric embedding in E^3 of unbounded regions with negative curvature*, Mat. Sb. **134** (1987), 119–134; English transl. in Math. USSR Sb. **62** (1989).
20. Zh. Kaĭdasov and E. V. Shikin, *On isometric embedding in E^3 of expanding strips on L-type manifolds*, Mat. Zametki **42** (1987), 842–853; English transl. in Math. Notes **42** (1987).
21. È. G. Poznyak, *Isometric embeddings of two-dimensional Riemannian metrics in Euclidean spaces*, Uspekhi Mat. Nauk **28** (1973), no. 5 (172), 47–76; English transl. in Russian Math. Surveys **28** (1973).
22. ______, *Geometric studies related to the equation $z_{xy} = \sin z$*, Itogi Nauki i Tekhniki: Problemy Geometrii, vol. 8, VINITI, Moscow, 1977, pp. 225–241; English transl. in J. Soviet Math. **13** (1980), no. 5.
23. ______, *Some new results on isometric embeddings of parts of the Lobachevsky plane in E^3*, All-Union Sci. Conf. Non-Euclidean Geometry: 150 Years of Lobachevskian Geometry (Kazan, 1976), VINITI, Moscow, 1977, pp. 73–78. (Russian)
24. B. L. Rozhdestvenskiĭ, *The system of quasilinear equations of the theory of surfaces*, Dokl. Akad. Nauk SSSR **143** (1962), 50–52; English transl. in Soviet Math. Dokl. **3** (1962).
25. G. Darboux, *Leçons sur la théorie générale des surfaces et les applications géométriques du calcul infinitésimal*. Part 3, Gauther-Villars, Paris, 1894.
26. R. Courant, *Partial differential equations*, Wiley, New York, 1962.

27. E. V. Shikin, *On the equations of isometric embeddings in three-dimensional Euclidean space of two-dimensional manifolds with negative curvature*, Mat. Zametki **31** (1982), 601–612; English transl. in Math. Notes **31** (1982).
28. V. F. Kagan, *Fundamentals of the theory of surfaces in tensor presentation*. Vol. 2, GITTL, Moscow, 1947. (Russian)
29. E. Beltrami, *Sulla superficie di rotazione che serve di tipo alle superficie pseudosferiche*, Giorn. Mat. Univ. Ital. (Napoli) **10** (1872), 147–159; reprinted in his *Opere Matematiche*, Vol. II, Ulrico Koepli, Milan, 1904, pp. 394–409.
30. A. V. Bäcklund, *Zur Theorie der Flächentransformationen*, Math. Ann. **19** (1882), 387–422.
31. D. Hilbert, *Grundlagen der Geometrie*, 7th ed., Teubner, Leipzig, 1930.
32. N. V. Efimov and È. G. Poznyak, *Generalization of Hilbert's theorem on surfaces with constant negative curvature*, Dokl. Akad. Nauk SSSR **137** (1961), 509–512; English transl. in Soviet Math. Dokl. **2** (1961).
33. L. Bianchi, *Lezioni di geometria differenziale*, 4th ed., Zanichelli, Bologna, 1927.
34. M. H. Amsler, *Des surfaces à courbure négative constante dans l'espace à trois dimensions et de leurs singularités*, Math. Ann. **130** (1955), 234–256.
35. R. Steuerwald, *Über Enneper'sche Flächen und Bäcklund'sche Transformation*, Abh. Bayer. Akad. Wiss. Math.-Nat. Abt. **40** (1936).
36. Ch. Wissler, *Globale Tschebyscheff-Netze auf Riemannschen Mannigfaltigkeiten und Fortsetzung von Flächen konstanter negativer Krümmung*, Comment. Math. Helv. **47** (1972), 348–372.
37. P. Tschebyscheff [Chebyshev], *Sur la coupe des vêtements*, Assoc. Française Avancement Sci., 7 Session, Paris, Séance du 28 Août 1879; see his *Oeuvres*, Vol. II, Acad. Imp. Sci., St Petersburg, 1907, p. 708.
38. A. V. Pogorelov, *Extrinsic geometry of convex surfaces*, Amer. Math. Soc., Providence, RI, 1973.
39. E. E. Levi, *Sur l'application des équations intégrales au problème de Riemann*, Nachr. Königl. Ges. Wiss. Göttingen Math.-Phys. Kl. (1908), 249–252; reprinted in his *Opere*, Vol. I, Edizione Cremonese, Rome, 1959, pp. 175–179.
40. A. V. Pogorelov, *An example of a two-dimensional Riemannian metric not admitting local realization in E^3*, Dokl. Akad. Nauk SSSR **198** (1971), 42–43; English transl. in Soviet Math. Dokl. **12** (1971).
41. È. G. Poznyak, *Examples of regular metrics on a sphere and on a disk unrealizable in the class of twice continuously differentiable surfaces*, Vestnik Moskov. Univ. Ser. I Mat. Mekh. **1960**, no. 2, 3–5. (Russian)
42. M. L. Gromov and V. A. Rokhlin, *Inclusions and embeddings in Riemannian geometry*, Uspekhi Mat. Nauk **25** (1970), no. 5 (155), 3–62; English transl. in Russian Math. Surveys **25** (1970).
43. I. N. Vekua, *Generalized analytic functions*, Addison-Wesley, Reading, MA, 1962.
44. R. S. Hamilton, *The inverse function theorem of Nash and Moser*, Bull. Amer. Math. Soc. (N.S.) **7** (1982), 65–222.
45. K. O. Friedrichs, *Symmetric positive linear differential equations*, Comm. Pure Appl. Math. **11** (1956), 333–418.

E. V. Shikin, 18 Matveevskaya, Kopr. 2, Apt. 39, Moscow 119517, Russia

Selected Titles in This Series

(Continued from the front of this publication)

145 **S. G. Dalalyan et al.,** Eight Papers Translated from the Russian
144 **S. D. Berman et al.,** Thirteen Papers Translated from the Russian
143 **V. A. Belonogov et al.,** Eight Papers Translated from the Russian
142 **M. B. Abalovich et al.,** Ten Papers Translated from the Russian
141 **H. Draškovičová et al.,** Ordered Sets and Lattices
140 **V. I. Bernik et al.,** Eleven Papers Translated from the Russian
139 **A. Ya. Aĭzenshtat et al.,** Nineteen Papers on Algebraic Semigroups
138 **I. V. Kovalishina and V. P. Potapov,** Seven Papers Translated from the Russian
137 **V. I. Arnol′d et al.,** Fourteen Papers Translated from the Russian
136 **L. A. Aksent′ev et al.,** Fourteen Papers Translated from the Russian
135 **S. N. Artemov et al.,** Six Papers in Logic
134 **A. Ya. Aĭzenshtat et al.,** Fourteen Papers Translated from the Russian
133 **R. R. Suncheleev et al.,** Thirteen Papers in Analysis
132 **I. G. Dmitriev et al.,** Thirteen Papers in Algebra
131 **V. A. Zmorovich et al.,** Ten Papers in Analysis
130 **M. M. Lavrent′ev, K. G. Reznitskaya, and V. G. Yakhno,** One-dimensional Inverse Problems of Mathematical Physics
129 **S. Ya. Khavinson,** Two Papers on Extremal Problems in Complex Analysis
128 **I. K. Zhuk et al.,** Thirteen Papers in Algebra and Number Theory
127 **P. L. Shabalin et al.,** Eleven Papers in Analysis
126 **S. A. Akhmedov et al.,** Eleven Papers on Differential Equations
125 **D. V. Anosov et al.,** Seven Papers in Applied Mathematics
124 **B. P. Allakhverdiev et al.,** Fifteen Papers on Functional Analysis
123 **V. G. Maz′ya et al.,** Elliptic Boundary Value Problems
122 **N. U. Arakelyan et al.,** Ten Papers on Complex Analysis
121 **V. D. Mazurov, Yu. I. Merzlyakov, and V. A. Churkin, Editors,** The Kourovka Notebook: Unsolved Problems in Group Theory
120 **M. G. Kreĭn and V. A. Jakubovič,** Four Papers on Ordinary Differential Equations
119 **V. A. Dem′janenko et al.,** Twelve Papers in Algebra
118 **Ju. V. Egorov et al.,** Sixteen Papers on Differential Equations
117 **S. V. Bočkarev et al.,** Eight Lectures Delivered at the International Congress of Mathematicians in Helsinki, 1978
116 **A. G. Kušnirenko, A. B. Katok, and V. M. Alekseev,** Three Papers on Dynamical Systems
115 **I. S. Belov et al.,** Twelve Papers in Analysis
114 **M. Š. Birman and M. Z. Solomjak,** Quantitative Analysis in Sobolev Imbedding Theorems and Applications to Spectral Theory
113 **A. F. Lavrik et al.,** Twelve Papers in Logic and Algebra
112 **D. A. Gudkov and G. A. Utkin,** Nine Papers on Hilbert's 16th Problem
111 **V. M. Adamjan et al.,** Nine Papers on Analysis
110 **M. S. Budjanu et al.,** Nine Papers on Analysis
109 **D. V. Anosov et al.,** Twenty Lectures Delivered at the International Congress of Mathematicians in Vancouver, 1974
108 **Ja. L. Geronimus and Gábor Szegő,** Two Papers on Special Functions
107 **A. P. Mišina and L. A. Skornjakov,** Abelian Groups and Modules

(See the AMS catalog for earlier titles)